LOGIC, SETS, AND NUMBERS

An Introduction to Abstract Mathematics

Frank Blume

With Contributions by Clark Scholz

Cover credit: © Igor Stevanovic | Dreamstime.com

ISBN-10: 197377936
ISBN-13: 978-1973779360

To
Ferdi and Babette

Acknowledgments

I am grateful to John Brown University for supporting my work on this book by granting my request for a sabbatical leave in the fall of 2015. Special thanks go to my colleagues Calvin Piston and Gregory Varner who kindly agreed to teach some of my courses during my absence. Furthermore, I also would like to thank my former student Clark Scholz for supplying all the main results in the section on *Constructing Multiplication.*

Preface

Logic, Sets, and Numbers is meant to be a textbook for a one-semester introductory college-level course. In my own teaching experience, the amount of material covered in this book is quite bit larger than the average student can be expected to process in a single three-hour course. But this excess in volume and topical variety is likely beneficial because it gives instructors using this text greater flexibility in course design and subject selection. Sections that can be considered optional are 1.6, 2.3, 2.4, 4.3, and 4.4; and optional as well are Chapters 5 and 6. Furthermore, sections and exercises that are marked with a ★ are typically a little more difficult than the material in the rest of the text.

In terms of text organization it is definitely worth noting that the problem of rigorously formulating mathematical statements is greatly emphasized throughout. For it is my experience that students frequently struggle with proofs not only because poofs require genuine creative insight but also because it frequently is not that easy to discern exactly what it is that one is trying to prove. This emphasis on rigor in statement formulation is helpful as well insofar as it causes the initial slope of the learning curve to be a little less steep.

There are no formal prerequisites for successfully studying this text, as all subjects are developed from the bottom up beginning with elementary formal logic. However, abstract mathematics has a reputation for being notoriously difficult, and the prior acquisition of a certain level of maturity in working with mathematical variables is therefore very much desirable.

That said, it only remains for me to express my hope that the reader may find some positive stimulation in the fruits of my labor and come to appreciate, in studying this text, the rational beauty and purity that abstract mathematics so prominently exhibits. The ability to think with unexcepting rigor is not acquired easily, but those who are properly dedicated to the task are likely to find the eventual increase in their cognitive competence to be well worth the effort.

FRANK BLUME, JULY 2017.

Contents

Chapter 1

Formal Logic

1.1 No Statement is a Statement

The fundamental object of study in the field of formal logic is the *statement*. Roughly speaking, a statement is a sentence or, more generally, a symbolic construct that can be said to be true or false with absolute certainty.

1.1.1 Example. Consider the following sentence:

> The author of this book is a human being.

On the surface it may seem that this sentence, according to the definition just provided, is clearly a statement because it clearly is true. But we mustn't forget that the author of this book, by being a college professor, is also a teacher— and more especially—a teacher of students who may be inclined not to agree. For in the eyes of some of these students it may be doubtful, to say the least, whether a living being who tortures other living beings by exposing their minds to the horrible contents that this book comprises can really be considered a person in the ordinary, compassionate sense. Put differently, these students may plausibly argue that the author of this book, by being relentless and mean, is also non-human.

The author and teacher himself, of course, will view the matter differently. In his perception, it is not he himself who causes the pain but rather the subject. So the culprit in his eyes is the quantification scheme, the equivalence class, or the proof by induction—the topic at hand—rather than its hapless presenter. In other words, the author and teacher in his own, more benevolent view, is a perfectly benign being whose only desire is always to help and support and who therefore can rightly be said to be human.

Consequently and in summary, we are led to conclude that the sentence above is actually *not* a statement because its *truth value*—true or false—is observer dependent and therefore uncertain.

In the light of this example, it should be quite obvious that a position of radical doubt can always be somehow adopted. And hence there really isn't such a thing as a sentence, expression, or sequence of symbols that can truly be said to be true or false with absolute certainty—except, of course, for the sentence that says that no sentence can ever be said to be true or false with absolute certainty. By implication, the only statement there is is the statement that no statement is ever a statement.

But be that as it may. Our purpose here is not to contemplate or solve a philosophical mystery, but rather to understand the intended meaning of the notion of a statement in a more practical, down-to-earth sense.

1.1.2 Example. The sentence, 'Dogs commonly have four legs', may rightly be considered a statement because it may rightly be considered to be true. Strictly speaking, of course, the word 'commonly' introduces a degree of vagueness that may cause us to hesitate. But for all practical purposes it is indeed the case that dogs *commonly*—that is, in absence of birth defects or crippling injuries— do indeed have four legs and that therefore the sentence making this claim is indeed a statement.

1.1.3 Example. More interesting from a mathematical point of view is the symbolic construct '$x + 2 = 4$'. In itself, this construct is not a statement because x is an unspecified variable that renders the value of $x + 2$ inherently uncertain. This deficit, however, can easily be overcome by adding on some further symbols and words. For instance, the sentences '$x + 2 = 4$ for all integers x', '$x + 2 = 4$ for some integer x', and '$x + 2 = 4$ whenever $x = 2$', are statements because they are false, true, and true in this order.

1.1.4 Example. Just for the fun of it and in order to deepen the reader's confusion, let us consider the following sentence:

This sentence is a statement.

Is this sentence a statement or is it not? If we said that it is not, then the assertion that the sentence makes about itself would be false. Thus the sentence would in fact be a statement because a sentence that is false has—by being false—a well defined truth value and hence is a statement. So the assumption that the sentence above is not a statement leads to the contradictory conclusion that it is in fact a statement. Consequently, we are forced to infer, or so it seems, that the sentence above is a statement because it cannot be a non-statement. But can we here agree? Can a sentence really be a statement if the quality of being a statement is the only quality that the sentence is asserted to have and if it is the sentence itself that does the asserting? It doesn't make much sense, does it? Let's take a closer look: the words 'This sentence' in the sentence above are meant to signify the sentence itself. Since this is so, it seems to follow undeniably that the sentence above, by being self-referential, is equivalent to the assertion that the sentence, 'This sentence is a statement', is a statement. However, this latter assertion is no longer self-referential because there is no longer any 'this' in it that refers to the whole of the assertion. After all, the 'This' in the sentence, 'This sentence is a statement', only refers to this sentence itself and not to the larger, embedding sentence which asserts this sentence to be a statement. Thus the sentence above cannot be a statement because if it were a statement, it would be a self-referential statement that is equivalent to a non-self-referential statement and therefore would be non-self-referential, which is absurd (or so at least it seems if we agree that self-referentiality is a quality that equivalent sentences always possess or don't possess in synchrony). By implication and in summary, the sentence above is neither a statement nor a non-statement but rather a non-(statement-or-non-statement). In other words, the solution to the puzzle appears to be that the puzzle is ill conceived to begin with. For the question of whether the sentence above is a statement wrongly presumes that every sentence must necessarily be identifiable as either a statement or a non-statement. So what is here revealed is the astonishing fact that there do exist sentences—such as the sentence, 'This sentence is a statement'—the truth value of which is neither certain nor uncertain but simply impolite to talk about.

Exercises

1.1.5. Which of the following sentences or symbolic constructs are statements?

a) For all integers x it is the case that $x^2 > 0$.

b) $x < x + 1$

c) $x < x + 1$ for all integers x.

d) $2 > 3$

1.1.6. Write a five-page essay concerning the very interesting question of whether it was justified, in the preceding 'argument', to infer the non-self-referential nature of the assertion that the sentence, 'This sentence is a statement', is a statement, given that the prior analysis had shown that this assertion, in turn, is in fact equivalent to the sentence, 'This sentence is a statement', and therefore self-referential—or so one might think.

1.1.7. Discuss whether the following sentences are statements (but don't expect the author to provide a solution in the back of the book because there may not really be one):

a) I am reading this sentence.

b) I am reading this statement.

1.1.8. What would be the negation of the sentence, 'This sentence is a statement', if this sentence were indeed a statement and therefore could be negated?

1.1.9. Perform an analysis analogous to the one provided in Example 1.1.4 for the sentence, 'This statement is a sentence' (provided that you feel driven to do so by a strong, unruly compulsion).

1.1.10. Contemplate the nature of the sentences listed below in the light of the assertion, made in Example 1.1.4, that there do exist sentences that are neither statements nor non-statements.

  a) One hour is a very long stretch of time.

  b) Blue is a very nice color.

  c) Unicorns commonly have very thick fur.

  d) Every sentence is either a statement or a non-statement.

1.1.11. Contrive a philosophical argument that sheds doubt on the assertion that the sentence, 'The sky is blue', is a true statement. (Note: you may assume that the sky in question is cloudless and is observed from the surface of Planet Earth during daytime when the sun is shining.)

1.2 Truth Functions

Having clarified the notion of a statement in Section 1.1 by totally confusing it, we now move on to study in the present section in what ways statements can be connected and how the truth value of the resulting compound statements can be determined from the truth values of the statements of which the compounds are made up. In order to facilitate this discussion and also in order to hide from our awareness the embarrassing fact that statements, strictly speaking, do not exist, we will commonly denote statements by black-box variables like P or Q that are assumed to carry the values T for 'true' or F for 'false' with absolute, illusory certainty.

A very natural and frequently occurring compound statement in which two separate statements are combined into one is the *conjunction* 'P and Q' which is commonly denoted by $P \wedge Q$ or simply PQ. By definition, $P \wedge Q$ is true if and only if both P and Q are true simultaneously. In a sense, of course, this definition is circular because it describes the connective 'and' by means of the connective 'and' itself (as in the phrase 'if *and* only if both P *and* Q are true'). But this problem is really not a problem at all if we think of the conjunction more abstractly as a *truth function* that assigns to every pair of truth values for P and Q the truth value of $P \wedge Q$. Taking this functional point of view, the conjunction is simply defined by the following *truth table*:

P	Q	$P \wedge Q$
T	T	T
T	F	F
F	T	F
F	F	F

In order to make the concept of a truth function appear less mysterious, it is helpful to recall the familiar concept of a multivariable function in calculus. Consider for instance the function $f(x,y) := x + y$. What we commonly mean when we write this defining equation is that to every pair of real numbers x and y there is assigned the real number $x + y$. Consequently, there is nothing odd about assigning to a given pair of truth values for P and Q the truth value of $P \wedge Q$ as specified in the table above.

As a matter of course, truth functions, just like multivariable functions in calculus, can depend on any number of variables—on one variable, two variables, three variables, or perhaps one hundred twenty-three variables—it doesn't matter. The conjunction is a function that depends on two variables, and the so-called *negation* is a function that depends on only one. The negation of P is denoted by $\neg P$ or $\bar{P}$ and is defined by the following truth table:

P	$\neg P$
T	F
F	T

1.2.1 Example. The truth function that assigns to any triple of truth values for the variables P, Q, and R, the truth value of $P \wedge (Q \wedge \neg R)$ or $P \wedge (Q \wedge \bar{R})$ is an example of a truth function that depends on three variables. Its truth table looks as follows:

P	Q	R	$P \wedge (Q \wedge \neg R)$
T	T	T	F
T	T	F	T
T	F	T	F
T	F	F	F
F	T	T	F
F	T	F	F
F	F	T	F
F	F	F	F

Having defined the conjunction and the negation, we are now going to turn to the nonexclusive 'or'—the so-called *alternation* or *disjunction*. The alternation of two statements P and Q is denoted by $P \vee Q$ and is defined as follows:

P	Q	$P \vee Q$
T	T	T
T	F	T
F	T	T
F	F	F

Note: the attribute 'nonexclusive' refers to the fact that $P \vee Q$ is construed to be true in particular whenever both P and Q are true. By contrast, the exclusive 'or' is true if and only if ((P is true and Q is false) or (P is false and Q is true)) (see Exercise 1.2.6).

The truth function that tends to cause beginning students of logic the most trouble when it comes to aligning its abstract functional definition with its intended intuitive meaning is the *material conditional* $P \Rightarrow Q$. Its defining truth table looks as follows:

P	Q	$P \Rightarrow Q$
T	T	T
T	F	F
F	T	T
F	F	T

Intuitively speaking, the word 'conditional' is suggestive of a causal connection in the sense that the truth of P is a sufficient condition for the truth of Q or, equivalently, that the truth of Q is a necessary condition for the truth of P. That is to say, we are naturally inclined to read the arrow in '$P \Rightarrow Q$' as being indicative of an if-then relation: if P is true, then so is Q because the truth of P is the reason for Q to be true as well. Unfortunately, this way of thinking is quite strongly misleading. To make this point clear, it may help to denote by P the statement, 'Grass is green', and by Q the statement, '$1 + 1 = 2$'. In this case, both P and Q are true and so is the material conditional $P \Rightarrow Q$, according to the truth table given above. But this is not to say, of course, that $1 + 1 = 2$ *because* grass is green. Causality is a metaphysical concept that simply cannot be captured by a crude truth-functional definition.

On the other hand, though, the notion of a necessary inference is clearly important in mathematics, and to understand whether and how this notion is linked to the definition of the material conditional given above is certainly desirable. So let us consider, as an example, the following assertions:

For all real numbers x it is the case that $1/x < 1$ if $x > 1$.   (1.1)

For all real numbers x it is the case that $x > 1 \Rightarrow 1/x < 1$.   (1.2)

In order to see that (1.1) and (1.2) are in fact equivalent, it is helpful to ask under what conditions (1.1) and (1.2) would be false. Clearly, (1.1) would be false if it were possible to find some real number x that is greater than 1 but doesn't satisfy the inequality $1/x < 1$. And (1.2) would be false if it were possible to find an x for which the material conditional $x > 1 \Rightarrow 1/x < 1$ is false, that is, for which the *antecedent* $x > 1$ is true and the *consequent* $1/x < 1$ is false. In other words, the assertion that (1.1) is false is equivalent to the assertion that (1.2) is false, and by implication, the assertion that (1.1) is true is equivalent to the assertion that (1.2) is true. Consequently, the 'if..., then'-statement, 'if $x > 1$, then $1/x < 1$'—which we naturally understand to be equivalent to the if-statement, '$1/x < 1$ if $x > 1$'—is properly represented by the material conditional $x > 1 \Rightarrow 1/x < 1$.

In the light of this conclusion, it is helpful to take another look at the definition of $P \Rightarrow Q$ in the truth table above and to examine the rows in that table one by one. If we identify $P \Rightarrow Q$ with the inferential statement 'if P then Q', as explained in the preceding example, then it is quite natural to assign to $P \Rightarrow Q$ the value T in the first row where both P and Q are true. For there is obviously no reason whatsoever to declare the inference 'if P then Q' to be false if both the antecedent and the consequent are true. Similarly, the truth value F that is assigned to $P \Rightarrow Q$ in the second row also makes perfect sense, because if the antecedent is true and the consequent false, then the inference 'if P then Q' is inadmissible. But what about the rows three and four? In both of these rows the antecedent is false, and the material conditional $P \Rightarrow Q$ is defined to be true. Why do we say that an inference is true if the antecedent is false? To answer this question it is instructive to consider the alternative that the value assigned to $P \Rightarrow Q$ was F in either one of the last two rows and then to look back at assertion (1.2) above. If $P \Rightarrow Q$ was construed to be false in row three, then the material conditional $x > 1 \Rightarrow 1/x < 1$ would be false for $x = -1$, and if it was construed to be false in row four, then the material conditional $x > 1 \Rightarrow 1/x < 1$ would be false for $x = 1$. In either case there would exist a value for x that falsifies the material conditional, and, by implication, assertion (1.2) would be false. In particular, (1.2) would no longer be equivalent to (1.1) because (1.1) is evidently true. In other words, if we wish statements (1.1) and (1.2) to be equivalent, then a material conditional must be construed to be true whenever the antecedent is false.

The last of the standard truth functions that mathematicians commonly use is the *material biconditional* which is denoted by $P \Leftrightarrow Q$ and is defined as follows:

P	Q	$P \Leftrightarrow Q$
T	T	T
T	F	F
F	T	F
F	F	T

Intuitively, the material biconditional conveys the notion of the equivalence of two statements because $P \Leftrightarrow Q$ is true if and only if the truth values of P and

Q are equal.

Exercises

1.2.2. Why did we not write $P \wedge Q \wedge \bar{R}$ or $(P \wedge Q) \wedge \bar{R}$ in Example 1.2.1? Would it have made a difference, and if so, what would that difference have been?

1.2.3. Construct a truth table for the function $\neg(P \wedge Q) \wedge R$.

1.2.4. What is the total number of truth functions that depend on one variable? What is the total number of truth functions that depend on two variables, or, in general, on n variables?

1.2.5. List all the truth functions that depend on two variables (i.e., write a truth table for each of them).

1.2.6. Construct a truth table for the exclusive 'or'.

1.2.7. Construct a truth table for $\neg(\bar{P} \wedge \bar{Q})$. What do you notice?

1.2.8. Which of the following statements are true?

a) Commonly, dogs have four legs and human beings have two arms.

b) Commonly, dogs have four legs or human beings have two arms.

c) Commonly, dogs have four legs and human beings have four arms.

d) Commonly, dogs have four legs or human beings have four arms.

1.2.9. Construct two separate truth tables for $\bar{Q} \Rightarrow \bar{P}$ and $\bar{P} \vee Q$. What do you notice?

1.2.10. Construct a truth table for $\bar{P} \Leftrightarrow \bar{Q}$. What do you notice?

1.2.11. Construct a truth table for $(P \Leftrightarrow Q) \Leftrightarrow (\bar{P} \Leftrightarrow \bar{Q})$. How is this table relevant to the observation made in Exercise 1.2.10?

1.2.12. On a certain island, each inhabitant is either a truth-teller or a liar. A truth-teller *always* tells the truth and a liar *always* lies. Linus and Lucy live on the island.

a) Suppose that Linus says, "If I am a truth-teller, then each person living on this island is either a truth-teller or a liar." Can you tell whether Linus is a truth-teller or a liar?

b) Suppose that Linus says, "If I am a truth-teller, then so is Lucy." Can you tell whether Linus and Lucy are truth-tellers or liars?

Note: this is a well known problem that can be found in slightly altered form for instance in [DG], p.29.

1.3 Symbols and Words

In everyday linguistic usage the logical schemata introduced in the preceding section may assume a variety of forms, that is, they may not necessarily always be represented by 'and', 'or', 'if..., then', or 'if and only if'. For instance, the sentence, 'Grumpy is stupid whereas Stupy is grumpy', is properly construed to be a conjunction in which 'whereas' takes on the role of 'and'. Similarly, the sentence, 'Grumpy is happy unless Stupy is grumpy', is to be interpreted as an alternation: 'Grumpy is happy or Stupy is grumpy'. Moreover, typical constructions that signify $P \Rightarrow Q$ are:

> if P, then Q,
>
> Q if P,
>
> Q whenever P,
>
> P only if Q,
>
> Q in case that P,
>
> not P unless Q.

Most of these constructions are easily recognized as adequate verbal renderings of $P \Rightarrow Q$, but perhaps the fourth and the last require a few words of explanation. If we say that P is true *only if* Q is true as well, then we are effectively saying that P cannot be true unless Q is true as well. In other words, if P is true, then so is Q. Consequently, the phrase 'if P, then Q' does indeed carry the same meaning as the phrase 'P only if Q'. Furthermore, the observation just made that the latter phrase is also captured in the statement that P cannot be true unless Q is true as well, leads us to infer that the phrase, 'P must be false unless Q is true'—or equivalently, 'not P unless Q'—is another, adequate rendering of the material conditional $P \Rightarrow Q$.

1.3.1 Example. Consider the following sentences:

> Grumpy is happy but Stupy is not. (1.3)
>
> Grumpy is happy even though Stupy is not. (1.4)

From a strictly formal point of view, both sentences are equivalent to the conjunction,

> Grumpy is happy and Stupy is not. (1.5)

But when it comes to the meaning that theses sentences convey, there are some subtle differences to be noted: the use of 'and' in (1.5) merely indicates the parallel truth of two independent facts—the happiness of Grumpy *and* the unhappiness of Stupy—whereas the use of 'but' in (1.3) suggests a contrasting of one fact with another. Thus the extended construction, 'Grumpy is happy but Stupy, *by contrast*, is not', seems decidedly more natural than the competing alternative, 'Grumpy is happy and Stupy, by contrast, is not'. Similarly, the 'even though'-construction in (1.4) quite clearly suggests an element of surprise or a noteworthy exception: commonly Grumpy and Stupy are happy or unhappy in tandem, but for once Grumpy is happy *even though* Stupy is not. Naturally, such differences at the level of semantics cannot be conveyed in the starkly reductive symbolisms of formal logic. That is to say, poverty of meaning is the price we have to pay for clarity of structure (and hence mathematics makes sense but philosophy does not).

1.3.2 Example. Following up on the last thought in the preceding example, it is instructive to note that the schemata of formal logic are not only indifferent to nuances of meaning but also devoid of the semantic ambiguities that ordinary speech and language are often beset with. Consider, for instance, the sentence,

> Grumpy will beat his chest (P) and Stupy will cry (Q) if Goofy is not at home (R).

The suggestion here is that Goofy's absence causes Grumpy and Stupy to behave somewhat erratically, that is, the suggestion is that $R \Rightarrow (P \wedge Q)$. But a possible alternative is that Grumpy is in the habit of routinely beating his chest and that only Stupy is affected by Goofy's absence. In other words, the sentence might also be construed to mean that $P \wedge (R \Rightarrow Q)$—which is not the same. Avoiding such ambiguities may require the use of careful and potentially cumbersome reformulations:

$R \Rightarrow (P \wedge Q)$: If Goofy is not at home, it will be the case that Grumpy beats his chest and that Stupy cries.

$P \wedge (R \Rightarrow Q)$: Grumpy will beat his chest, and if Goofy is not at home, Stupy will cry.

Note: various similar examples that further highlight the difficulties encountered in translating words into symbols can be found in [Q], pp.53–60.

Exercises

1.3.3. Write each of the following statements in symbolic form, using the truth functions defined in the preceding section.

a) Grumpy has a long nose but Stupy doesn't.

b) Grumpy has a long nose even though his head is quite small.

c) Grumpy is grumpy while Stupy is stupid.

d) If Grumpy is happy, then so is Stupy.

e) Grumpy is happy only if Stupy is too.

f) Grumpy is happy if Stupy loses his nose.

g) Grumpy cries provided that Stupy doesn't pet him.

h) Grumpy grows fur whenever Stupy is gone for too long.

i) Grumpy gets lost unless he gets up.

j) Grumpy gets up unless he gets lost.

k) If Grumpy eats too much then it always is the case that he gets depressed and that Stupy scolds him.

l) If Grumpy eats too much then he gets depressed, and Stupy scolds him.

m) If Grumpy eats too much then he gets depressed, and Stupy commonly scolds Grumpy.

n) Grumpy goes for a walk if and only if Stupy takes off his socks.

1.3.4. Discuss the differences at the level of semantics between i) and j) in Exercise 1.3.3.

1.4   Truth-Value Analysis

If a mathematician wishes to convince another mathematician that a certain theorem or proposition is true, he or she must offer a proof. In other words, proofs are essential to mathematics, and so are the rules of logic that proofs must obey. As a simple example, let us consider the following proposition:

For all real numbers x it is the case that $x > 1$ if and only if $0 < 1/x < 1$.

How do we prove it? To begin with, we need to identify the logical form in which the proposition is stated: the clause, '$x > 1$ if and only if $0 < 1/x < 1$', signifies that we are dealing with a material biconditional. In other words, using the notation introduced in Section 1.2 and denoting the statements $x > 1$ and $0 < 1/x < 1$ by $P(x)$ and $Q(x)$, respectively, the proposition above can be rewritten as follows:

For all real numbers x it is the case that $P(x) \Leftrightarrow Q(x)$.

On the other hand though, the formulation '$P(x)$ if and only if $Q(x)$' also signifies a conjunction of two material conditionals: '$P(x)$ only if $Q(x)$' and '$P(x)$ if $Q(x)$', or, equivalently,

$$(P(x) \Rightarrow Q(x)) \wedge (Q(x) \Rightarrow P(x)).$$

Apparently then, in order to establish that $P(x) \Leftrightarrow Q(x)$ is true for all real numbers x, we need to show that both $P(x) \Rightarrow Q(x)$ and $Q(x) \Rightarrow P(x)$ are true for all x. That is to say, we need to show that the assumption $x > 1$

implies that $0 < 1/x < 1$ and that, conversely, the assumption $0 < 1/x < 1$ implies that $x > 1$.

Since both of these inferences are trivially valid, we need not be concerned with them in any greater detail. But what we do need to look at a bit more closely is the underlying assertion that a material biconditional is *equivalent* to the conjunction of two ordinary material conditionals, or, more explicitly and formally, that the schema $P \Leftrightarrow Q$ is *equivalent* to the schema $(P \Rightarrow Q) \wedge (Q \Rightarrow P)$ (we are leaving out the argument '(x)' because it is no longer of any interest). In order to understand exactly what is meant by saying that these two schemata are equivalent, it is helpful to consider the following table:

P	Q	$P \Rightarrow Q$	$Q \Rightarrow P$	$(P \Rightarrow Q) \wedge (Q \Rightarrow P)$	$P \Leftrightarrow Q$
T	T	T	T	T	T
T	F	F	T	F	F
F	T	T	F	F	F
F	F	T	T	T	T

Since the last two columns are completely identical, the truth function given by $P \Leftrightarrow Q$ is exactly the same as the truth function given by $(P \Rightarrow Q) \wedge (Q \Rightarrow P)$. In other words, the material biconditional

$$(P \Leftrightarrow Q) \Leftrightarrow ((P \Rightarrow Q) \wedge (Q \Rightarrow P)) \tag{1.6}$$

is true regardless of what truth values we happen to assign to P and Q (by the definition of the material biconditional that was stated in Section 1.2). In the light of this observation, it is useful to introduce the following definition:

1.4.1 Definition.   a) A truth-functional schema S is said to be *valid* or *tautological* if S is true for all assignments of truth values to the statement variables that S contains. (Note: tautological schemata are naturally referred to as *tautologies*.)

b) Two truth-functional schemata S_1 and S_2 are said to be *equivalent*, if the material biconditional $S_1 \Leftrightarrow S_2$ is valid.

c) A truth-functional schema S_1 is said to *imply* a schema S_2, if the material conditional $S_1 \Rightarrow S_2$ is valid.

d) A truth-functional schema S is said to be *consistent*, if it is true for at least one assignment of truth values to the statement variables that S contains. That is to say, S is consistent if $\neg S$ is not valid. (Note: if S is valid, then S is consistent as well.)

e) A truth-functional schema S is said to be *inconsistent* or *contradictory*, if it is not consistent, that is, if $\neg S$ is valid.

1.4.2 Example. If we set

$$S_1 :\Leftrightarrow (P \Leftrightarrow Q)$$

(i.e., S_1 is *by definition* equivalent to $P \Leftrightarrow Q$) and

$$S_2 :\Leftrightarrow ((P \Rightarrow Q) \wedge (Q \Rightarrow P)),$$

then the schema (1.6) assumes the form $S_1 \Leftrightarrow S_2$, and since the truth table above shows this schema to be valid, it follows that S_1 and S_2 are equivalent according to Definition 1.4.1a,b.

In order to establish the validity or consistency of a given schema, we can always use a truth table analysis as demonstrated above, but since the number of rows in such tables grows exponentially with the number of variables on which a schema depends, a truth table analysis can quickly become very laborious. A better method is frequently the so-called *truth-value analysis*, in the process of which we assign truth values to the variables in a given schema sequentially—beginning with the first and ending with the last—and then simplify the resulting schemata, step by step, as far as possible.

1.4.3 Example. We wish to show that schema (1.6) is valid by means of a truth-value analysis. To do so, we will show that this schema is true regardless of whether P is true or false. Assigning first the value T to P, the schema assumes the form

$$(T \Leftrightarrow Q) \Leftrightarrow ((T \Rightarrow Q) \wedge (Q \Rightarrow T)). \qquad (1.7)$$

Since $T \Leftrightarrow Q$ is true if Q is true and false if Q is false, it follows that $T \Leftrightarrow Q$ is equivalent to Q, and so is the schema $T \Rightarrow Q$ for exactly the same reason. Furthermore, the schema $Q \Rightarrow T$ is true because a material conditional with a true consequent is true (by the definition of the material conditional that was stated on p.6). Taking these simplifications into account, the schema (1.7) reduces to

$$Q \Leftrightarrow (Q \wedge T).$$

Since $Q \wedge T$ is true if Q is true and false if Q is false, we may replace $Q \wedge T$ simply with Q. This leaves us with the schema

$$Q \Leftrightarrow Q$$

which is evidently true. Having thus arrived at the desired conclusion in the case where P is true, it only remains to consider the case where P is false. The corresponding schema, along with the subsequent simplifications, looks as follows:

$$(F \Leftrightarrow Q) \Leftrightarrow ((F \Rightarrow Q) \wedge (Q \Rightarrow F))$$
$$\bar{Q} \Leftrightarrow (T \wedge \bar{Q}) \quad \text{(see Exercise 1.4.8)}$$
$$\bar{Q} \Leftrightarrow \bar{Q}$$
$$T.$$

Consequently, schema (1.6) is valid.

1.4.4 Example. Let us consider statement i) of Exercise 1.3.3:

> Grumpy gets lost unless he gets up.

Instinctively, we may be inclined to read this statement as a conditional: Grumpy gets lost *if* he does not get up. So if we denote by Q the statement, 'Grumpy gets lost', and by P the statement, 'Grumpy does not get up', then the schema that captures the statement above is $P \Rightarrow Q$. Alternatively, though, we also can read this statement as an alternation: Grumpy gets lost, *or*—if he doesn't get lost—he gets up. Since the symbolic form of this alternation is $Q \vee \bar{P}$—or, equivalently, $\bar{P} \vee Q$ (because the alternation is easly seen to be commutative)—we are led to conjecture that $P \Rightarrow Q$ is equivalent to $\bar{P} \vee Q$, that is, that the schema

$$(P \Rightarrow Q) \Leftrightarrow (\bar{P} \vee Q) \tag{1.8}$$

is a tautology. To establish this claim, we perform a truth-value analysis: if P is true, then the corresponding simplification yields

$$(T \Rightarrow Q) \Leftrightarrow (F \vee Q)$$
$$Q \Leftrightarrow Q$$
$$T,$$

and if P is false, we find that

$$(F \Rightarrow Q) \Leftrightarrow (T \vee Q)$$
$$T \Leftrightarrow T$$
$$T,$$

as desired. In the light of this result, it may be tempting to conclude that the statement

> Grumpy gets lost unless he gets up.

has exactly the same meaning as the permuted version

> Grumpy gets up unless he gets lost.

After all and as we just noted, $Q \vee \bar{P}$ is certainly equivalent to $\bar{P} \vee Q$. But here we need to be careful, for insofar as 'meaning' here refers to the *intuitive* meaning conveyed by human language, the two statements are not exactly alike. For whereas the former statement suggests that Grumpy gets lost if he stays in bed too long, the latter seems to imply that sometimes Grumpy gets lost while still in bed and therefore forgets to get up. So in this latter case Grumpy gets up if, antecedently, he does not get lost while still in bed. Thus the causal relations in the two cases are reversed: in the former case it is the not-getting-up that brings about the getting-lost, whereas in the latter, by contrast, it is the not-getting-lost that effects the getting-up. Symbolically, these two cases correspond to the schemata $P \Rightarrow Q$ and $\bar{Q} \Rightarrow \bar{P}$, respectively. However, since causality is far too subtle and too intangible a notion to be meaningfully reflected at a purely formal, truth-functional level (see the discussion on p.6), these two schemata ought to be—as truth-functions—completely identical. That is to say, the schema

$$(P \Rightarrow Q) \Leftrightarrow (\bar{Q} \Rightarrow \bar{P})$$

ought to be a tautology. In order to show that this indeed is so, we observe that the equivalence of $P \Rightarrow Q$ and $\bar{P} \vee Q$ (established above) implies, by way of substitution, that $\bar{Q} \Rightarrow \bar{P}$ is equivalent to $\neg\bar{Q} \vee \bar{P}$ (simply replace P with $\bar{Q}$ and Q with $\bar{P}$) which in turn is equivalent to $\bar{P} \vee Q$ (because the alternation is commutative and because $\neg\bar{Q} \Leftrightarrow Q$ is a tautology). Thus it follows, as desired, that $P \Rightarrow Q$ is equivalent to $\bar{Q} \Rightarrow \bar{P}$ because both of these schemata are equivalent to $\bar{P} \vee Q$ (see also Exercise 1.4.10).

A useful list of tautologies, which also contains the tautologies discussed in Examples 1.4.3 and 1.4.4, is given in the following proposition:

1.4.5 Proposition. *Each of the following schemata is a tautology:*

    a) $\neg(P \wedge Q) \Leftrightarrow (\bar{P} \vee \bar{Q})$ *(first law of DeMorgan),*

    b) $\neg(P \vee Q) \Leftrightarrow (\bar{P} \wedge \bar{Q})$ *(second law of DeMorgan),*

    c) $(P \Rightarrow Q) \Leftrightarrow (\bar{P} \vee Q)$,

    d) $\neg(P \Rightarrow Q) \Leftrightarrow (P \wedge \bar{Q})$,

    e) $(P \Rightarrow Q) \Leftrightarrow (\bar{Q} \Rightarrow \bar{P})$,

    f) $(P \wedge (Q \wedge R)) \Leftrightarrow ((P \wedge Q) \wedge R)$ *(first associative law),*

    g) $(P \vee (Q \vee R)) \Leftrightarrow ((P \vee Q) \vee R)$ *(second associative law),*

    h) $(P \vee (Q \wedge R)) \Leftrightarrow ((P \vee Q) \wedge (P \vee R))$ *(first distributive law),*

    i) $(P \wedge (Q \vee R)) \Leftrightarrow ((P \wedge Q) \vee (P \wedge R))$ *(second distributive law),*

    j) $(P \Leftrightarrow Q) \Leftrightarrow ((P \Rightarrow Q) \wedge (Q \Rightarrow P))$,

    k) $((P \Leftrightarrow Q) \wedge (Q \Leftrightarrow R)) \Leftrightarrow ((P \Rightarrow Q) \wedge (Q \Rightarrow R) \wedge (R \Rightarrow P))$.

Proof. We will only prove h) and leave it to the reader to demonstrate the validity of a sub-selection of the remaining statements in Exercise 1.4.11. Assuming that P is true, the schema given in h) simplifies as follows:

$$(T \vee (Q \wedge R)) \Leftrightarrow ((T \vee Q) \wedge (T \vee R))$$
$$T \Leftrightarrow (T \wedge T)$$
$$T.$$

Furthermore, if P is false, then we find that

$$(F \lor (Q \land R)) \Leftrightarrow ((F \lor Q) \land (F \lor R))$$
$$(Q \land R) \Leftrightarrow (Q \land R)$$
$$T,$$

as desired.   □

The next and last proposition in this section shows that all the standard truth functions introduced in Section 1.2 can be represented using only the negation and the conjunction (see Exercise 1.4.13).

1.4.6 Proposition. *Each of the following schemata is a tautology:*

a) $(P \lor Q) \Leftrightarrow \neg(\bar{P} \land \bar{Q})$,

b) $(P \Rightarrow Q) \Leftrightarrow \neg(P \land \bar{Q})$,

c) $(P \Leftrightarrow Q) \Leftrightarrow (\neg(P \land \bar{Q}) \land \neg(Q \land \bar{P}))$.

1.4.7 Example. On the surface it may seem that the sufficiency claim of Proposition 1.4.6 is purely theoretical in nature and therefore totally useless in practical terms. But consider for instance a situation in which one person feels compelled to assert—in the presence of another—the following causal relation:

I want to marry you *because* I love you.   (1.9)

Setting
$$P :\Leftrightarrow \text{'I love you'}$$
and
$$Q :\Leftrightarrow \text{'I want to marry you'},$$

it is tempting to think that (1.9) is equivalent to $P \Rightarrow Q$, but here we need to be careful because a material conditional is true in particular whenever the antecedent is false. So presently the truth of $P \Rightarrow Q$ would be guaranteed if P were false, or equivalently, if $\bar{P}$—'I do *not* love you'—were true. Since this seems odd, we are led to realize that the intended meaning of (1.9) is more accurately conveyed by the schema

$P \land (P \Rightarrow Q)$—'I love you, and if I love you, then I want to marry you',

which in turn is equivalent to

$$P \land \neg(P \land \bar{Q}) \quad \text{(by Proposition 1.4.6b)}.$$

Consequently, a structurally purified rephrasing of (1.9) that correctly reflects the sufficiency of negation and conjunction in the spirit of Proposition 1.4.6— and that also sounds more poignant and romantic—can be stated as follows:

I love you, and it is not the case that I love you and do not want to marry you.

For clarity we wish to add, that the person making this claim must be careful not to insert another 'I' following the second 'and'. For the altered claim, 'I love you, and it is not the case that I love you and *I* do not want to marry you', would surely be decidedly unorthodox. Furthermore, using the first law of DeMorgan and the second distributive law (as stated in Proposition 1.4.5a,i), we find that

$$P \wedge \neg(P \wedge \bar{Q}) \Leftrightarrow P \wedge (\bar{P} \vee \neg\bar{Q})$$
$$\Leftrightarrow P \wedge (\bar{P} \vee Q)$$
$$\Leftrightarrow (P \wedge \bar{P}) \vee (P \wedge Q)$$
$$\Leftrightarrow F \vee (P \wedge Q)$$
$$\Leftrightarrow P \wedge Q.$$

So in the final analysis, the *because*-construction in (1.9) is here revealed to be a simple conjunction:

I love you, and I want to marry you.

Without undue exaggeration it may be said that this conclusion—arrived at algebraically and wholly unexpectedly—powerfully demonstrates the astonishing depth of practical applicability that formal logic is endowed with.

Exercises

1.4.8. Explain why $Q \Rightarrow F$ is equivalent to $\bar{Q}$.

1.4.9. Perform a truth-value analysis to show that the schema $(P \Leftrightarrow Q) \Leftrightarrow (P \Rightarrow Q)$ is consistent but not valid, and then confirm this result by means of a truth table.

1.4.10. Use a truth-value analysis to confirm that the schema $(P \Rightarrow Q) \Leftrightarrow (\bar{Q} \Rightarrow \bar{P})$ is a tautology.

1.4.11. Prove Proposition 1.4.5a,b,d,f,k. To do so, you may use truth-value analyses or apply previously established results.

1.4.12. Generalize Proposition 1.4.5k to the case of n statement variables.

1.4.13. Use truth-value analyses or previously established results to prove Proposition 1.4.6.

1.4.14. Show that—in analogy to Proposition 1.4.6—all standard truth function can be expressed in terms of the alternation and the negation.

1.5   Quantification

Let us consider again the statement with which we began our discussion in Section 1.4:

For all real numbers x it is the case that $x > 1$ if and only if $0 < 1/x < 1$.

The latter part of this sentence can be written, as we saw, in the symbolic form $P(x) \Leftrightarrow Q(x)$ if we set $P(x) :\Leftrightarrow x > 1$ and $Q(x) :\Leftrightarrow 0 < 1/x < 1$. But what about the first part—the phrase, 'For all real numbers x it is the case that'? From a formal point of view it is desirable to represent this part in symbolic form as well, especially because *categorical statements* that assert the universal validity of a certain property—as indicated by the words 'For all'—are extremely common in mathematics. Thus it is expedient to introduce the *quantifier* $\forall$, which is read as 'for all'. Using this quantifier, the statement above can be rewritten as follows:

$$\forall_x \text{ if } x \text{ is a real number, then } P(x) \Leftrightarrow Q(x). \qquad (1.10)$$

Remark. Since this statement appeals to the notion of a real number, it seems appropriate to point out that, strictly speaking, at this point in the course, we do not know what a real number is because we do not even know what a positive integer is. After all, the Peano Axioms that establish the set of counting numbers as an object of rigorous mathematical study have not been stated as yet, and therefore, we do not know how to count. But then again, this sort of purist approach can easily become self-defeating. For our purpose ought to be to try to understand how logic is foundational to mathematical thought, and mathematical thought in empty space, without any proper object of thought, is somewhat difficult to practice. Consequently, we will be bold! We will assume that we know what we don't know and agree—with fear and trembling—that the following sets are well defined entities:

- the set $\mathbb{N}$ of positive counting numbers or *natural* numbers,
- the set $\mathbb{Z}$ of positive and negative integers,
- the set $\mathbb{Q}$ of rational numbers,
- the set $\mathbb{R}$ of real numbers.

Furthermore, in order to have some added flexibility in working with these sets, we also wish to introduce the standard notation for finite subsets of $\mathbb{R}$ and intervals that are contained in $\mathbb{R}$: if $x_1, \ldots, x_n \in \mathbb{R}$ are given real numbers, then the set containing these numbers is denoted by

$$\{x_1, \ldots, x_n\},$$

and if a and b are numbers in $\mathbb{R}$ such that $a \leq b$, then we denote by...

- (a, b) the *open interval* of all values $x \in \mathbb{R}$ that are strictly greater than a and strictly less than b, i.e., $a < x < b$.
- $[a, b]$ the *closed interval* of all values $x \in \mathbb{R}$ that are greater than or equal to a and less than or equal to b, i.e., $a \leq x \leq b$.
- $[a, b)$ the *half-open interval* of all values $x \in \mathbb{R}$ that are greater than or equal to a and strictly less than b, i.e., $a \leq x < b$.
- $(a, b]$ the *half-open interval* of all values $x \in \mathbb{R}$ that are strictly greater than a and less than or equal to b, i.e., $a < x \leq b$.

Equivalently, we can also use *set-builder notation* and write

$$(a, b) := \{x \in \mathbb{R} \mid a < x < b\},$$
$$[a, b] := \{x \in \mathbb{R} \mid a \leq x \leq b\},$$
$$[a, b) := \{x \in \mathbb{R} \mid a \leq x < b\},$$
$$(a, b] := \{x \in \mathbb{R} \mid a < x \leq b\}.$$

1.5.1 Example. Using the notation introduced in the preceding remark, statement (1.10) can be written in the form

$$\forall_x \; x \in \mathbb{R} \Rightarrow (P(x) \Leftrightarrow Q(x))$$

or, equivalently,

$$\forall_{x \in \mathbb{R}} \; P(x) \Leftrightarrow Q(x).$$

Apart from assertions of universality we also encounter very frequently assertions of existence for which we therefore introduce the existence quantifier $\exists$.

1.5.2 Example. Consider the following statement:

There exists a real number x such that $x > 1$ and $x^2 - x - 1 = 0$.

Setting $P(x) :\Leftrightarrow x > 1$ and $Q(x) :\Leftrightarrow x^2 - x - 1 = 0$, this statement can be written in quantified form as follows:

$$\exists_x \; x \in \mathbb{R} \wedge P(x) \wedge Q(x). \tag{1.11}$$

As in the case of the for-all quantifier, the *universe* from which x is taken can also be transfered to the argument of the quantifier itself. That is to say, instead of statement (1.11) we may write

$$\exists_{x \in \mathbb{R}} \; P(x) \wedge Q(x).$$

However, in this particular case we may also consider the universe to be the set of all real numbers greater than 1—because $P(x) \Leftrightarrow x > 1$. This yields

$$\exists_{x \in (1, \infty)} \; Q(x),$$

where $(1, \infty)$ denotes the set of all real numbers greater than 1, that is,

$$(1, \infty) := \{x \in \mathbb{R} \mid x > 1\}.$$

The general forms of the categorical statements that we encountered in the preceding examples are

$$\forall_{x \in \mathcal{U}} \ S(x)$$

and

$$\exists_{x \in \mathcal{U}} \ S(x),$$

where $\mathcal{U}$ is a given universe and $S(x)$ is a so called *term schema* that depends on x (such as $P(x) \Leftrightarrow Q(x)$). As we saw, the removal of the universe from the quantifier itself yields a material conditional in the for-all case and a conjunction in the existence case:

$$\boxed{\forall_{x \in \mathcal{U}} \ S(x) \Leftrightarrow \forall_x \ x \in \mathcal{U} \Rightarrow S(x)}$$

and

$$\boxed{\exists_{x \in \mathcal{U}} \ S(x) \Leftrightarrow \exists_x \ x \in \mathcal{U} \wedge S(x).}$$

Remark. If the universe $\mathcal{U}$ is empty, then the statement $x \in \mathcal{U}$ is false for all x, and therefore, the statement $\forall_x \ x \in \mathcal{U} \Rightarrow S(x)$ is true (because the antecedent $x \in \mathcal{U}$ is false), and the statement $\exists_x \ x \in \mathcal{U} \wedge S(x)$ is false. Consequently, the statements $\forall_{x \in \mathcal{U}} \ S(x)$ and $\exists_{x \in \mathcal{U}} \ S(x)$ are respectively true and false as well.

1.5.3 Example. We wish to write each of the following sentences as a quantification schema:

a) Either all animals with fur have four legs, and some animals have fur, or all animals with only two legs have no fur.

b) All animals with fur have four legs, or the existence of furless animals implies that some animals have two legs but no fur.

c) Either all animals are furless whenever they have four legs, or some animals are furless while others have fur but only two legs.

d) Some animals are furless unless they have four legs while others are furless but equipped with two legs.

Denoting by $\mathcal{U}$ the universe of all animals and setting

$$P(x) :\Leftrightarrow \text{'}x \text{ has fur'},$$
$$Q(x) :\Leftrightarrow \text{'}x \text{ has two legs'},$$
$$R(x) :\Leftrightarrow \text{'}x \text{ has four legs'},$$

the quantified schematic renderings of a)-d) are as follows:

a) $((\forall_{x \in \mathcal{U}} P(x) \Rightarrow R(x)) \wedge \exists_{x \in \mathcal{U}} P(x)) \vee \forall_{x \in \mathcal{U}} Q(x) \Rightarrow \bar{P}(x),$

b) $(\forall_{x \in \mathcal{U}} P(x) \Rightarrow R(x)) \vee ((\exists_{x \in \mathcal{U}} \bar{P}(x)) \Rightarrow \exists_{x \in \mathcal{U}} Q(x) \wedge \bar{P}(x)),$

c) $\left(\forall_{x\in\mathcal{U}}\, R(x) \Rightarrow \bar{P}(x)\right) \vee \left(\left(\exists_{x\in\mathcal{U}}\, \bar{P}(x)\right) \wedge \exists_{x\in\mathcal{U}}\, P(x) \wedge Q(x)\right),$

d) $\left(\exists_{x\in\mathcal{U}}\, \bar{P}(x) \vee R(x)\right) \wedge \exists_{x\in\mathcal{U}}\, \bar{P}(x) \wedge Q(x).$

Unfortunately, term schemata that are *monadic* in the sense that they depend on only one variable (such as x) are still too inflexible to really be useful in mathematics. That is to say, in order to cover the full range of statements (propositions, theorems, etc.) encountered in mathematics, term schemata must be allowed to be *polyadic*. A simple example of a statement that requires the use of a *dyadic* schema is the assertion that every nonnegative real number has a square root. For what we mean by saying that every nonnegative x has a square root is evidently that for every such x there exists a y such that $y^2 = x$. Consequently, the correct quantification schema is

$$\forall_{x\in\mathbb{R}}\left(x \geq 0 \Rightarrow \exists_{y\in\mathbb{R}}(y \geq 0 \wedge x = y^2)\right). \tag{1.12}$$

To see the formal structure of this statement more clearly, we set

$$P(x) :\Leftrightarrow x \geq 0$$

and

$$Q(x, y) :\Leftrightarrow y \geq 0 \wedge x = y^2.$$

This yields

$$\forall_{x\in\mathbb{R}}\left(P(x) \Rightarrow \exists_{y\in\mathbb{R}}\, Q(x, y)\right),$$

where the term schemata $P(x)$ and $Q(x, y)$ are monadic and dyadic, respectively.

Having witnessed, in the preceding paragraph, how the claim that every nonnegative real number has a square root can be written in schematic, quantified form, the uninitiated reader may naturally wonder why anyone would care. Why is it important to express a well known, well understood, and perfectly elementary mathematical statement in such a denuded logical form? The answer is that mathematics, as current-day mathematicians commonly practice it, would more or less cease to exist if it were to be divorced from the absolute logical rigor that we are presently in the process of introducing.

Mathematics is a creative activity of the human mind whose general purpose and theme it is to rigorously derive novel rational truths from *a priori* given, axiomatic assumptions. But as it is, there simply is no way to ever derive anything from anything, with unexcepting rigor, if one doesn't know and cannot clearly state exactly what it is that one is trying to derive. For instance, it would be very difficult indeed to convincingly *prove* the existence of square roots, if one cannot clearly say what precisely is meant by this existence assertion. So before the world of modern mathematics can be creatively explored, its constituting logical language must be acquired and thoroughly mastered.

This is not to say by any means that mathematicians commonly forego the use of ordinary English words and compose all their scholarly writings of fearfully cumbersome formal-logic notations. But it is to say that a mathematician who was unable to do so—that is, to express him- or herself essentially wordlessly—would simply be entirely incompetent. This judgment may sound harsh, but it is accurate regardless. That said, our guiding purpose in the examples that follow will be and ought to be to clearly understand the precise logical structure of commonly occurring, mathematical statements and then to translate these statements into a word-free quantified form.

1.5.4 Example. Given three positive integers m, n, and p, we wish to write the following statement as a quantification schema:

$$p \text{ is the greatest common divisor of } n \text{ and } m \text{ (i.e., } p = \gcd(m,n)). \quad (1.13)$$

To say that p is a divisor of m means that there exists a $k \in \mathbb{N}$ such that $m = kp$. Consequently, the assertion that p is a *common* divisor of m *and* n is equivalent to the schema

$$\exists_{k \in \mathbb{N}} \, m = kp \wedge \exists_{k \in \mathbb{N}} \, n = kp. \quad (1.14)$$

Alternatively, we can also write

$$\exists_{k \in \mathbb{N}} \, m = kp \wedge \exists_{i \in \mathbb{N}} \, n = ip,$$

but since each existence statement in the conjunction (1.14) is completely self-contained, there really is no need to introduce a second variable i. Using a second variable may be visually preferable, but formally necessary it is not. Furthermore, the added assertion that p is the *greatest* common divisor of m and n means that any arbitrary common divisor of m and n will always be less than or equal to p. That is to say, for all positive integers q it is the case that the assumed validity of (1.14)—with q in place of p—implies that $q \leq p$. Consequently, in setting

$$P(m,n,p) :\Leftrightarrow \exists_{k \in \mathbb{N}} \, m = kp \wedge \exists_{k \in \mathbb{N}} \, n = kp,$$

we find that (1.13) is equivalent to

$$P(m,n,p) \wedge \forall_{q \in \mathbb{N}} \, (P(m,n,q) \Rightarrow q \leq p).$$

Introducing the notation

$$p|m :\Leftrightarrow \exists_{k \in \mathbb{N}} \, m = kp,$$

we may also write

$$p = \gcd(m,n) \Leftrightarrow (p|m \wedge p|n \wedge \forall_{q \in \mathbb{N}} \, ((q|m \wedge q|n) \Rightarrow q \leq p)).$$

1.5.5 Example. We wish to write the following statement as a quantification schema:

> The points $(1, -1)$ and $(-2, 2)$ are equidistant from all points that are on the line described by the equation $y = x + 1$.

Using the theorem of Pythagoras, it is easy to see (and probably well known to the reader) that the distance between two given points (x_1, y_1) and (x_2, y_2) in an xy-coordinate system is

$$\sqrt{(x_2 - x_1)^2 + (y_2 - y_1)^2}.$$

Thus the statement above is equivalent to

$$\forall_{x,y \in \mathbb{R}} \, y = x + 1 \Rightarrow \sqrt{(x-1)^2 + (y+1)^2} = \sqrt{(x+2)^2 + (y-2)^2}.$$

1.5.6 Example. Using standard optimization techniques (as explained, for instance, in [B], Chapter 12) or elementary vector algebra, it is not difficult to establish the following geometric proposition:

> The shortest distance between a point (x_0, y_0) and a line given by the equation $y = mx + b$ is
>
> $$D(x_0, y_0, m, b) := \frac{|mx_0 + b - y_0|}{\sqrt{1 + m^2}}.$$

Implicit in this statement there are two separate assertions:

(1) There is a point (x, y) on the line given by the equation $y = mx + b$ that has distance $D(x_0, y_0, m, b)$ from (x_0, y_0).

(2) No point on the line in question has a distance from (x_0, y_0) that is less than $D(x_0, y_0, m, b)$. In other words, all points on the line have a distance from (x_0, y_0) that is greater than or equal to $D(x_0, y_0, m, b)$.

Since (1) is equivalent to

$$\exists_{x,y \in \mathbb{R}} \left(y = mx + b \wedge \sqrt{(x - x_0)^2 + (y - y_0)^2} = D(x_0, y_0, m, b) \right),$$

and since (2) is equivalent to

$$\forall_{x,y \in \mathbb{R}} \left(y = mx + b \Rightarrow \sqrt{(x - x_0)^2 + (y - y_0)^2} \geq D(x_0, y_0, m, b) \right),$$

it follows that the statement above is equivalent to

$$\exists_{x,y \in \mathbb{R}} \left(y = mx + b \wedge E(x, y) \right) \wedge \forall_{x,y \in \mathbb{R}} \left(y = mx + b \Rightarrow G(x, y) \right),$$

where
$$E(x, y) :\Leftrightarrow \sqrt{(x - x_0)^2 + (y - y_0)^2} = D(x_0, y_0, m, b)$$
and
$$G(x, y) :\Leftrightarrow \sqrt{(x - x_0)^2 + (y - y_0)^2} \geq D(x_0, y_0, m, b).$$

To continue this discussion, let us consider the additional statement that there is *exactly* one point on the line given by the equation $y = mx + b$ the distance of which from (x_0, y_0) is equal to $D(x_0, y_0, m, b)$. Generally speaking, a given object is the unique possessor of a certain property if and only if all other objects that also possess this property are identical with this one given object. Thus, in setting

$$P(x, y) :\Leftrightarrow y = mx + b \wedge E(x, y),$$

the uniqueness claim presently in question can be written in the form

$$\exists_{x,y \in \mathbb{R}} \left(P(x, y) \wedge \forall_{s,t \in \mathbb{R}} \left(P(s, t) \Rightarrow (x = s \wedge y = t) \right) \right) \qquad (1.15)$$

or, equivalently, in the form

$$\exists_{x,y \in \mathbb{R}} P(x, y) \wedge \forall_{s,t \in \mathbb{R}} \forall_{v,w \in \mathbb{R}} \left((P(s, t) \wedge P(v, w)) \Rightarrow (s = v \wedge t = w) \right). \qquad (1.16)$$

In order to prove that (1.15) and (1.16) are indeed equivalent, we will show that (1.15) $\Rightarrow$ (1.16) and (1.16) $\Rightarrow$ (1.15) (because, by Proposition 1.4.5j, $P \Leftrightarrow Q$ is equivalent to $(P \Rightarrow Q) \wedge (Q \Rightarrow P)$). To establish the first of these two conditionals, we assume that (1.15) is true, that is, we assume that there are numbers $x, y \in \mathbb{R}$ such that

$$P(x, y) \wedge \forall_{s,t \in \mathbb{R}} \left(P(s, t) \Rightarrow (x = s \wedge y = t) \right). \qquad (1.17)$$

Given this assumption, we obviously only need to show that

$$\forall_{s,t \in \mathbb{R}} \forall_{v,w \in \mathbb{R}} \left((P(s, t) \wedge P(v, w)) \Rightarrow (s = v \wedge t = w) \right).$$

So let $s, t, v, w \in \mathbb{R}$ such that

$$P(s, t) \wedge P(v, w)$$

is true. Then (1.17) implies that $x = s$ and $y = t$ and also that $x = v$ and $y = w$. Consequently, $s = v$ and $t = w$, as desired.

Assuming now conversely that (1.16) is true, it follows that there are $x, y \in \mathbb{R}$ such that $P(x, y)$ is true. Thus we only need to show that

$$\forall_{s,t \in \mathbb{R}} \left(P(s, t) \Rightarrow (x = s \wedge y = t) \right).$$

So let $s, t \in \mathbb{R}$ such that $P(s, t)$ is true. Then $P(x, y) \wedge P(s, t)$ is true as well, and the assumption

$$\forall_{s,t \in \mathbb{R}} \forall_{v,w \in \mathbb{R}} \left((P(s, t) \wedge P(v, w)) \Rightarrow (s = v \wedge t = w) \right)$$

in (1.16) therefore implies that $x = s$ and $y = t$, as desired.

Having thus established the equivalence of (1.15) and (1.16), we can further inquire whether this equivalence is purely schematic in kind or wether instead it is also dependent upon the specific form of the statements $x = s \wedge y = t$ and $s = v \wedge t = w$ that form the consequents in (1.15) and (1.16). To test the former hypothesis, we define

$$Q(x, y, s, t) :\Leftrightarrow x = s \wedge y = t.$$

Given this definition, (1.15) assumes the form

$$\exists_{x,y \in \mathbb{R}} \left(P(x, y) \wedge \forall_{s,t \in \mathbb{R}} \left(P(s, t) \Rightarrow Q(x, y, s, t) \right) \right) \tag{1.18}$$

and (1.16) assumes the form

$$\exists_{x,y \in \mathbb{R}} P(x, y) \wedge \forall_{s,t \in \mathbb{R}} \forall_{v,w \in \mathbb{R}} \left((P(s, t) \wedge P(v, w)) \Rightarrow Q(s, t, v, w) \right). \tag{1.19}$$

If these statements were equivalent in a purely schematic sense, then they would be equivalent as well if

$$P(x, y) :\Leftrightarrow x \geq 0 \wedge y \geq 0$$

and

$$Q(x, y, s, t) :\Leftrightarrow s \geq x \wedge t \geq y.$$

However, this is not the case because with these definitions for P and Q, the statement

$$P(x, y) \wedge \forall_{s,t \in \mathbb{R}} \left(P(s, t) \Rightarrow Q(x, y, s, t) \right)$$

is true for $x = 0$ and $y = 0$, but the statement

$$\forall_{s,t \in \mathbb{R}} \forall_{v,w \in \mathbb{R}} \left((P(s, t) \wedge P(v, w)) \Rightarrow Q(s, t, v, w) \right)$$

is false because $(P(1, 1) \wedge P(0, 0)) \Rightarrow Q(1, 1, 0, 0)$ is evidently false. In other words, (1.18) does not in general imply (1.19). Conversely, however, it is the case and can be shown quite easily that (1.19) implies (1.18) (see Exercise 1.5.15).

1.5.7 Example. We wish to write the following statememt—which is known as the *Well-Ordering Principle*—in quantified form:

Every nonempty subset of $\mathbb{N}$ has a minimal element.

Denoting the empty set by the symbol $\emptyset$, as is customary, this statement effectively asserts that any subset M of $\mathbb{N}$ (written as '$M \subset \mathbb{N}$') that satisfies the assumption $M \neq \emptyset$ contains an element that is less than or equal to all the elements contained in M. That is to say, the quantified rendering of the statement above is

$$\forall_{M \subset \mathbb{N}} M \neq \emptyset \Rightarrow \exists_{n \in M} \forall_{m \in M} n \leq m.$$

1.5.8 Example. Consider the following statement concerning a given subset S of, say, the real numbers $\mathbb{R}$ (i.e., $S \subset \mathbb{R}$):

> S has a minimal element (i.e., a minimum) and also a maximal element (i.e., a maximum).

Given the characterization, in the preceding example, of a minimal element (or minimum) of a set of positive integers as a number contained in that set that is less than or equal to all other numbers in that set, we readily find that the quantification schema expressing the statement above is

$$(\exists_{y \in S} \forall_{x \in S} \, y \le x) \wedge (\exists_{y \in S} \forall_{x \in S} \, y \ge x). \tag{1.20}$$

For clarity we wish to point out that the use of the variable y for *both* the minimum *and* the maximum of S does not at all imply that the minimum and the maximum of S are equal. Both of the existence statements that make up the conjunction (1.20) are perfectly self-contained and therefore disjoint from each other. The first existence statement says that S has a minimum and the second that S has a maximum and that's it. If we wish, we can of course also write

$$(\exists_{y \in S} \forall_{x \in S} \, y \le x) \wedge (\exists_{z \in S} \forall_{x \in S} \, z \ge x),$$

but the use of z instead of y in the second existence statement is purely a matter of taste. It doesn't alter the content of the statement in the least and is therefore strictly speaking irrelevant.

For the final set of examples in this section we will assume that the reader is familiar with the elementary notion of a real-valued function that is defined on an interval in $\mathbb{R}$. Readers who are not familiar with this notion may skip ahead to the next section.

1.5.9 Example. We wish to find a function f that is defined on the interval $[0, 1]$, takes values in $[0, 1]$, and satisfies the following statement:

$$\forall_{y \in [0,1]} \exists_{x \in [0,1]} \, f(x) = y. \tag{1.21}$$

What does this statement tell us about f? Well, it says that every potential output value $y \in [0, 1]$ is in fact an actual output value, that is, it is equal to $f(x)$ for some x in the domain $[0, 1]$. Probably the simplest function that satisfies this condition is the linear straight-line function $f(x) := x$, because if $f(x) = x$ for all $x \in [0, 1]$, then for any given value $y \in [0, 1]$, we simply set $x := y$ and find that $f(x) = x = y$, as desired.

1.5.10 Example. We wish to find a function f which—as in Example 1.5.9—is defined on the interval $[0, 1]$ and takes values in $[0, 1]$, but which instead of (1.21), satisfies the following alternative statement:

$$\exists_{x_1, x_2 \in [0,1]} \, f(x_1) = f(x_2) \wedge x_1 \ne x_2. \tag{1.22}$$

The function $f(x) = x$ from Example 1.5.9 does not satisfy this condition because if $x_1 \neq x_2$ and $f(x) = x$, then $f(x_1) = x_1 \neq x_2 = f(x_2)$. So for this particular function f, the conditions $f(x_1) = f(x_2)$ and $x_1 \neq x_2$ are mutually exclusive. However, if we switch from a straight-line function to a parabola, the statement above can very easily be satisfied: if

$$f(x) := x(1 - x),$$

then $f(x) \in [0, 1]$ for all $x \in [0, 1]$ and for $x_1 := 0$ and $x_2 := 1$ we have $f(x_1) = 0 = f(x_2)$. In fact, if we set $x_2 := 1 - x_1$ for any given value $x_1 \in [0, 1]$ that is different from $1/2$, then $x_1 \neq x_2$ and $f(x_2) = x_2(1 - x_2) = (1 - x_1)x_1 = f(x_1)$. Consistent with these observations, the graph of f is shown in Figure 1.1.

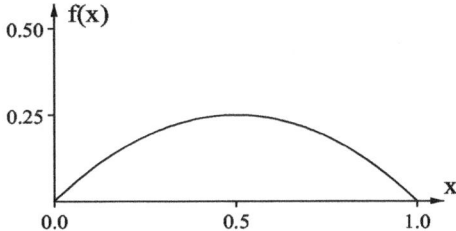

Figure 1.1: graph of $f(x) = x(1 - x)$.

Remark. In solving the problem posed in the next example we will find it helpful that we solved Example 1.5.10 above the way we solved it, by means of the function $f(x) = x(1 - x)$. However, if we had been interested only in Example 1.5.10 and not in working also Example 1.5.11 afterwards, we could have solved the problem in the former example by means of a function that is much simpler than the one we actually used. Thus the purpose of Exercise 1.5.19 below is precisely to find that simpler alternative function.

1.5.11 Example. We wish to find a function f defined on $[0, 1]$ that satisfies the properties (1.21) and (1.22) in conjunction. That is to say, the property that f is required to possess is given by the following statement:

$$\left(\forall_{y \in [0,1]} \exists_{x \in [0,1]} f(x) = y\right) \wedge \left(\exists_{x_1, x_2 \in [0,1]} f(x_1) = f(x_2) \wedge x_1 \neq x_2\right).$$

The function $f(x) = x(1 - x)$ from Example 1.5.10 doesn't work because it doesn't reach all the way up to 1. That is to say, it doesn't satisfy property (1.21) because there are values $y \in [0, 1]$ for which there are no values $x \in [0, 1]$ with $f(x) = y$. In fact, in completing the square, we notice that

$$f(x) = x(1 - x) = x - x^2 = \frac{1}{4} - \left(x - \frac{1}{2}\right)^2$$

and that, by implication, the maximal value of f on $[0,1]$ is $f(1/2) = 1/4$. Consequently, if we multiply f by the factor 4, then the maximal value of f on $[0,1]$ will be 1, and property (1.21) will be satisfied. So a function that does the trick is

$$f(x) := 4x(1-x).$$

This function satisfies (1.22) because we still have $f(0) = 0 = f(1)$ (as in Example 1.5.10), and it also satisfies (1.21) because if y is an arbitrary value in $[0,1]$, then, in solving the quadratic equation $y = x(1-x)$ for x, we readily find that $x = (1 \pm \sqrt{1-y})/2$, and therefore,

$$f(x) = f\left((1 \pm \sqrt{1-y})/2\right) = y.$$

Notice also that $x = (1 \pm \sqrt{1-y})/2$ is contained in $[0,1]$, as required in (1.21), because if $y \in [0,1]$, then $1 \pm \sqrt{1-y} \in [0,2]$, and hence $x \in [0,1]$.

Exercises

1.5.12. Show that the schema $(P \Rightarrow Q) \Rightarrow ((P \wedge Q) \Leftrightarrow P)$ is a tautology and explain how this observation is relevant to the reduction of the universe from $\mathbb{R}$ to $(1, \infty)$ in Example 1.5.2.

1.5.13. Write each of the following sentences as a quantification schema:

a) If no animal had four legs, an animal's furlessness would always cause it to be two-legged and long-nosed.

b) For an animal to have four legs it is necessary that it also has fur, and furless animals cannot exist in absence of four-legged animals with fur.

c) All four-legged animals either have fur or a long nose, and some two-legged animals have a long nose but no fur.

d) Among the animals with a long nose there are some that have fur and others that don't, but no animal with a long nose will ever be two-legged.

1.5.14. Would it be okay to leave out the parentheses in statement (1.12)? Explain your answer.

1.5.15. Explain why (1.19) implies (1.18).

1.5.16. Write each of the following statements as a quantification schema:

a) The shortest distance from the circle given by the equation $(x-2)^2 + y^2 = 1$ to the origin $(0,0)$ is 1.

b) If (x_1, y_1) and (x_2, y_2) are contained in the disc of radius 1 centered at the origin $(0,0)$ of an xy-plane, then the distance between (x_1, y_1) and (x_2, y_2) must be less than or equal to 2.

c) 60 is the least common multiple of 12 and 10.

d) If m and n are positive integers (i.e., $m, n \in \mathbb{N}$), then m and n have a least common multiple (in $\mathbb{N}$).

e) The line given by the equation $y = -(x - 1/\sqrt{2}) + 1/\sqrt{2}$ is tangent to the circle described by the equation $x^2 + y^2 = 1$.

f) p is a prime number (i.e., p is an integer greater than one that is divisible only by one and by itself).

1.5.17. Let $P(n) :\Leftrightarrow \forall_{m \in \mathbb{N}}((\exists_{j \in \mathbb{N}}\ jm = n) \Rightarrow (m = 1 \vee m = n))$. Write each of the statements listed below as an English sentence. The sentence should be as brief and as poignant as possible. Note: the first statement is false, the second true, and the truth value of the third is currently still unknown.

a) $\forall_{n \in \mathbb{N}} ((\exists_{k \in \mathbb{N}}\ n = 2k + 1) \Rightarrow P(n))$

b) $\forall_{n \in \mathbb{N}} (n > 2 \Rightarrow (\forall_{x \in \mathbb{N}}\forall_{y \in \mathbb{N}}\forall_{z \in \mathbb{N}} z^n \neq y^n + x^n))$

c) $\forall_{n \in \mathbb{N}}((\exists_{k \in \mathbb{N}}(k > 1 \wedge n = 2k)) \Rightarrow \exists_{p,q \in \mathbb{N}}(p > 1 \wedge q > 1 \wedge n = p + q \wedge P(p) \wedge P(q)))$

1.5.18. Write each of the following statements in quantified form:

a) Any set $S \subset \mathbb{R}$ that has a minimum and a maximum contains exactly one element if and only if its minimum is equal to its maximum.

b) Let S be a nonempty subset of $\mathbb{R}$. If there is a number $M \in \mathbb{R}$ such that $M \geq s$ for all $s \in S$ then the set of all real numbers that are greater than or equal to all elements in S contains a minimal element.

1.5.19. Find the function referred to in the remark on p.27.

1.5.20. For each of the statements given below draw the graph of a function f that is defined on $[0, 1]$, takes values in $[0, 1]$, and satisfies the property that the statement specifies.

a) $(\exists_{y \in [0,1]} \forall_{x \in [0,1]}\ f(x) \neq y) \wedge \forall_{x_1, x_2 \in [0,1]} (f(x_1) = f(x_2) \Rightarrow x_1 = x_2)$,

b) $(\forall_{x \in [0,1]} \exists_{x_0 \in [0,1]}\ f(x) < f(x_0)) \wedge (\forall_{x \in [0,1]} \exists_{x_0 \in [0,1]}\ f(x) > f(x_0))$,

c) $(\forall_{y \in [0,1]} \exists_{x \in [0,1]}\ f(x) = y) \wedge \forall_{x_1, x_2 \in [0,1]} (x_1 < x_2 \Rightarrow f(x_1) > f(x_2))$,

d) $\exists_{y \in [0,1]} ((\forall_{x \in [0,1]}\ f(x) \neq y) \wedge (\forall_{z \in [0,1]} (y \neq z \Rightarrow \exists_{x \in [0,1]}\ f(x) = z)))$.

1.5.21. ★ Sketch the graph of a function f that is defined on $[-1, 1]$, takes values in $[0, 1]$ and satisfies the following property:

$$\forall_{y \in [0,1]} \exists_{x_1, x_2 \in [-1,1]} (f(x_1) = f(x_2) = y \wedge x_1 \neq x_2 \wedge P(f, x_1, x_2, y)),$$

where

$$P(f, x_1, x_2, y) :\Leftrightarrow \forall_{x \in [-1,1]} (f(x) = y \Rightarrow (x = x_1 \vee x = x_2))$$

1.6   A Case Study: Sequences

To continue our discussion of quantification schemata, we will now take a somewhat closer look at the notion of an infinite *sequence* because this notion will
be important to us in constructing the reals from the rationals in Section 6.3.
Strictly speaking, a real or rational sequence is a function from $\mathbb{N}$ to $\mathbb{R}$ or $\mathbb{Q}$,
but since the notion of a function (which probably is well familiar to most
readers) has not as yet been rigorously introduced, we will think of a sequence
more informally as an infinite ordered listing of real or rational numbers a_n:

$$(a_n)_{n\in\mathbb{N}} = (a_1, a_2, a_3, \dots).$$

1.6.1 Example. If $a_n = 1/n$ for all $n \in \mathbb{N}$, then

$$(a_n)_{n\in\mathbb{N}} = (1, 1/2, 1/3, 1/4, \dots),$$

1.6.2 Example. If $b_n = 1 - 1/n$ for all $n \in \mathbb{N}$, then

$$(b_n)_{n\in\mathbb{N}} = (0, 1/2, 2/3, 3/4, \dots).$$

1.6.3 Example. If $c_n = (-1)^n$ for all $n \in \mathbb{N}$, then

$$(c_n)_{n\in\mathbb{N}} = (-1, 1, -1, 1, \dots).$$

As we examine the values a_n and b_n, in Examples 1.6.1 and 1.6.2, for
increasingly large values of n, we readily notice that a_n approaches zero and
that b_n approaches one. Using standard calculus lingo, we therefore say that
the *limit* of the sequence $(a_n)_{n\in\mathbb{N}}$ is equal to zero and that the limit of $(b_n)_{n\in\mathbb{N}}$
is equal to one. Thus we write

$$\lim_{n\to\infty} a_n = \lim_{n\to\infty} \frac{1}{n} = 0$$

and

$$\lim_{n\to\infty} b_n = \lim_{n\to\infty} \left(1 - \frac{1}{n}\right) = 1.$$

Moreover, we say that the limit of the sequence $(c_n)_{n\in\mathbb{N}}$ in Example 1.6.3 does
not exist because, as n tends to infinity, the values $c_n = (-1)^n$ continue to
alternate between -1 and 1, and there is thus no single limiting value that
they eventually assume or approach.

Given these simple observations, there arises the natural question as to
what exactly it means to say—in general—that the limit of a sequence $(a_n)_{n\in\mathbb{N}}$
exists and is equal to L for some $L \in \mathbb{R}$. Historically, it took a long time for
this question to fully emerge to sight and be answered properly. In fact, it took
the better part of two centuries, from the creation of the calculus in the second
half of the 17th century, to the 'arithmetization' of the calculus in the 19th, for

this question to be adequately answered. But from our contemporary vantage point—with the benefit of hindsight—the matter can be sorted out quite easily. For to say that an infinite sequence of values $a_1, a_2, a_3, \ldots$ approaches a value L simply means that any open interval centered at L—no matter how small— always contains all but finitely many of the sequence values a_n. That is to say, for any such interval I, we can always find an integer $N \in \mathbb{N}$ such that $a_n \in I$ for all $n > N$. The first one million values a_n may not be anywhere near the interval I, but once the index n has grown sufficiently large, the values a_n will all—without exception—be contained in I. Furthermore, since every open interval centered at L is of the form $(L - \varepsilon, L + \varepsilon)$ for some positive real number ε, it follows that the assertion that the limit of the sequence $(a_n)_{n \in \mathbb{N}}$ exists and is equal to L is equivalent to the following quantification schema:

$$\forall_{\varepsilon>0} \exists_{N \in \mathbb{N}} \forall_{n \in \mathbb{N}} \, (n > N \Rightarrow a_n \in (L - \varepsilon, L + \varepsilon)).$$

Finally, since the condition $a_n \in (L-\varepsilon, L+\varepsilon)$ is equivalent to the assertion that the inequality $|a_n - L| < \varepsilon$ be satisfied, we arrive at the following definition:

1.6.4 Definition. A sequence $(a_n)_{n \in \mathbb{N}}$ of real or rational numbers is said to be *convergent* if

$$\exists_{L \in \mathbb{R}} \forall_{\varepsilon>0} \exists_{N \in \mathbb{N}} \forall_{n \in \mathbb{N}} \, (n > N \Rightarrow |a_n - L| < \varepsilon). \tag{1.23}$$

Furthermore, if $(a_n)_{n \in \mathbb{N}}$ is convergent and L is a number in $\mathbb{R}$ such that

$$\forall_{\varepsilon>0} \exists_{N \in \mathbb{N}} \forall_{n \in \mathbb{N}} \, (n > N \Rightarrow |a_n - L| < \varepsilon), \tag{1.24}$$

then we say that L is the *limit* of $(a_n)_{n \in \mathbb{N}}$, and we write

$$\lim_{n \to \infty} a_n = L.$$

Since our main concern at present is the study of quantification schemata, we will focus our attention in this section on the logical structure of the defining statement (1.24) and not on the mathematical properties of convergent sequences that can be derived from it. But in reference to the phrase "L is *the* limit of $(a_n)_{n \in \mathbb{N}}$," which we used in Definition 1.6.4 above, we need to point out that, in speaking of "*the* limit" of a certain sequence, we are strongly suggesting that the limit of a sequence—if it exists—is unique. In fact, if a sequence could have more than one limit, the notation $\lim_{n \to \infty} a_n$ would simply be meaningless because we wouldn't know, in any particular example, which of the various limits of $(a_n)_{n \in \mathbb{N}}$ this limit notation is supposed to refer to. A formal proof of the uniqueness of limits will be provided in Section 2.3. Furthermore, in order to give the reader an initial impression of how Definition 1.6.4 allows us to establish the convergence of a given sequence by means of a concise, rigorous proof, we will consider the following example:

1.6.5 Example. We wish to show that the sequence

$$a_n := \frac{\sqrt{n}}{\sqrt{n}+2}$$

converges to $L = 1$ as n tends to infinity. Since

$$|a_n - 1| = \frac{2}{\sqrt{n}+2} < \frac{2}{\sqrt{n}},$$

we may infer—in reference to (1.24)—that the inequality $|a_n - 1| < \varepsilon$, for any given $\varepsilon > 0$, is satisfied whenever $2/\sqrt{n} < \varepsilon$, or equivalently, whenever $n > 4/\varepsilon^2$. So if—for a given $\varepsilon > 0$—we choose N to be an integer that is larger than $4/\varepsilon^2$, then the assumption $n > N$ implies that $|a_n - 1| < \varepsilon$, and therefore,

$$\lim_{n\to\infty} \frac{\sqrt{n}}{\sqrt{n}+2} = 1,$$

as desired.

1.6.6 Example. To better understand the logical nature of statement (1.24), let us assume that we are given a real or rational sequence $(a_n)_{n\in\mathbb{N}}$ that satisfies the following variant of (1.24):

$$\exists_{N\in\mathbb{N}} \, \forall_{\varepsilon>0} \, \forall_{n\in\mathbb{N}} \, (n > N \Rightarrow |a_n - L| < \varepsilon). \tag{1.25}$$

How is this statement different from (1.24) and what does it tell us about the sequence $(a_n)_{n\in\mathbb{N}}$? Obviously, the formal difference between (1.24) and (1.25) is that the first two quantifiers are interchanged. But what does it mean for the sequence $(a_n)_{n\in\mathbb{N}}$ if there exists one $N \in \mathbb{N}$ so that *for all $\varepsilon > 0$ and all $n > N$ it is the case that $|a_n - L| < \varepsilon$? To answer this question, we simply pick an $n > N$ and inquire what it means to say that for this particular n and *for all $\varepsilon > 0$ we have $|a_n - L| < \varepsilon$. How can a non-negative number—the absolute value $|a_n - L|$—be smaller than all positive values ε? Clearly the answer is that $|a_n - L|$ must be equal to zero, or equivalently, that a_n must be equal to L. In other words, for all $n \in \mathbb{N}$ that are greater than N it must be the case that $a_n = L$. Consequently, the assertion made by statement (1.25) about the sequence $(a_n)_{n\in\mathbb{N}}$ is that this sequence is constant for sufficiently large indices n (i.e., indices greater than N). This observation shows that statement (1.25) is actually stronger than (1.24) because any sequence that satisfies (1.25), that is, any sequence that eventually assumes the constant value L, obviously converges to L, but not every sequence that converges to L is necessarily constant from some point onwards (after all, the sequences in Examples 1.6.1, 1.6.2, and 1.6.5 are all convergent but none of them is eventually constant).

To proceed, we will examine some further variants of statement (1.24), but instead of determining what types of sequences might satisfy these statements, we will consider only the special case where

$$a_n = 1/n$$

and

$$L = 0.$$

That is to say, we will consider variants of the statement

$$\forall_{\varepsilon > 0} \exists_{N \in \mathbb{N}} \forall_{n \in \mathbb{N}} \ (n > N \Rightarrow 1/n < \varepsilon), \tag{1.26}$$

which correctly asserts that $\lim_{n \to \infty} 1/n = 0$, and simply ask the question in each case whether the altered statement is true or false.

1.6.7 Example. The statement

$$\exists_{\varepsilon > 0} \exists_{N \in \mathbb{N}} \forall_{n \in \mathbb{N}} \ (n > N \Rightarrow 1/n < \varepsilon), \tag{1.27}$$

which differs from (1.26) only in the kind of the first quantifier is true because if we choose ε to be a number larger than 1 (e.g., $\varepsilon := 2$), then the consequent $1/n < \varepsilon$ is true for all $n \in \mathbb{N}$, and therefore the material conditional $n > N \Rightarrow 1/n < \varepsilon$ is true as well (because in general $P \Rightarrow Q$ is true whenever Q is true). Notice also that this argument is valid regardless of the value we choose for N. So the existence quantifier for N is in this case irrelevant.

1.6.8 Example. The statement

$$\forall_{\varepsilon > 0} \forall_{N \in \mathbb{N}} \exists_{n \in \mathbb{N}} \ (n > N \Rightarrow 1/n < \varepsilon)$$

is true because if we set $n := 1$, then the antecedent $n > N$ is false for all $N \in \mathbb{N}$, and therefore, the material conditional $n > N \Rightarrow 1/n < \varepsilon$ is true (because $P \Rightarrow Q$ is true whenever P is false). Furthermore and as a note at the side, the value of ε is irrelevant.

1.6.9 Example. The statement

$$\forall_{\varepsilon > 0} \forall_{N \in \mathbb{N}} \ ((\forall_{n \in \mathbb{N}} \ n > N) \Rightarrow (\forall_{n \in \mathbb{N}} \ 1/n < \varepsilon))$$

is true because the antecedent $\forall_{n \in \mathbb{N}} \ n > N$ is false for all $N \in \mathbb{N}$. After all, there is no positive integer N that is less than all positive integers n.

1.6.10 Example. Consider the statement

$$\exists_{\varepsilon > 0} \forall_{N \in \mathbb{N}} \ ((\forall_{n \in \mathbb{N}} \ n \geq N) \Rightarrow (\forall_{n \in \mathbb{N}} \ 1/n < \varepsilon)).$$

In this case the antecedent $\forall_{n \in \mathbb{N}} \ n \geq N$ is not false for all $N \in \mathbb{N}$ because it is true for $N = 1$. But if we set $\varepsilon := 2$, then the consequent $\forall_{n \in \mathbb{N}} \ 1/n < \varepsilon$ is true, and therefore, the statement as a whole is true as well.

Exercises

1.6.11. Use Definition 1.6.4 to prove that the sequence $a_n := (2n+3)/(n+1)$ converges to $L = 2$.

1.6.12. For what real or rational sequences $(a_n)_{n\in\mathbb{N}}$ is the following statement satisfied?

$$\forall_{N\in\mathbb{N}} \, \forall_{\varepsilon>0} \, \forall_{n\in\mathbb{N}} \, (n > N \Rightarrow |a_n - L| < \varepsilon).$$

Hint: the tempting answer that the only sequence that satisfies this statement is the constant sequence $a_n = L$ is almost correct but not quite.

1.6.13. Determine the truth value—true or false—of each of the statements listed below. Explain your answer in each case.

a) $\forall_{\varepsilon>0} \forall_{N\in\mathbb{N}} \forall_{n\in\mathbb{N}} \, (n > N \Rightarrow 1/n < \varepsilon)$,

b) $\exists_{N\in\mathbb{N}} \forall_{\varepsilon>0} \exists_{n\in\mathbb{N}} \, (n > N \Rightarrow 1/n < \varepsilon)$,

c) $\exists_{\varepsilon>0} \forall_{N\in\mathbb{N}} \forall_{n\in\mathbb{N}} \, (n > N \Rightarrow 1/n < \varepsilon)$,

d) $\forall_{\varepsilon>0} \forall_{n\in\mathbb{N}} \exists_{N\in\mathbb{N}} \, (n > N \Rightarrow 1/n < \varepsilon)$,

e) $\forall_{\varepsilon>0} \exists_{n\in\mathbb{N}} \forall_{N\in\mathbb{N}} \, (n > N \Rightarrow 1/n < \varepsilon)$,

f) $\exists_{n\in\mathbb{N}} \forall_{\varepsilon>0} \forall_{N\in\mathbb{N}} \, (n > N \Rightarrow 1/n < \varepsilon)$,

g) $\exists_{n\in\mathbb{N}} \forall_{\varepsilon>0} \forall_{N\in\mathbb{N}} \, (n \geq N \Rightarrow 1/n < \varepsilon)$,

h) $\forall_{n\in\mathbb{N}} \forall_{\varepsilon>0} \exists_{N\in\mathbb{N}} \, (n \geq N \Rightarrow 1/n < \varepsilon)$.

1.6.14. Determine the truth value of each of the following statements:

a) $\forall_{\varepsilon>0} \exists_{N\in\mathbb{N}} ((\forall_{n\in\mathbb{N}} \, n > N) \Rightarrow (\forall_{n\in\mathbb{N}} 1/n < \varepsilon))$,

b) $\forall_{\varepsilon>0} \forall_{N\in\mathbb{N}} ((\forall_{n\in\mathbb{N}} \, n \geq N) \Rightarrow (\exists_{n\in\mathbb{N}} 1/n < \varepsilon))$,

c) $\forall_{\varepsilon>0} \exists_{N\in\mathbb{N}} ((\forall_{n\in\mathbb{N}} \, n \geq N) \Rightarrow (\forall_{n\in\mathbb{N}} 1/n < \varepsilon))$,

d) $\forall_{\varepsilon>0} \forall_{N\in\mathbb{N}} ((\forall_{n\in\mathbb{N}} \, n \geq N) \Rightarrow (\forall_{n\in\mathbb{N}} 1/n < \varepsilon))$,

e) $\exists_{\varepsilon>0} \forall_{N\in\mathbb{N}} ((\forall_{n\in\mathbb{N}} \, n \geq N) \Rightarrow (\exists_{n\in\mathbb{N}} 1/n < \varepsilon))$,

f) $\exists_{\varepsilon>0} \exists_{N\in\mathbb{N}} ((\forall_{n\in\mathbb{N}} \, n \geq N) \Rightarrow (\forall_{n\in\mathbb{N}} 1/n < \varepsilon))$,

g) $\exists_{\varepsilon>0} \forall_{N\in\mathbb{N}} ((\forall_{n\in\mathbb{N}} \, n \geq N) \Rightarrow (\forall_{n\in\mathbb{N}} 1/n > \varepsilon))$,

h) $\exists_{\varepsilon>0} \exists_{N\in\mathbb{N}} ((\forall_{n\in\mathbb{N}} \, n \geq N) \Rightarrow (\forall_{n\in\mathbb{N}} 1/n > \varepsilon))$.

1.7 Negations, Orderings, and Passages

In Section 1.2, p.6, we examined the following statement:

For all real numbers x it is the case that $1/x < 1$ if $x > 1$.

Our purpose in Section 1.2 was to understand the definition of the material conditional and in order to do so we asked the question under what conditions the statement above would be false. And our intuitively plausible—and correct—answer was that the statement "would be false if it were possible to find some real number x that is greater than 1 but doesn't satisfy the inequality $1/x < 1$." In other words, the negation of

$$\forall_{x \in \mathbb{R}}\ x > 1 \Rightarrow 1/x < 1 \tag{1.28}$$

is

$$\exists_{x \in \mathbb{R}}\ x > 1 \land 1/x \geq 1. \tag{1.29}$$

Setting

$$S(x) :\Leftrightarrow (x > 1 \Rightarrow 1/x < 1),$$

we may use Proposition 1.4.5b,c to infer that

$$\neg S(x) \Leftrightarrow \neg(x \leq 1 \lor 1/x < 1) \Leftrightarrow (x > 1 \land 1/x \geq 1).$$

So the general negation pattern, suggested by (1.29) and (1.28), is

$$\boxed{\neg \forall_{x \in \mathcal{U}}\ S(x) \Leftrightarrow \exists_{x \in \mathcal{U}}\ \neg S(x).} \tag{1.30}$$

Clearly, this rule makes perfect sense. For to say that it is not the case that $S(x)$ is true for all $x \in \mathcal{U}$ is surely equivalent to the assertion that there exists some $x \in \mathcal{U}$ for which $S(x)$ is false, or equivalently, for which $\neg S(x)$ is true.

In order to derive the general rule for the negation of an existence statement of the form $\exists_{x \in \mathcal{U}}\ S(x)$ from (1.30), we argue as follows: since

$$(P \Leftrightarrow Q) \Leftrightarrow (\bar{P} \Leftrightarrow \bar{Q})$$

is easily seen to be a tautology (the reader may want to verify this claim by means of a truth-value analysis) and since $\neg \bar{R} \Leftrightarrow R$, it follows that (1.30) is equivalent to

$$\forall_{x \in \mathcal{U}}\ S(x) \Leftrightarrow \neg \exists_{x \in \mathcal{U}}\ \neg S(x).$$

Since this statement is true for all monadic term schemata $S(x)$, it is true as well for $\neg S(x)$. Thus, we may substitute $\neg S(x)$ for $S(x)$ and use the symmetry of the material biconditional (that is, the evident fact that $(P \Leftrightarrow Q) \Leftrightarrow (Q \Leftrightarrow P)$ is a tautology) to infer that

$$\boxed{\neg \exists_{x \in \mathcal{U}}\ S(x) \Leftrightarrow \forall_{x \in \mathcal{U}}\ \neg S(x).} \tag{1.31}$$

1.7.1 Example. In Example 1.6.3 we asserted that there is no limiting value L to which the sequence $c_n := (-1)^n$ converges. What exactly did we mean by making this claim? In order to answer this question, we need to recall that—according to (1.23)—the assertion that $(c_n)_{n\in\mathbb{N}}$ *is* convergent is equivalent to the following statement:

$$\exists_{L\in\mathbb{R}} \forall_{\varepsilon>0} \exists_{N\in\mathbb{N}} \forall_{n\in\mathbb{N}} (n > N \Rightarrow |c_n - L| < \varepsilon).$$

So in denoting this statement by P, we find that the claim that $(c_n)_{n\in\mathbb{N}}$ is *not* convergent can be equivalently transformed as follows:

$$\bar{P} \Leftrightarrow \neg\exists_{L\in\mathbb{R}} \forall_{\varepsilon>0} \exists_{N\in\mathbb{N}} \forall_{n\in\mathbb{N}} (n > N \Rightarrow |c_n - L| < \varepsilon)$$

$$\Leftrightarrow \forall_{L\in\mathbb{R}} \neg\forall_{\varepsilon>0} \exists_{N\in\mathbb{N}} \forall_{n\in\mathbb{N}} (n > N \Rightarrow |c_n - L| < \varepsilon) \quad \text{(by (1.31))}$$

$$\Leftrightarrow \forall_{L\in\mathbb{R}} \exists_{\varepsilon>0} \neg\exists_{N\in\mathbb{N}} \forall_{n\in\mathbb{N}} (n > N \Rightarrow |c_n - L| < \varepsilon) \quad \text{(by (1.30))}$$

$$\Leftrightarrow \forall_{L\in\mathbb{R}} \exists_{\varepsilon>0} \forall_{N\in\mathbb{N}} \neg\forall_{n\in\mathbb{N}} (n > N \Rightarrow |c_n - L| < \varepsilon) \quad \text{(by (1.31))}$$

$$\Leftrightarrow \forall_{L\in\mathbb{R}} \exists_{\varepsilon>0} \forall_{N\in\mathbb{N}} \exists_{n\in\mathbb{N}} \neg(n > N \Rightarrow |c_n - L| < \varepsilon) \quad \text{(by (1.30))}$$

$$\Leftrightarrow \forall_{L\in\mathbb{R}} \exists_{\varepsilon>0} \forall_{N\in\mathbb{N}} \exists_{n\in\mathbb{N}} \neg(n \leq N \vee |c_n - L| < \varepsilon) \quad \text{(by (1.8))}$$

$$\Leftrightarrow \forall_{L\in\mathbb{R}} \exists_{\varepsilon>0} \forall_{N\in\mathbb{N}} \exists_{n\in\mathbb{N}} (n > N \wedge |c_n - L| \geq \varepsilon) \quad \text{(by Proposition 1.4.5b).}$$

So in order to show that indeed the squence $c_n = (-1)^n$ is not convergent, we need to establish that

$$\forall_{L\in\mathbb{R}} \exists_{\varepsilon>0} \forall_{N\in\mathbb{N}} \exists_{n\in\mathbb{N}} (n > N \wedge |(-1)^n - L| \geq \varepsilon). \tag{1.32}$$

To do so, we observe to begin with that for any number $L \in \mathbb{R}$ it is the case that

$$|(-1)^n - L| = \begin{cases} |1 + |L|| = 1 + |L| > 1 & \text{if } n \text{ is even and } L < 0, \\ |-1 - L| = 1 + L \geq 1 & \text{if } n \text{ is odd and } L \geq 0. \end{cases}$$

Consequently, if L is an arbitrary real number and $\varepsilon := 1$, then for any given $N \in \mathbb{N}$ we may define

$$n := \begin{cases} 2N & \text{if } L < 0, \\ 2N + 1 & \text{if } L \geq 0 \end{cases}$$

and conclude that $n > N$ and $|(-1)^n - L| \geq 1 = \varepsilon$. This shows, as desired, that (1.32) is indeed valid.

Apart from rules concerning negations of quantifications, we also need to be aware of rules that pertain to changes in the order and placement quantifiers. For instance, in Example 1.6.6 we saw that the interchange of the first two quantifiers in (1.24) yields a more restrictive statement in the sense that any sequence that satisfies (1.25) also satisfies (1.24) but not vice versa. Consistent with this observation we state the following proposition:

1.7.2 Proposition. *For any dyadic term schema $S(x, y)$ and any universes $\mathcal{U}$ and $\mathcal{V}$, the following statements are valid:*

a) $\forall_{x\in\mathcal{U}} \forall_{y\in\mathcal{V}} S(x,y) \Leftrightarrow \forall_{y\in\mathcal{V}} \forall_{x\in\mathcal{U}} S(x,y)$,

b) $\exists_{x\in\mathcal{U}} \exists_{y\in\mathcal{V}} S(x,y) \Leftrightarrow \exists_{y\in\mathcal{V}} \exists_{x\in\mathcal{U}} S(x,y)$,

c) $\exists_{x\in\mathcal{U}} \forall_{y\in\mathcal{V}} S(x,y) \Rightarrow \forall_{y\in\mathcal{V}} \exists_{x\in\mathcal{U}} S(x,y)$.

Proof. The assertions a) and b) are elementary rules of thought that we accept to be true simply because they are self-evident. To establish c), we assume that the antecedent $\exists_{x\in\mathcal{U}} \forall_{y\in\mathcal{V}} S(x,y)$ is true. That is to say, we assume that there exists an element $x_0 \in \mathcal{U}$ such that $S(x_0,y)$ is true for all $y \in \mathcal{V}$, (Note: the index '0' is here attached to x in order to highlight the fact that x_0 is one specific element in $\mathcal{U}$.) Given this assumption, we need to show that for every $y \in \mathcal{V}$ there exists an element $x_y \in \mathcal{U}$ such that $S(x_y,y)$ is true. (Note: here we attach the index 'y' to x in order to emphasize that x may dependent on y.) So let y be an arbitrary element in $\mathcal{V}$. Setting $x_y := x_0$ (so that x_y is actually not dependent on y), it follows that $S(x_y,y)$ is true, as desired, because $S(x_0,y)$ is true by assumption. $\square$

Remark. If $\mathcal{U} = \mathcal{V}$, then we commonly write $\forall_{x,y\in\mathcal{U}}$ and $\exists_{x,y\in\mathcal{U}}$ instead of $\forall_{x\in\mathcal{U}} \forall_{y\in\mathcal{U}}$ and $\exists_{x\in\mathcal{U}} \exists_{y\in\mathcal{U}}$—and we already did so in various examples in Section 1.5. The formal justifications for these reductions in the number of the quantifiers are the commutation rules a) and b) in Proposition 1.7.2.

1.7.3 Example. In order to make fully apparent that c) in Proposition 1.7.2 is indeed only a one-way implication and not an equivalence (a fact that we already observed in the context of Example 1.6.6, as noted above), we set $S(x,y) :\Leftrightarrow x \leq y$ for all $x,y \in \mathbb{R}$. Then the statement $\forall_{y\in\mathbb{R}} \exists_{x\in\mathbb{R}} S(x,y)$ is true because for every $y \in \mathbb{R}$ there clearly exists an $x \in \mathbb{R}$ such that $x \leq y$ (simply set $x := y$). But the statement $\exists_{x\in\mathbb{R}} \forall_{y\in\mathbb{R}} S(x,y)$, by contrast, is false because there obviously does not exist any real number x that is less than or equal to all real numbers. Thus the statement $\forall_{y\in\mathcal{V}} \exists_{x\in\mathcal{U}} S(x,y)$ does not in general imply that $\exists_{x\in\mathcal{U}} \forall_{y\in\mathcal{V}} S(x,y)$.

To further explore the nature of quantification schemata, we consider again statement (1.12) from Section 1.5:

$$\forall_{x\in\mathbb{R}} \left(x \geq 0 \Rightarrow \exists_{y\in\mathbb{R}} (y \geq 0 \wedge x = y^2) \right). \tag{1.33}$$

A natural question that we may ask is whether this statement is equivalent to

$$\forall_{x\in\mathbb{R}} \exists_{y\in\mathbb{R}} \left(x \geq 0 \Rightarrow (y \geq 0 \wedge x = y^2) \right). \tag{1.34}$$

As it turns out, the affirmative answer to this question is provided by the following *rules of passage:*

1.7.4 Proposition. *If $\mathcal{U}$ is a nonempty universe, R a statement, and $P(x)$ a monadic term schema (i.e., a statement depending on x), then...*

a) $\forall_{x\in\mathcal{U}}(R\wedge P(x))$ *is equivalent to* $R\wedge\forall_{x\in\mathcal{U}}P(x)$,

b) $\forall_{x\in\mathcal{U}}(R\vee P(x))$ *is equivalent to* $R\vee\forall_{x\in\mathcal{U}}P(x)$,

c) $\exists_{x\in\mathcal{U}}(R\wedge P(x))$ *is equivalent to* $R\wedge\exists_{x\in\mathcal{U}}P(x)$,

d) $\exists_{x\in\mathcal{U}}(R\vee P(x))$ *is equivalent to* $R\vee\exists_{x\in\mathcal{U}}P(x)$,

e) $\forall_{x\in\mathcal{U}}(R\Rightarrow P(x))$ *is equivalent to* $R\Rightarrow\forall_{x\in\mathcal{U}}P(x)$,

f) $\exists_{x\in\mathcal{U}}(R\Rightarrow P(x))$ *is equivalent to* $R\Rightarrow\exists_{x\in\mathcal{U}}P(x)$,

g) $\forall_{x\in\mathcal{U}}(P(x)\Rightarrow R)$ *is equivalent to* $(\exists_{x\in\mathcal{U}}P(x))\Rightarrow R$,

h) $\exists_{x\in\mathcal{U}}(P(x)\Rightarrow R)$ *is equivalent to* $(\forall_{x\in\mathcal{U}}P(x))\Rightarrow R$.

Proof. In order to prove a), we observe that $\forall_{x\in\mathcal{U}}(R\wedge P(x))$ and $R\wedge\forall_{x\in\mathcal{U}}P(x)$ are both equivalent to $\forall_{x\in\mathcal{U}}P(x)$ whenever R is true. If R is false, then $R\wedge\forall_{x\in\mathcal{U}}P(x)$ is false, and since $\mathcal{U}$ is nonempty, $\forall_{x\in\mathcal{U}}(R\wedge P(x))$ is false as well (see the remark below). Thus $\forall_{x\in\mathcal{U}}(R\wedge P(x))$ and $R\wedge\forall_{x\in\mathcal{U}}P(x)$ are indeed equivalent whenever $\mathcal{U}$ is nonempty. Since the proofs of b), c), and d) are very similar to the proof of a), they will be left as exercises to the reader (see Exercise 1.7.10). Furthermore, statements e) and g) follow from b) and d) respectively as follows:

$$\forall_{x\in\mathcal{U}}(R\Rightarrow P(x))\leftrightarrow\forall_{x\in\mathcal{U}}(\bar{R}\vee P(x))\Leftrightarrow(\bar{R}\vee\forall_{x\in\mathcal{U}}P(x))\Leftrightarrow(R\Rightarrow\forall_{x\in\mathcal{U}}P(x))$$

and

$$\forall_{x\in\mathcal{U}}(P(x)\Rightarrow R)\Leftrightarrow\forall_{x\in\mathcal{U}}(\bar{P}(x)\vee R)\Leftrightarrow((\forall_{x\in\mathcal{U}}\bar{P}(x))\vee R)$$
$$\Leftrightarrow((\neg\exists_{x\in\mathcal{U}}P(x))\vee R)\Leftrightarrow((\exists_{x\in\mathcal{U}}P(x))\Rightarrow R).$$

The analogous derivations of f) and h) from b) and d) will be left as exercises to the reader (see Exercise 1.7.11). □

Remark. If $\mathcal{U}$ is empty, then statement a) of Proposition 1.7.4 is no longer valid because if R is false, then $R\wedge\forall_{x\in\mathcal{U}}P(x)$ is false but $\forall_{x\in\mathcal{U}}(R\wedge P(x))$ is true (as explained in the remark on p.20). Furthermore, statements d), f), and h) are no longer valid either, but the remaining statements b), c), e), and g) remain valid even if the universe $\mathcal{U}$ is empty (see Exercise 1.7.12).

Returning now to the question posed above, we observe that Proposition 1.7.4f implies that

$$x\geq 0\Rightarrow\exists_{y\in\mathbb{R}}(y\geq 0\wedge x=y^2)$$

is equivalent to

$$\exists_{y\in\mathbb{R}}\left(x\geq 0\Rightarrow(y\geq 0\wedge x=y^2)\right),$$

and therefore, (1.33) is indeed equivalent to (1.34).

1.7.5 Example. In order to see why, for example, in Proposition 1.7.4h the tempting statement, '$\exists_{x\in\mathcal{U}}(P(x) \Rightarrow R)$ is equivalent to $(\exists_{x\in\mathcal{U}}P(x)) \Rightarrow R$', is *not* correct, let us assume that R is false. Since, in general, $Q \Rightarrow F$ is equivalent to $\bar{Q}$, it follows that in this case $\exists_{x\in\mathcal{U}}(P(x) \Rightarrow R)$ is equivalent to $\exists_{x\in\mathcal{U}}\bar{P}(x)$ and $(\exists_{x\in\mathcal{U}}P(x)) \Rightarrow R$ is equivalent to $\neg\exists_{x\in\mathcal{U}}P(x)$ which in turn is equivalent to $\forall_{x\in\mathcal{U}}\bar{P}(x)$. Consequently, $\exists_{x\in\mathcal{U}}(P(x) \Rightarrow R)$ does not in general imply $\exists_{x\in\mathcal{U}}P(x) \Rightarrow R$ because $\exists_{x\in\mathcal{U}}\bar{P}(x)$ obviously does not in general imply $\forall_{x\in\mathcal{U}}\bar{P}(x)$.

1.7.6 Example. We wish to re-examine the statement

$$\forall_{\varepsilon>0}\forall_{N\in\mathbb{N}}\left((\forall_{n\in\mathbb{N}}\,n > N) \Rightarrow (\forall_{n\in\mathbb{N}}\,1/n < \varepsilon)\right)$$

from Example 1.6.9 in the light of Proposition 1.7.4. Using property g) with $R :\Leftrightarrow \forall_{n\in\mathbb{N}}\,1/n < \varepsilon$, it follows that this statement is equivalent to

$$\forall_{\varepsilon>0}\left((\exists_{N\in\mathbb{N}}\forall_{n\in\mathbb{N}}\,n > N) \Rightarrow (\forall_{n\in\mathbb{N}}\,1/n < \varepsilon)\right),$$

and using property e), this latter statement in turn is equivalent to

$$(\exists_{N\in\mathbb{N}}\forall_{n\in\mathbb{N}}\,n > N) \Rightarrow (\forall_{\varepsilon>0}\forall_{n\in\mathbb{N}}\,1/n < \varepsilon).$$

Thus we arrive at the same conclusion as in Example 1.6.9, namely that the statement as a whole is true, because the antecedent $\exists_{N\in\mathbb{N}}\forall_{n\in\mathbb{N}}\,n > N$ is obviously false.

In the next proposition below there is no reference to a universe $\mathcal{U}$ because all statements are valid regardless of whether the universe from which the elements x are drawn is empty or nonempty.

1.7.7 Proposition. *If $P(x)$ and $Q(x)$ are monadic term schemata, then...*

a) $\forall_x P(x) \wedge \forall_x Q(x)$ *is equivalent to* $\forall_x(P(x) \wedge Q(x))$,

b) $\forall_x P(x) \vee \forall_x Q(x)$ *implies* $\forall_x(P(x) \vee Q(x))$,

c) $\exists_x(P(x) \vee Q(x))$ *is equivalent to* $\exists_x P(x) \vee \exists_x Q(x)$,

d) $\exists_x(P(x) \wedge Q(x))$ *implies* $\exists_x P(x) \wedge \exists_x Q(x)$,

e) $\forall_x P(x) \Rightarrow \exists_x Q(x)$ *is equivalent to* $\exists_x(P(x) \Rightarrow Q(x))$,

f) $\exists_x P(x) \Rightarrow \forall_x Q(x)$ *implies* $\forall_x(P(x) \Rightarrow Q(x))$,

g) $\forall_x(P(x) \Leftrightarrow Q(x))$ *implies* $\forall_x P(x) \Leftrightarrow \forall_x Q(x)$,

h) $\forall_x(P(x) \Leftrightarrow Q(x))$ *implies* $\exists_x P(x) \Leftrightarrow \exists_x Q(x)$.

Proof. Statement a) is self-evident because $P(x)$ and $Q(x)$ are true for all x if and only if $P(x)$ is true for all x and $Q(x)$ is true for all x. Concerning b), we observe that the assumption $\forall_x P(x) \vee \forall_x Q(x)$ implies that for any given x either $P(x)$ or $Q(x)$ is true. To prove c) we argue as follows:

$$\exists_x (P(x) \vee Q(x)) \Leftrightarrow \neg \forall_x (\bar{P}(x) \wedge \bar{Q}(x)) \Leftrightarrow \neg(\forall_x \bar{P}(x) \wedge \forall_x \bar{Q}(x)) \quad \text{(by a))}$$
$$\Leftrightarrow \neg \forall_x \bar{P}(x) \vee \neg \forall_x \bar{Q}(x) \Leftrightarrow \exists_x P(x) \vee \exists_x Q(x).$$

The proofs of d) and e) are left as exercises to the reader (see Exercise 1.7.17), and to establish f), we observe that

$$\exists_x P(x) \Rightarrow \forall_x Q(x) \Leftrightarrow \neg (\exists_x P(x)) \vee \forall_x Q(x) \Leftrightarrow \forall_x \bar{P}(x) \vee \forall_x Q(x)$$
$$\Rightarrow \forall_x (\bar{P}(x) \vee Q(x)) \quad \text{(by b))}$$
$$\Leftrightarrow \forall_x (P(x) \Rightarrow Q(x)),$$

as desired. The proof of g) is left as an exercise to the reader (see Exercise 1.7.17), and concerning h) we notice that the assumption $\forall_x (P(x) \Leftrightarrow Q(x))$ is equivalent to the assertion that the truth values of $P(x)$ and $Q(x)$ are the same for all x. Consequently, the truth values of $\exists_x P(x)$ and $\exists_x Q(x)$ must be equal as well, and therefore, $\exists_x P(x) \Leftrightarrow \exists_x Q(x)$ must be true. $\qquad \square$

1.7.8 Example. We wish to show that the reverse implication in Proposition 1.7.7b is in general not valid. To do so, we only need to provide one specific counterexample, and to this end in turn we set $P(x) :\Leftrightarrow x \geq 0$ and $Q(x) :\Leftrightarrow x < 0$ for all $x \in \mathbb{R}$. Then $\forall_{x \in \mathbb{R}}(P(x) \vee Q(x))$ is true, and $\forall_x P(x) \vee \forall_x Q(x)$ is false. Therefore, the former statement does not in general imply the latter.

Exercises

1.7.9. Negate statements a), b), and g) in both Exercise 1.6.13 and Exercise 1.6.14.

1.7.10. Prove Proposition 1.7.4b,c,d.

1.7.11. Use d) in Proposition 1.7.4 to prove f) and h).

1.7.12. Explain why statements d), f), and h) in Proposition 1.7.4 are not valid if the universe $\mathcal{U}$ is empty and explain as well why statements b), c), e), and g) remain valid regardless.

1.7.13. Discuss how the statement $\forall_{x \in \mathcal{U}}(R \Leftrightarrow P(x))$ is related to the statement $R \Leftrightarrow \forall_{x \in \mathcal{U}} P(x)$.

1.7.14. Discuss how the statement $\exists_{x \in \mathcal{U}}(R \Leftrightarrow P(x))$ is related to the statement $R \Leftrightarrow \exists_{x \in \mathcal{U}} P(x)$.

1.7.15. Show that $(\exists_{x \in \mathcal{U}} P(x)) \Rightarrow R$ implies $\exists_{x \in \mathcal{U}} (P(x) \Rightarrow R)$ in any nonempty universe $\mathcal{U}$.

1.7.16. Explain, in reference to Proposition 1.7.4g, why the statement, '$\forall_{x \in \mathcal{U}} (P(x) \Rightarrow R)$ is equivalent to $(\forall_{x \in \mathcal{U}} P(x)) \Rightarrow R$', is not correct.

1.7.17. Prove Proposition 1.7.7d,e,g.

1.7.18. Show that the reverse implications in Proposition 1.7.7d,f,g,h are not valid.

1.7.19. Consider the following statements involving two monadic term schemata $P(x)$ and $Q(x)$:

(1) $\forall_x (P(x) \Rightarrow \forall_y Q(y))$

(2) $\forall_x (P(x) \Rightarrow Q(x))$

(3) $\forall_x P(x) \Rightarrow \forall_y Q(y)$

Prove that...

   a) (1) implies (2),

   b) (2) does not imply (1),

   c) (2) implies (3),

   d) (3) does not imply (2),

   e) (1) implies (3),

   f) (3) does not imply (1).

1.7.20. Prove that $\exists_x (P(x) \Leftrightarrow Q(x))$ does not imply and is not implied by $\exists_x P(x) \Leftrightarrow \exists_x Q(x)$.

Chapter 2

Proofs

2.1 Proofs by Contradiction

If we wish to establish, for example, a categorical assertion of the form

$$\forall_x (P(x) \Rightarrow Q(x)), \tag{2.1}$$

it is frequently useful to employ a *proof by contradiction*. To do so, we assume that (2.1) is false and then attempt to show that this assumption yields a contradiction. That is to say, in setting $R :\Leftrightarrow \exists_x (P(x) \wedge \bar{Q}(x))$ (because $\neg\forall_x (P(x) \Rightarrow Q(x)) \Leftrightarrow \exists_x \neg(\bar{P}(x) \vee Q(x)) \Leftrightarrow \exists_x (P(x) \wedge \bar{Q}(x))$) we try to demonstrate that the assumption that R is true implies a false statement. The rigorous logical justification for this approach is the fact that a material conditional with a false consequent is equivalent to the negation of the antecedent. In other words, if $R \Rightarrow F$ is true then R must be false and $\bar{R}$ must be true, as desired (because $\bar{R}$ is equivalent to (2.1)).

2.1.1 Example. We wish to prove the following statement:

$$\sqrt{2} \text{ is not a rational number.} \tag{2.2}$$

In order to link this statement to (2.1), we set

$$P(x) :\Leftrightarrow \exists_{n,m \in \mathbb{N}} \, x = \frac{m}{n}$$

and

$$Q(x) :\Leftrightarrow x \neq \sqrt{2}.$$

Since $P(x)$ is equivalent to the assertion that x is a positive rational number, it follows that (2.2) is equivalent to

$$\forall_{x \in \mathbb{R}} \, P(x) \Rightarrow Q(x).$$

Concerning the question as to why it might be advantageous to prove this statement by way of contradiction, we wish to point out that the negation $\exists_{x \in \mathbb{R}} P(x) \wedge \bar{Q}(x)$ is more accessible and tangible a statement than the original universal assertion that all positive rational numbers are different from $\sqrt{2}$. For if we assume that there exists an $x \in \mathbb{R}$ such that $P(x)$ and $\bar{Q}(x)$ are true—or equivalently, that $m/n = x = \sqrt{2}$ for some $m, n \in \mathbb{N}$—then we actually have something concrete to work with, namely the equation $m/n = \sqrt{2}$. As a matter of course, there cannot exist a general rule that tells us unfailingly when to use a proof by contradiction and when to use a proof that is direct. After all and as we said before (p.21), mathematics is a creative activity of the mind and is, as such, inherently free and full of surprises. On the other hand, though, there also are present in mathematical thought certain recurrent themes and elements that experience acquaints us with and that we can learn to use effectively with patience and practice. That said, we now return to our proof-by-contradiction assumption:

$$\frac{m}{n} = \sqrt{2}.$$

Since any fraction (of natural numbers) can be reduced so that the only common divisor of the numerator and the denominator is 1, we may infer that there are integers $p, q \in \mathbb{N}$ such that

$$\gcd(p, q) = 1,$$

and

$$\frac{p}{q} = \frac{m}{n} = \sqrt{2}.$$

Rewriting this latter equation, we find that $p^2 = 2q^2$ and that, by implication, p^2 is even. Consequently, p is even as well, because if p were odd, then so would be p^2. Hence there is a $k \in \mathbb{N}$ such that $p = 2k$, and therefore, $4k^2 = p^2 = 2q^2$. Thus q^2 is even and so is q by implication. This shows that 2 divides both p and q and that therefore the greatest common divisor of p and q must be greater than one. In other words, the assumption that (2.2) is false has led us to correctly infer the false conclusion that there are integers $p, q \in \mathbb{N}$ such that $1 = \gcd(p, q) > 1$. Consequently, the assumption must be false, and statement (2.2) must therefore be true.

To understand the logic of this argument in more detail, it is helpful to set $R(m, n) :\Leftrightarrow \sqrt{2} = m/n$ and $S(p, q) :\Leftrightarrow \gcd(p, q) = 1$ for all $m, n, p, q \in \mathbb{N}$. For given these definitions, the above argument shows that $R(p, q)$ implies $\bar{S}(p, q)$ for all $p, q \in \mathbb{N}$ and that for all $m, n \in \mathbb{N}$ the assumption $R(m, n)$ implies that there exist $p, q \in \mathbb{N}$ such that $R(p, q) \wedge S(p, q)$ is true. So in setting

$$A :\Leftrightarrow \forall_{n,m \in \mathbb{N}}(R(m, n) \Rightarrow \exists_{p,q \in \mathbb{N}}(R(p, q) \wedge S(p, q)))$$

and
$$B :\Leftrightarrow \forall_{p,q \in \mathbb{N}}(R(p,q) \Rightarrow \bar{S}(p,q)),$$
it follows that $A \wedge B$ is true. So if we can show that the material conditional
$$(A \wedge B) \Rightarrow \forall_{m,n \in \mathbb{N}} \bar{R}(m,n)$$
is true, then it further follows that $\forall_{m,n \in \mathbb{N}} \bar{R}(m,n)$ is true as well, as desired, because a material conditional with a true antecedent is true only if the consequent is true too. Given this observation, we proceed to assume that there exist integers $m, n \in \mathbb{N}$ such that $\bar{R}(m,n)$ is false, or equivalently, such that $R(m,n)$ is true. Then, according to A, we can find $p, q \in \mathbb{N}$ such that $R(p,q) \wedge S(p,q)$ is true, and according to B, it therefore is the case that $S(p,q)$ is false. Consequently, the assumption that $R(m,n)$ is true must have been false because otherwise there would exist $p, q \in \mathbb{N}$ for which $S(p,q)$ is true and false simultaneously. Thus $\bar{R}(m,n)$ is true for all $m, n \in \mathbb{N}$, as claimed.

2.1.2 Example. Another famous fact that is typically established via a proof by contradiction concerns the infinity of the set of all prime numbers:

$$\text{The set of prime numbers is infinite.} \tag{2.3}$$

In order to give a proof by contradiction, we assume that statement (2.3) is false, that is, we assume that the set of prime numbers is finite (where a prime number, by definition, is a positive integer greater than one that is divisible only by one and itself). In that case there are integers $n, p_1, \ldots, p_n \in \mathbb{N}$ such that $\{p_1, \ldots, p_n\}$ is the set of prime numbers. Since every integer greater than one is divisible by a prime number (because every integer greater than one can in fact be written as a product of primes), we may infer that

$$m := p_1 \cdots p_n + 1$$

is divisible by p_k for some $k \in \{1, \ldots, n\}$. Since $p_1 \cdots p_n$ is evidently divisible by p_k as well, it follows that p_k divides $m - p_1 \cdots p_n$, and since this latter number is equal to 1 (by the definition of m), we arrive at the false conclusion that p_k divides 1. (Note: the conclusion is false because p_k, as any prime number, is greater than one and therefore cannot be a divisor of one.) Thus the assumption of the finiteness of the set of all prime numbers must have been false, and the set of prime numbers therefore must indeed be infinite.

Remark. We did not explicitly use quantification notation in the preceding proof, but implicitly we made use of the fact that the assumption of finiteness with regard to the set of prime numbers is equivalent to the following quantification schema:

$$\exists_{n \in \mathbb{N}} \exists_{p_1, \ldots, p_n \in \mathbb{N}} \forall_{p \in \mathbb{N}} \left(p \text{ prime} \Rightarrow \exists_{k \in \{1, \ldots, n\}} p = p_k \right),$$

where
$$p \text{ prime} \Leftrightarrow (p > 1 \wedge \forall_{j \in \mathbb{N}} \forall_{i \in \mathbb{N}} \; ij = p \Rightarrow (i = p \vee j = p)).$$

Exercises

2.1.3. Prove that $\sqrt{p}$ is irrational whenever p is a prime number.

2.1.4. Prove that the following statements are equivalent for any given number $p \in \mathbb{N}$:

(1) $\forall_{j \in \mathbb{N}} \forall_{i \in \mathbb{N}} \ ij = p \Rightarrow (i = p \vee j = p)$

(2) $\forall_{j \in \mathbb{N}} (\exists_{i \in \mathbb{N}} \ ij = p) \Rightarrow (j = 1 \vee j = p)$

2.1.5. Prove that the sum $x + y$ of an irrational number x and a rational number y is irrational.

2.1.6. Prove that $\sqrt{2} + \sqrt{3}$ is irrational.

2.1.7. Prove that $\sqrt{p} + \sqrt{q}$ is irrational for all primes p and q.

2.1.8. Prove that the n-th root of 2 is irrational for all integers $n > 1$.

2.2 Proofs by Contraposition

In our proof of the irrationality of $\sqrt{2}$ we made use of the fact that an integer p is even if and only if p^2 is even. A rigorous proof of this fact is most conveniently given by means of a so-called *proof by contraposition* which is based on the fact that $P \Rightarrow Q$ is equivalent to $\bar{Q} \Rightarrow \bar{P}$ (by Proposition 1.4.5e). So instead of proving that P implies Q we prove, by way of contraposition, that the negation of Q implies the negation of P.

2.2.1 Example. We wish to prove the following statement:

An integer $p \in \mathbb{N}$ is even if and only if p^2 is even.

To begin with, we observe that the statement here in question is equivalent to

$$\forall_{p \in \mathbb{N}} \ p \text{ even} \Leftrightarrow p^2 \text{ even}.$$

Using the fact that $P \Leftrightarrow Q$ is equivalent to $(P \Rightarrow Q) \wedge (Q \Rightarrow P)$ (by Proposition 1.4.5j), as well as the above mentioned fact that $P \Rightarrow Q$ is equivalent to $\bar{Q} \Rightarrow \bar{P}$, we find that

$$\left(\forall_{p \in \mathbb{N}} \ p \text{ even} \Leftrightarrow p^2 \text{ even} \right)$$
$$\Leftrightarrow \left(\forall_{p \in \mathbb{N}} \ (p \text{ even} \Rightarrow p^2 \text{ even}) \wedge (p^2 \text{ even} \Rightarrow p \text{ even}) \right) \qquad (2.4)$$
$$\Leftrightarrow \left(\forall_{p \in \mathbb{N}} \ (p \text{ even} \Rightarrow p^2 \text{ even}) \wedge (p \text{ odd} \Rightarrow p^2 \text{ odd}) \right).$$

In order to prove the last of these statements, we first assume that p is even. In that case there exists a $k \in \mathbb{N}$ such that $p = 2k$. Hence $p^2 = 4k^2$ is even as well. If on the other hand p is odd, then there exists a $k \in \mathbb{N}$ such that $p = 2k - 1$, and therefore, $p^2 = 4k^2 - 4k + 1$ is also odd.

Remark. Setting

$$P(p) :\Leftrightarrow ((\exists_{k \in \mathbb{N}} \, p = 2k) \Rightarrow (\exists_{k \in \mathbb{N}} \, p^2 = 2k))$$

and

$$Q(p) :\Leftrightarrow ((\exists_{k \in \mathbb{N}} \, p^2 = 2k - 1) \Rightarrow (\exists_{k \in \mathbb{N}} \, p = 2k - 1)),$$

the last statement in (2.4) is equivalent to $\forall_{p \in \mathbb{N}} \, (P(p) \wedge Q(p))$.

2.2.2 Example. Taking yet another look at our earlier proof of the fact that $\sqrt{2}$ is not a rational number, and setting

$$P(p, q) :\Leftrightarrow \sqrt{2} = p/q$$

and

$$Q(p, q) :\Leftrightarrow \gcd(p, q) = 1,$$

we may say that in essence we showed that $P(p, q)$ implies $\bar{Q}(p, q)$ for all $p, q \in \mathbb{N}$ (because $P(p, q)$ implies that both p and q are divisible by 2). That is to say, we demonstrated that

$$\forall_{p, q \in \mathbb{N}} \, P(p, q) \Rightarrow \bar{Q}(p, q). \tag{2.5}$$

Since the assertion that any fraction n/m can always be reduced to p/q with $\gcd(p, q) = 1$ is equivalent to the statement

$$\forall_{m, n \in \mathbb{N}} \exists_{p, q \in \mathbb{N}} (m/n = p/q \wedge Q(p, q)) \Leftrightarrow: R,$$

it follows that our overall argument is properly expressed in the following material conditional:

$$(R \wedge \forall_{p, q \in \mathbb{N}} (P(p, q) \Rightarrow \bar{Q}(p, q))) \Rightarrow \forall_{m, n \in \mathbb{N}} \bar{P}(m, n). \tag{2.6}$$

To establish the validity of this argument, we will use a proof by contraposition, that is, we will show that

$$\exists_{m, n \in \mathbb{N}} P(m, n) \Rightarrow \left(\bar{R} \vee \neg \forall_{p, q \in \mathbb{N}} (P(p, q) \Rightarrow \bar{Q}(p, q)) \right). \tag{2.7}$$

So let us assume that $P(m, n)$ is true for some $m, n \in \mathbb{N}$. If $\bar{R}$ is true, then the consequent in (2.7) is true as well, and we are done. If, on the other hand, $\bar{R}$ is false, then R is true, and we can find $p, q \in \mathbb{N}$ such that $m/n = p/q \wedge Q(p, q)$. Hence $p/q = \sqrt{2}$ (because $P(m, n)$ is true and because $m/n = p/q$), and therefore $P(p, q)$ is true. Consequently, the statement

$$\exists_{p, q \in \mathbb{N}} P(p, q) \wedge Q(p, q)$$

is true as well. Furthermore, since this latter statement is equivalent to the negation of (2.5), we may infer that the consequent in (2.7) is true, as desired.

For clarity we wish to add that the proof above only establishes the truth of the material conditional (2.6) as a material conditional. That is to say, our argument only shows that the consequent is true under the assumption that the antecedent is true as well, but it does not establish that the antecedent is in fact true. The truth of the antecedent was established in Example 2.1.1 but not in the present proof by contraposition.

Exercises

2.2.3. Why is it permissible to use the same letter k in each of the existence statements in the remark on p.46?

2.2.4. Prove that the sum of two integers $n, m \in \mathbb{N}$ is even if and only if either both n and m are even or both n and m are odd.

2.2.5. Prove that the product of two integers $n, m \in \mathbb{N}$ is odd if and only if both n and m are odd.

2.3 More about Sequences

In order to provide the reader with additional examples of elementary proofs and to prepare our construction of the real number system in Chapter 6, we will discuss in this section some further basic properties of (real or rational) sequences. A simple but very useful tool that will facilitate this discussion is the so-called *triangle inequality:*

2.3.1 Proposition. *For all real or rational numbers x and y it is the case that*

$$|x + y| \leq |x| + |y|. \tag{2.8}$$

Proof. If $x, y \geq 0$, then $|x + y| = x + y = |x| + |y|$ and (2.8) is therefore valid. If $x, y < 0$, then $|x + y| = -x - y = |x| + |y|$, and again (2.8) is valid. If $x \geq 0$, $y < 0$, and $x + y \geq 0$, then $|x + y| = x + y = |x| + y < |x| + |y|$, as desired. If $x \geq 0$, $y < 0$, and $x + y < 0$, then $|x + y| = -x - y = -x + |y| \leq |x| + |y|$, as desired. Finally, the validity of (2.8) for the case $x < 0$ and $y \geq 0$ follows by symmetry (see Exercise 2.3.8) from the case $x \geq 0$ and $y < 0$ (which we just discussed) because $|x + y| = |y + x|$ and $|x| + |y| = |y| + |x|$. $\qquad\square$

2.3.2 Corollary. *For all real or rational numbers x and y the following inequalities are valid:*

a) $|x| - |y| \leq |x - y|$

b) $||x| - |y|| \leq |x - y|$

Proof. **a)** According to Proposition 2.3.1, it is the case that

$$|x| = |x - y + y| \leq |x - y| + |y|,$$

and therefore, $|x| - |y| \leq |x - y|$, as desired.

b) Using a) with x in place of y and y in place of x, we find that

$$-(|x| - |y|) = |y| - |x| \leq |y - x| = |-(x - y)| = |x - y|.$$

Since also $|x| - |y| \leq |x - y|$, it follows that $||x| - |y|| \leq |x - y|$ (because, in general, $a \leq b$ and $-a \leq b$ if and only if $|a| \leq b$).   □

Following up on our discussion on p.31, concerning the uniqueness of limits, our first objective in this section is to prove precisely that property:

2.3.3 Proposition. *The limit of a convergent sequence is unique.*

Proof. We need to show that any two limits K and L of a real or rational sequence $(a_n)_{n \in \mathbb{N}}$ are equal. In other words, we need to show that $|L - K| < \varepsilon$ for all $\varepsilon > 0$. So let $\varepsilon > 0$. Since K and L are limits of $(a_n)_{n \in \mathbb{N}}$, we can find numbers $M, N \in \mathbb{N}$ such that

$$\forall_{n \in \mathbb{N}} \, n > M \Rightarrow |a_n - K| < \frac{\varepsilon}{2}$$

and

$$\forall_{n \in \mathbb{N}} \, n > N \Rightarrow |a_n - L| < \frac{\varepsilon}{2}.$$

So if we choose a number $n \in \mathbb{N}$ that is greater than both M and N (e.g, $n = M + N$), then the triangle inequality implies that

$$|L - K| = |a_n - K + L - a_n| \leq |a_n - K| + |L - a_n|$$
$$= |a_n - K| + |a_n - L| < \frac{\varepsilon}{2} + \frac{\varepsilon}{2} = \varepsilon,$$

as desired.   □

2.3.4 Definition. A real or rational sequence $(a_n)_{n \in \mathbb{N}}$ is said to be *bounded* if there exists a number $M \in \mathbb{N}$ (or $\mathbb{Q}$ or $\mathbb{R}$) such that $|a_n| \leq M$ for all $n \in \mathbb{N}$.

2.3.5 Proposition. *A real or rational sequence that is convergent is necessarily bounded as well.*

Proof. If $(a_n)_{n \in \mathbb{N}}$ is a convergent sequence, then there exists an $L \in \mathbb{R}$ such that

$$\forall_{\varepsilon > 0} \exists_{N \in \mathbb{N}} \forall_{n \in \mathbb{N}} \, n > N \Rightarrow |a_n - L| < \varepsilon.$$

So in setting $\varepsilon := 1$, we can find an $N \in \mathbb{N}$ such that $|a_n - L| \leq 1$ for all $n > N$. Using Corollary 2.3.2a, it follows that $|a_n| - |L| \leq 1$, and therefore,

$|a_n| \leq 1 + |L|$ for all $n > N$. Consequently, if we define M to be the maximum of the values $|a_1|, \ldots, |a_N|, 1 + |L|$, that is,

$$M := \max\{|a_1|, \ldots, |a_N|, 1 + |L|\},$$

then $|a_n| \leq M$ for all $n \in \mathbb{N}$, and therefore, $(a_n)_{n \in \mathbb{N}}$ is bounded. $\square$

Naturally, the question that arises here is whether the reverse conclusion that every bounded sequence is convergent is valid as well. As it turns out, the answer is "no" because the sequence $a_n = (-1)^n$ is bounded but not convergent: it is bounded because $|a_n| = |(-1)^n| = 1 \leq 1$ for all $n \in \mathbb{N}$, and it is not convergent by the argument given in Example 1.7.1.

2.3.6 Definition. A real or rational sequence $(a_n)_{n \in \mathbb{N}}$ is said to be a *Cauchy sequence* if

$$\forall_{\varepsilon > 0} \, \exists_{N \in \mathbb{N}} \, \forall_{m,n \in \mathbb{N}} \, (m > N \wedge n > N) \Rightarrow |a_n - a_m| < \varepsilon.$$

Intuitively, this definition says that a sequence is a Cauchy sequence if for any given margin $\varepsilon > 0$ it is the case that any two numbers a_m and a_n in the sequence will differ from each other by less than that margin whenever the indices m and n are sufficiently large. Consequently, we may be led to suspect that the notion of a Cauchy sequence is very closely linked to the notion of a convergent sequence because as the numbers in a convergent sequence get closer and closer to a certain limiting value L they also, of course, get closer and closer to each other. Somewhat surprisingly, though, the question of whether these two notions are indeed equivalent depends for its answer on the number system that the sequence values inhabit. That is to say, a convergent sequence is always a Cauchy sequence, but a Cauchy sequence, in general, is a convergent sequence only if the sequence and its limit are allowed to be real numbers. To understand this dependence of the convergence property on the underlying number system more clearly, let us consider, for instance, the number

$$\sqrt{2} = 1.4142...$$

which we showed to be irrational in Example 2.1.1. Setting

$$a_1 := 1.4,$$
$$a_2 := 1.41,$$
$$a_3 := 1.414,$$
$$a_4 := 1.4142,$$

$$\vdots$$

it follows that $(a_n)_{n \in \mathbb{N}}$ is a sequence of *rational* numbers that converges to the *irrational* limit $L = \sqrt{2}$. In other words, $(a_n)_{n \in \mathbb{N}}$ is not convergent in $\mathbb{Q}$

but only in $\mathbb{R}$ because its limit is not rational. Furthermore, since $(a_n)_{n\in\mathbb{N}}$ is *convergent* in $\mathbb{R}$ it is—as we just said and as we will show in the proof of Theorem 2.3.7 below—a Cauchy sequence as well. So there are rational Cauchy sequences that do not converge to rational numbers. In this sense we say that $\mathbb{Q}$ is *not complete*, and the task of completing $\mathbb{Q}$ will lead us in Chapter 5 to actually define the set of real numbers $\mathbb{R}$ by way of the notion of a Cauchy sequence of rational numbers. So in order to fully understand this matter the reader will have to be patient and wait until this subject is fully developed in the closing chapter of this book.

2.3.7 Theorem. *A real or rational sequence that is convergent is necessarily a Cauchy sequence as well.*

Proof. Assume that $(a_n)_{n\in\mathbb{N}}$ is a real or rational sequence that converges to a limiting value L in $\mathbb{R}$ or $\mathbb{Q}$. Then, for a given $\varepsilon > 0$, there exists an $N \in \mathbb{N}$ such that

$$|a_n - L| < \frac{\varepsilon}{2}$$

for all $n > N$. Consequently, in using the triangle inequality as stated in Proposition 2.3.1, we find that for all $m, n > N$ it is the case that

$$|a_n - a_m| = |a_n - L - (a_m - L)| \le |a_n - L| + |a_m - L| < \frac{\varepsilon}{2} + \frac{\varepsilon}{2} = \varepsilon,$$

as desired (see Definition 2.3.6). $\qquad\square$

Exercises

2.3.8. What exactly does it mean to say in the proof of Proposition 2.3.1 that "the validity of (2.8) for the case $x < 0$ and $y \ge 0$ follows by symmetry from the case $x \ge 0$ and $y < 0$... because $|x+y| = |y+x|$ and $|x|+|y| = |y|+|x|$"?

2.3.9. Prove that $|x + y + z| \le |x| + |y| + |z|$ for all $x, y, z \in \mathbb{R}$.

2.3.10. Determine the truth value of each of the following statements, given that $a_n := 1/n$ for all $n \in \mathbb{N}$:

a) $\forall_{\varepsilon>0} \exists_{N\in\mathbb{N}} \forall_{m\in\mathbb{N}} \forall_{n\in\mathbb{N}} ((m > N \wedge n > N) \Rightarrow |a_n - a_m| < \varepsilon)$,

b) $\exists_{N\in\mathbb{N}} \forall_{\varepsilon>0} \forall_{m\in\mathbb{N}} \forall_{n\in\mathbb{N}} ((m > N \wedge n > N) \Rightarrow |a_n - a_m| < \varepsilon)$,

c) $\forall_{\varepsilon>0} \forall_{n\in\mathbb{N}} \forall_{m\in\mathbb{N}} \exists_{N\in\mathbb{N}} ((m > N \wedge n > N) \Rightarrow |a_n - a_m| < \varepsilon)$,

d) $\forall_{\varepsilon>0} \exists_{N\in\mathbb{N}} \exists_{m\in\mathbb{N}} \exists_{n\in\mathbb{N}} ((m > N \wedge n > N) \Rightarrow |a_n - a_m| < \varepsilon)$,

e) $\forall_{\varepsilon>0} \forall_{N\in\mathbb{N}} \forall_{m\in\mathbb{N}} \forall_{n\in\mathbb{N}} ((m > N \wedge n > N) \Rightarrow |a_n - a_m| < \varepsilon)$,

f) $\exists_{\varepsilon>0} \forall_{N\in\mathbb{N}} \forall_{m\in\mathbb{N}} \forall_{n\in\mathbb{N}} ((m > N \wedge n > N) \Rightarrow |a_n - a_m| < \varepsilon)$.

2.3.11. Find a sequence of irrational numbers that converges to a rational limit.

2.3.12. Assume that $(a_n)_{n\in\mathbb{N}}$ is a convergent sequence such that $a_n \in \mathbb{N}$ for all $n \in \mathbb{N}$. Show that there exists an $N \in \mathbb{N}$ such that $a_n = a_N$ for all $n \ge N$.

2.4 ★ A Case Study: Periodic Functions

For the material covered in this final section of Chapter 2 we appeal again to the elementary notion of a real-valued function defined on $\mathbb{R}$ or a subset of $\mathbb{R}$ (as we did previously in the concluding examples of Section 1.5). Readers that are altogether unfamiliar with this notion may move ahead to Chapter 3.

More precisely, what we will be dealing with is the more specific notion of a *periodic* function which is commonly encountered in trigonometry. For instance, the sine function $f(t) := \sin(t)$ is a periodic function defined on $\mathbb{R}$ whose period is 2π (as shown in Figure 2.1). When we ask ourselves what

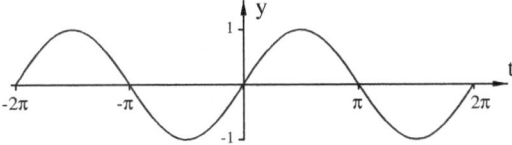

Figure 2.1: graph of $f(t) = y = \sin(t)$.

exactly we mean by saying that the sine function is periodic, the natural answer that occurs to us is that the sine function repeats itself in intervals of length 2π. That is to say, the value of f at t is always the same as the value of f at $t + 2\pi$:

$$\forall_{t \in \mathbb{R}} \, \sin(t) = \sin(t + 2\pi).$$

Given this observation, it is natural to agree that in general a real-valued function f defined on $\mathbb{R}$ is periodic if there exists a value $T > 0$ such that $f(t) = f(t + T)$ for all $t \in \mathbb{R}$. But what if the function f is not defined on all of $\mathbb{R}$ but only on a subset D of $\mathbb{R}$ (such as, for instance the tangent and cotangent functions from trigonometry)? In that case we need to make sure that the value $t + T$ is in fact contained in D whenever t is contained in D. For otherwise the expression $f(t+T)$ is meaningless. Thus we are led to introduce the following definition:

2.4.1 Definition. (Periodicity, First Version) A real-valued function f that is defined on a subset D of $\mathbb{R}$ is said to be *periodic* if

$$\exists_{T>0} \forall_{t \in D} \, (t + T \in D \wedge f(t + T) = f(t)).$$

If we further inquire as to what it means for us to say that the *period* of the sine function is 2π, we readily notice that the equation $\sin(t) = \sin(t+2\pi)$ does not suffice to identify 2π as the period of sine because this equation is satisfied as well for any integer multiple of 2π: $\sin(t) = \sin(t + 2\pi n)$ for all $n \in \mathbb{Z}$ and all $t \in \mathbb{R}$. So what we really mean by saying that the period of sine is 2π is that $T = 2\pi$ is the *smallest* positive number that satisfies the equation

$\sin(t) = \sin(t + T)$ for all $t \in \mathbb{R}$. By implication, there now arises the added question as to whether for every function f that is periodic in the sense of Definition 2.4.1 there does exist such a smallest value $T > 0$? Unfortunately, the answer here is "no." For the constant function $f(t) := 1$, defined on $\mathbb{R}$, is evidently periodic but does not possess a well defined period. After all, if f is constant, then the equation $f(t + T) = f(t)$ is satisfied for all $t \in \mathbb{R}$ and *all* $T > 0$. So there cannot be a smallest positive T if all positive T are possible choices.

2.4.2 Definition. Let f be a real-valued function defined on a subset D of $\mathbb{R}$ and let

$$P(T) :\Leftrightarrow \forall_{t \in D} \, (t + T \in D \land f(t + T) = f(t))$$

for all $T > 0$. Then we say that f has a *well defined period* if

$$\exists_{T>0} \, (P(T) \land \forall_{S>0} \, (P(S) \Rightarrow S \geq T)) \, .$$

In this case we say that the number T, thus identified, is the *period* of f.

In order to make the meaning of this definition more apparent, we wish to point out that the statement $\exists_{T>0} P(T)$ simply says that f is periodic and that the added assertion—$\forall_{S>0} \, (P(S) \Rightarrow S \geq T)$—has the purpose to identify $S = T$ as the smallest positive value for which $P(S)$ is true. And indeed, if the truth of $P(S)$ always implies that S is greater than or equal to T and if $P(T)$ is true, then surely $S = T$ is the smallest value for which $P(S)$ is true. Furthermore, the minimality of T also implies that T, as the period of f, is in fact unique.

2.4.3 Proposition. *If f is a real-valued function defined on a subset D of $\mathbb{R}$ and if f has a well defined period, then the period of f is unique.*

Proof. We need to show that any two periods T and U of f are equal. So let us assume that T and U are positive numbers that are both periods of f in the sense of Definition 2.4.2. Then it is the case that the statements

$$P(T) \land \forall_{S>0} \, (P(S) \Rightarrow S \geq T) \tag{2.9}$$

and

$$P(U) \land \forall_{S>0} \, (P(S) \Rightarrow S \geq U) \tag{2.10}$$

are both true. Setting $S := T$ in (2.10) and $S := U$ in (2.9), it follows that $T \geq U$ according to (2.10) and $U \geq T$ according to (2.9). Thus $T = U$, as desired. $\qquad\qquad\square$

Remark. The uniqueness property established in Proposition 2.4.3 allows us to meaningfully speak of the period of f whenever f does in fact have a well defined period. That is to say, if periods were not unique, we would have to specify which of the various periods of f we are referring to whenever we speak of the period of f. As it turns out, this need to establish the uniqueness of a certain property is very frequently encountered in mathematics, and it already came our way once before when we established the uniqueness of the limit of a sequence in Proposition 2.3.3.

Moving on now to the main purpose of this section, we observe that the statement "f is periodic but does not have a well defined period" is equivalent to the following schema (see Exercise 2.4.8):

$$\exists_{T>0} \, P(T) \wedge \forall_{U>0} \, (P(U) \Rightarrow \exists_{S>0} \, (P(S) \wedge S < U)) \qquad (2.11)$$

To help the reader better understand the structure and meaning of this statement, it may be helpful to point out that the first part—$\exists_{T>0} \, P(T)$—guarantees that f is periodic and that the second part—$\forall_{U>0} \, (P(U) \Rightarrow \exists_{S>0} \, (P(S) \wedge S < U))$—asserts that whenever the periodicity condition $P(U)$ is satisfied for some value $U > 0$, then it is satisfied as well for some value $S > 0$ that is smaller than U. So if for every value $U > 0$ for which P is true there is always a smaller value $S > 0$ for which P is true, then perhaps the true meaning of (2.11) is that P is true for values $S > 0$ that are arbitrarily small. In other words, we may here be led to conjecture that (2.11) is in fact equivalent to

$$\exists_{T>0} \, P(T) \wedge \forall_{U>0} \, \exists_{S>0} \, (P(S) \wedge S < U). \qquad (2.12)$$

But is it actually possible to rigorously prove this conjecture? Well, since $Q \Rightarrow (R \Rightarrow Q)$ is a tautology, we readily find that (2.12) implies (2.11) (simply replace Q by $P(U)$ and R by $\exists_{S>0}(P(S) \wedge S < U)$), but what about the reverse conclusion? If we wish to show that (2.11) implies (2.12), we need move beyond a mere comparison of the overall logical structures of these statements to an examination of mathematical content, because $(R \Rightarrow Q) \Rightarrow Q$ is not a tautology (the schema is false if R and Q are both false). On a more intuitive level the formal nonequivalence of (2.11) and (2.12) can be understood as follows: assume that f is a periodic function for which there exists a $U_0 > 0$ such that the statement $\exists_{S>0}(P(S) \wedge S < U_0)$ is false. Then (2.12) is false, but statement (2.11) may still be true, because $P(U_0)$ may be false.

In order to explore the relationship between (2.11) and (2.12) at the level of mathematical content, it is best to simplify our assumptions by setting $D := \mathbb{R}$ and thereby guaranteeing that the condition $t + T \in D$ (which is part of $P(T)$) is satisfied by default. If we wish to show that under this additional assumption (2.11) implies (2.12), then we have to demonstrate that (2.12) is true whenever (2.11) is true. In this context, the following proposition is very helpful, because it shows that the assumption $D = \mathbb{R}$ allows us to replace (2.11) by a slightly stronger statement.

2.4.4 Lemma. *If $D = \mathbb{R}$ and if* (2.11) *is true, then*

$$\forall_{U>0} \, (P(U) \Rightarrow \exists_{S>0} \, (P(S) \wedge S \leq U/2)). \qquad (2.13)$$

Proof. Assume that $D = \mathbb{R}$ and that (2.11) is true. We need to show that for any $U > 0$ the statement $\exists_{S>0}(P(S) \wedge S \leq U/2)$ is true whenever $P(U)$ is true. So suppose that $P(U)$ is true for some $U > 0$. Then (2.11) allows us to infer that there exists an $S_0 > 0$ such that $P(S_0)$ is true and $S_0 < U$. Depending on the value of S_0 we now consider two cases:

Case 1: $S_0 \leq U/2$.
Setting $S := S_0$, it follows that $P(S)$ is true (because $P(S_0)$ is true) and therefore, $\exists_{S>0}(P(S) \wedge S \leq U/2)$ is true as well, as desired.

Case 2: $S_0 > U/2$.
Setting $S := U - S_0$, it follows that $S < U/2$ and since $D = \mathbb{R}$, it also is true that $t + S \in D$ for all $t \in D$. Furthermore, for a given $t \in D = \mathbb{R}$, the assumption that both $P(S_0)$ and $P(U)$ are true allows us to conclude that

$$f(t + S) = f(t + U - S_0) = f(t - S_0) = f(t - S_0 + S_0) = f(t).$$

Consequently, $P(S)$ is true, and the statement $\exists_{S>0}(P(S) \wedge S \leq U/2)$ is therefore true as well. So in both cases we have arrived at the same desired conclusion, and the proof of the lemma is thus complete. $\square$

2.4.5 Proposition. *If $D = \mathbb{R}$, then (2.11) and (2.12) are equivalent.*

Proof. Since we already know that (2.12) implies (2.11), we only need to show that (2.11) implies (2.12). So let us assume that (2.11) is true. Then the first part of (2.12), namely the statement $\exists_{T>0}P(T)$, is true because it is the first part of the conjunction that constitutes (2.11). Consequently, we only need to show that $\forall_{U>0}\exists_{S>0}(P(S) \wedge Q(S,U))$. So let $U > 0$. Since $\exists_{T>0}P(T)$ is true, we may infer that there exists a $T_0 > 0$ such that $P(T_0)$ is true. Since U is greater than zero, there exists an $n \in \mathbb{N}$ such that $T_0/2^n < U$ (this can be rigorously proven, but since we do not have available, at this point in the course, a rigorous definition of $\mathbb{N}$ and $\mathbb{R}$, we will be content to consider this assertion to be self-evident). Furthermore, according to Lemma 2.4.4, there exists an $S_1 > 0$ such that $P(S_1)$ is true and $S_1 \leq T_0/2$. Applying again Lemma 2.4.4, we may infer the existence of a number $S_2 > 0$ such that $P(S_2)$ is true and $S_2 \leq S_1/2 \leq T_0/2^2$. Continuing in this manner, we may conclude that there exist numbers $S_3, \ldots, S_n > 0$ such that $P(S_k)$ is true and $S_k \leq S_{k-1}/2 \leq T_0/2^k$ for all $k \in \{3, \ldots, n\}$. Setting $S := S_n$, it follows that $P(S)$ is true and $S \leq T_0/2^n < U$. Consequently, the statement $\exists_{S>0}(P(S) \wedge Q(S,U))$ is true as desired. $\square$

As we look more closely at the proof of Lemma 2.4.4, we notice that the derivation of the equation $f(t + S) = f(t)$ was made possible by the fact that the value $t - S_0$ was contained in the domain D of f (as D was assumed to be all of $\mathbb{R}$). If $t - S_0$ had not been in D, the expression $f(t - S_0)$ would have been meaningless. This observation leads us to realize that the statement of

Proposition 2.4.5 remains valid as well if Definition 2.4.1 is replaced by the following, slightly altered definition (see Exercise 2.4.9):

2.4.6 Definition. (Periodicity, Second Version) A real-valued function f that is defined on a subset D of $\mathbb{R}$ is said to be *periodic* if

$$\exists_{T>0} \forall_{t \in D} \, (t \pm T \in D \wedge f(t+T) = f(t)).$$

Since Proposition 2.4.5 shows that (2.11) and (2.12) are equivalent whenever the requirement $t+T \in D$ is satisfied by default (as in the case when D is assumed to be all of $\mathbb{R}$), it follows that (2.11) and (2.12) can be non-equivalent only if the condition $\forall_{t \in D} \, t+T \in D$ is non-trivial. In order to be able to focus our attention exclusively on this latter condition, it is helpful and natural to consider a case in which the equation $f(t + T) = f(t)$ is trivially satisfied. To do so, we set $f(t) := 1$. For if f is constant, then, clearly, $f(t)$ equals $f(t+T)$ whenever both t and $t + T$ are in the domain D of f. As we now attempt to construct a domain D for which (2.11) is true and (2.12) is false, we first observe that (2.12) implies in particular that for any positive number U there are distinct points in D that are separated by a distance less than U. Given this observation, it may occur to us to set D equal to $\mathbb{Z}$, because the distance of any two elements in $\mathbb{Z}$ is always greater than or equal to 1. Unfortunately, a moment's reflection also reveals that for $D = \mathbb{Z}$ and $f(t) := 1$ we have a well defined period in the sense of Definition 2.4.2, namely $T = 1$.

Spurned on by this failure and with our reputation as problem solvers on the line we press on and envision D to be a set that has a hole but also contains points that are arbitrarily close to each other. For instance, we may set D equal to $\{0\} \cup (1, \infty)$ (the set containing 0 and all real numbers greater than 1) or perhaps $\{0\} \cup [1, \infty)$, but the problem with the latter choice is of course that again we have the well defined period $T = 1$.

2.4.7 Proposition. If $D = \{0\} \cup (1, \infty)$ and if $f(t) := 1$ for all $t \in D$, then f satisfies statement (2.11) but not (2.12). In other words, (2.11) does not in general imply (2.12).

Proof. To begin with we observe that the definition of D trivially implies that $2 + t \in D$ for all $t \in D$. Consequently, the statement $\exists_{T>0} P(T)$ is true. In order to show that $\forall_{U>0} (P(U) \Rightarrow \exists_{S>0} (P(S) \wedge S < U))$ is true as well (see (2.11)), we assume that U is a positive number for which $P(U)$ is true. Then, given that $0 \in D$, we may infer that $U = 0 + U \in D$. In other words, U must be an element of $(1, \infty)$ and therefore greater than 1. Setting $S := (U+1)/2$, it follows that $1 < S < U$ and, by implication, that $P(S)$ is true. Consequently, $\exists_{S>0} (P(S) \wedge S < U)$ is true as well, and the material conditional $P(U) \Rightarrow \exists_{S>0} (P(S) \wedge S < U)$ has thus been shown to be true for all $U > 0$, as desired. In other words, (2.11) is indeed true.

In order to demonstrate that (2.12) is false, it is sufficient to show that the statement $\exists_{S>0} (P(S) \wedge S < U)$ is false for $U = 1$. That is to say, it is sufficient to show that $\exists_{S>0} (P(S) \wedge S < 1)$ is false or, equivalently, that $\forall_{S>0} (P(S) \Rightarrow S \geq 1)$ is true. So let us assume that $P(S)$ is true for a given $S > 0$. Then, as above, we may conclude that S is an element of $(1, \infty)$ and therefore greater than 1. So the statement $\forall_{S>0} (P(S) \Rightarrow S \geq 1)$ is true, as desired. $\square$

Exercises

2.4.8. Explain why statement (2.11) is equivalent to the assertion that f is periodic but does not have a well defined period.

2.4.9. Show that for any real-valued function f, defined on D, that is periodic in the sense of Definition 2.4.6, the statements (2.11) and (2.12) are equivalent. Note: the statement $P(T)$ is here to be re-construed as $\forall_{t \in D} (t \pm T \in D \wedge f(t + T) = f(t))$.

2.4.10. Assume that f is a real-valued function, defined on $\mathbb{R}$, that has a well-defined period $T > 0$ in the sense of Definition 2.4.2. Explain why a number $S > 0$ will satisfy the periodicity condition $\forall_{t \in \mathbb{R}} f(t + S) = f(t)$ if and only if there exists an $n \in \mathbb{N}$ such that $S = nT$.

2.4.11. Find a real-valued function f that is defined on $\mathbb{R}$, is non-periodic in the sense of Definition 2.4.2, and satisfies the condition

$$\forall_{\varepsilon>0} \forall_{t \in \mathbb{R}} \exists_{T>0} (f(t) = f(t + T) \wedge T < \varepsilon).$$

2.4.12. Find a real-valued function f that is defined on $\mathbb{R}$, is non-periodic in the sense of Definition 2.4.2, and satisfies the condition

$$\forall_{t \in \mathbb{R}} \exists_{T>0} f(t) = f(t + T)$$

but not the condition

$$\forall_{\varepsilon>0} \forall_{t \in \mathbb{R}} \exists_{T>0} (f(t) = f(t + T) \wedge T < \varepsilon).$$

Chapter 3

Sets, Relations, and Functions

3.1 What in the World Is a Set?

> Future generations will look at set theory as a sickness, from which mathematicians will have recovered. (Henri Poincaré, 1908)

> No one shall be able to drive us from the paradise that Cantor created for us. (David Hilbert, 1925)

Mathematics, at bottom, is the study of sets. But what in the world is a set? Well, naively we might say that a set is a collection of elements which we somehow can recognize as distinct, discrete objects. In other words, sets are collections and collections are sets and elements are objects and objects are elements. It doesn't make much sense, but never mind, it's not philosophy but merely mathematics.

3.1.1 Definition. (Georg Cantor, 1845–1918) By a *set* we mean any collection M into a whole of definite, distinct objects m (which are called the *elements* of M) of our perception or of our thought.

Unfortunately, this definition isn't just confusing philosophically but also logically completely contradictory. For if Cantor's definition is to have any meaning at all, then sets themselves must surely be considered to be "distinct objects of our perception or of our thought." Consequently, the set of all sets ought to be a well defined entity. Denoting this set by S, it would follow that $S \in S$, because S, by being a set itself, must be an element of the set of all sets. Having thus shown that inherent in Cantor's definition is the notion of self-containment, we may regard the set

$$M := \{A \mid A \text{ is a set, and } A \notin A\}$$

to be well defined. Furthermore, since self-containment is a property that sets, as we just argued, must be allowed to have, there are now two possibilities: $M \in M$ or $M \notin M$. In the former case the definition of M implies that $M \notin M$ and in the latter it implies that $M \in M$. Thus we are left in either case with the contradictory conclusion $M \in M$ and $M \notin M$.

In order to resolve this antinomy, Bertrand Russell and Alfred North White-head elaborated a comprehensive axiomatic system in a famous monumental work—the *Principia Mathematica*—that in essence was designed to bring within its explanatory reach the entirety of all known mathematics. In particular, the system in question was intended to be consistent and complete. However, in 1931, the Austrian logician Kurt Gödel was able to prove that completeness is a property that cannot be achieved. For according to Gödel's famous incompleteness theorems, any *finite* axiomatic system that is sufficiently complex to accommodate the notion of a positive counting number will necessarily allow for the formulation of statements that cannot be proven true or false within the propositional framework that the system provides.

Such logical perplexities are bound to seem decidedly discouraging. But in the light of the Socratic dictum that true wisdom is knowledge of one's ignorance, we may as well rejoice in our lack of comprehension in this case because it is precisely in acknowledging this lack that we display the deepest kind of wisdom there can be. So what we ought to do is gladly revel in our ignorance and start exploring sets despite the fact that we do not really have a clue what all of this at bottom is about.

On a more serious note, we also wish to mention that there does exist in current-day mathematics an axiomatic definition of the notion of a set that is more adequate by far than the original definition proposed by Georg Cantor. However, a detailed discussion of this system is clearly not within the scope of the present exposition. But readers with an interest in this topic may consult for instance the introduction to set theory by Just and Weese in [JW]. So given that we cannot here provide a fully rigorous account, we will have to be content, as Cantor was, to consider a set to be a collection of objects of our perception or thought. This cannot be completely satisfactory, but it can teach us all the same how, starting from sets, the whole of modern mathematics can be built. As a matter of course, we won't be able to survey in this text all known mathematics, but the truth of this claim—that mathematics is grounded in sets—will be demonstrable in various specific examples and will thereby be shown to be globally plausible.

3.2 Set Notation

As we already saw in Section 1.5 (on p.18f), the presence of a set is commonly indicated by a pair of curly brackets. For instance, the set containing the numbers 1, 2, and 4 is written as

$$\{1, 2, 4\},$$

or, in *set-builder notation*, as

$$\{x \mid x = 1 \vee x = 2 \vee x = 4\}.$$

In words the latter option would be described as the set of all x that are equal to 1, 2, or 4. More generally, set-builder notation allows us to denote the set of all x in a given universe $\mathcal{U}$ that satisfy a certain statement $P(x)$ by

$$\{x \in \mathcal{U} \mid P(x)\}. \tag{3.1}$$

3.2.1 Example. In the universe of the real numbers, the half-open interval $(0, 1]$ is the set of all x for which the statement $P(x) :\Leftrightarrow (x > 0 \wedge x \leq 1)$ is true, that is
$$(0, 1] = \{x \in \mathbb{R} \mid x > 0 \wedge x \leq 1\}.$$

3.2.2 Example. In the Cartesian coordinate plane, that is, in the universe of all ordered pairs of real numbers (a precise definition of an ordered pair *as a set* will be given in Section 3.5), the line L that is described by the equation $y = 2x + 1$ is the set of all pairs (x, y) that satisfy the equation $y = 2x + 1$. That is to say,
$$L = \{(x, y) \mid x, y \in \mathbb{R} \wedge y = 2x + 1\}.$$

Furthermore, if we wish to express the fact that a certain x is an *element* of a set S, we write
$$x \in S,$$
and otherwise, if x is not contained in S, we write

$$x \notin S.$$

3.2.3 Example. For the set $S := \{1, 2, 4\}$ it is the case that $1 \in S$ and $3 \notin S$.

Exercises

3.2.4. Write each of the following sets in set-builder notation.

a) The unit circle in the Cartesian plane.
b) The interval $(2, 5]$.

c) The set of all prime numbers.

d) The set of all odd positive integers.

e) The x-axis in the Cartesian plane.

f) The parabola described by the equation $y = x^2$.

g) The set of all positive integers that are squares.

h) The points on the line segment that connects $(1, 2)$ with $(3, -1)$ (and includes these endpoints).

3.2.5. Which of the following sets contain the element $x = 1$?

a) The interval $[0, 1)$.

b) The interval $[0, 1]$.

c) The set of all prime numbers.

d) The unit circle in the Cartesian plane.

3.3 Relations Between Sets

The following two definitions establish the most basic relations between sets—the relations of equality and inclusion.

3.3.1 Definition. Two sets A and B are said to be *equal* if

$$\forall_x \, x \in A \Leftrightarrow x \in B,$$

and in this case we write $A = B$.

3.3.2 Definition. A set A is said to be a *subset* of a set B if

$$\forall_x \, x \in A \Rightarrow x \in B,$$

and in this case we write $A \subset B$.

3.3.3 Proposition. *The statements below are true for all sets A, B, and C.*

a) $A = A$.

b) *If $A = B$ then $B = A$.*

c) *If $A = B$ and $B = C$, then $A = C$.*

d) *If $A = B$, then $A \subset B$.*

e) $A \subset A$.

f) *If $A \subset B$ and $B \subset C$, then $A \subset C$.*

g) $A = B$ *if and only if $A \subset B$ and $B \subset A$.*

Proof. We will prove properties c, f, and g and leave the proofs of the remaining properties as exercises to the reader (see Exercise 3.3.12).

c) We need to show that $\forall_x\, x \in A \Leftrightarrow x \in C$, and we know that $\forall_x\, x \in A \Leftrightarrow x \in B$ and $\forall_x\, x \in B \Leftrightarrow x \in C$ (because $A = B$ and $B = C$). Since $((P \Leftrightarrow Q) \wedge (Q \Leftrightarrow R)) \Rightarrow (P \Leftrightarrow R)$ is a tautology, we may infer that for any given object x it is the case that $x \in A \Leftrightarrow x \in C$, as desired, because the conjunction $(x \in A \Leftrightarrow x \in B) \wedge (x \in B \Leftrightarrow x \in C)$ is true by assumption (where $P :\Leftrightarrow x \in A$, $Q :\Leftrightarrow x \in B$, and $R :\Leftrightarrow x \in C$).

f) We need to show that $\forall_x\, x \in A \Rightarrow x \in C$, given that $\forall_x\, x \in A \Rightarrow x \in B$ and $\forall_x\, x \in B \Rightarrow x \in C$. Since $((P \Rightarrow Q) \wedge (Q \Rightarrow R)) \Rightarrow (P \Rightarrow R)$ is a tautology, it follows that for any given object x the truth of the conjunction $(x \in A \Rightarrow x \in B) \wedge (x \in B \Rightarrow x \in C)$ implies the truth of the material conditional $x \in A \Rightarrow x \in C$, as desired.

g) Since $(P \Leftrightarrow Q) \Leftrightarrow ((P \Rightarrow Q) \wedge (Q \Rightarrow P))$ is a tautology (by Proposition 1.4.5j), we find that

$$
\begin{aligned}
A = B &\Leftrightarrow \forall_x\, (x \in A \Leftrightarrow x \in B) \\
&\Leftrightarrow \forall_x\, ((x \in A \Rightarrow x \in B) \wedge (x \in B \Rightarrow x \in A)) \\
&\Leftrightarrow \forall_x\, (x \in A \Rightarrow x \in B) \wedge \forall_x\, (x \in B \Rightarrow x \in A) \\
&\qquad \text{(by Proposition 1.7.7a)} \\
&\Leftrightarrow A \subset B \wedge B \subset A,
\end{aligned}
$$

as desired. $\square$

3.3.4 Definition. A set A is said to be *empty* if

$$\forall_x\, x \notin A.$$

3.3.5 Proposition. *The following statements are true for all sets A and B:*

a) *If A is empty, then $A \subset B$.*

b) *If A and B are both empty, then $A = B$.*

Proof. **a)** According to Definition 3.3.2, we need to show that $\forall_x (x \in A \Rightarrow x \in B)$. Since A is empty, the antecedent $x \in A$ is false for all x, and therefore, the material conditional $x \in A \Rightarrow x \in B$ is true for all x.

b) According to a), the assumption that both A and B are empty implies that $A \subset B$ and $B \subset A$. Hence $A = B$ by Proposition 3.3.3g. $\square$

According to Proposition 3.3.5, all empty sets are equal. Consequently, if we accept the axiomatic assertion that empty sets exist, then it is permissible to speak of a set that contains no elements as *the* empty set. By convention, this unique empty set is denoted by the symbol '$\emptyset$'.

3.3.6 Definition. If S is a set, then the *power set* of S is

$$\mathcal{P}(S) := \{M \mid M \subset S\}.$$

In words: $\mathcal{P}(S)$ is the set of all subsets of S.

3.3.7 Example. If $S = \{0, 1\}$, then $\mathcal{P}(S) = \{\emptyset, \{0\}, \{1\}, \{0, 1\}\}$.

3.3.8 Proposition. *If S is a finite set with n elements, then the number of elements in $\mathcal{P}(S)$ is 2^n.*

Proof. We will postpone a fully rigorous proof of this proposition until there is available to us the technique known as 'mathematical induction' (see Proposition 5.1.8 on p.137). At this point, therefore, we will have to be content to offer an argument that is easy to follow and certainly strongly persuasive but also a little informal. So let us assume that S is a finite set with n distinct elements $x_1, \ldots, x_n$ and that M is a subset of S, that is, $S = \{x_1, \ldots, x_n\}$ and $M \in \mathcal{P}(S)$. Given this assumption, we set

$$y_i(M) := \begin{cases} 1 & \text{if } x_i \in M \\ 0 & \text{if } x_i \notin M \end{cases}$$

and

$$y(M) := (y_1(M), \ldots, y_n(M)).$$

The purpose of this definition is to associate each element M of $\mathcal{P}(S)$ with exactly one binary sequence of length n: the i-th entry in $y(M)$ is one if x_i is an element of M and otherwise it is zero. For given this association, it is clear that the number of elements in $\mathcal{P}(M)$ is equal to the number of binary sequences of length n, and since the latter number is equal to 2^n (one factor 2 for each of the n entries in the sequence), it follows that the number of elements in $\mathcal{P}(S)$ is equal to 2^n as well. $\qquad\qquad\square$

3.3.9 Proposition. *The following statement is true for all sets A and B: $\mathcal{P}(A) \subset \mathcal{P}(B)$ if and only if $A \subset B$.*

Proof. '$\Rightarrow$' If $\mathcal{P}(A) \subset \mathcal{P}(B)$, then $A \in \mathcal{P}(B)$ because $A \in \mathcal{P}(A)$ (by Proposition 3.3.3e). Consequently, A is a subset of B.
'$\Leftarrow$' If $A \subset B$ then every subset of A is also a subset of B (by Proposition 3.3.3f). In other words, every element of $\mathcal{P}(A)$ is also an element of $\mathcal{P}(B)$, and therefore, $\mathcal{P}(A) \subset \mathcal{P}(B)$. $\qquad\qquad\square$

3.3.10 Example. We wish to decide whether the following statements are true or false:

a) $\emptyset \in \emptyset$

b) $\emptyset \subset \emptyset$

c) $\emptyset \in \{\emptyset\}$

d) $\emptyset \in \mathcal{P}(\emptyset)$

Statement a) is false because there are no elements in the empty set. By contrast, b) is true because the empty set is a subset of every set (by Proposition 3.3.5a). Furthermore, c) is true as well (because the empty set is indeed an element of the set whose only element is the empty set), and so is d) because the empty set is a subset of the empty set and therefore, by definition, an element of $\mathcal{P}(\emptyset)$.

3.3.11 Example. We wish to decide whether the following statements are true for all sets S:

a) $\emptyset \in \mathcal{P}(S)$

b) $S \in \mathcal{P}(S)$

c) $S \subset \mathcal{P}(S)$

d) $\{S\} \subset \mathcal{P}(S)$

Statement a) is true because the empty set is a subset of S and therefore, by definition, an element of $\mathcal{P}(S)$. Statement b) is true as well because S is a subset of S (by Proposition 3.3.3e) and thus an element of $\mathcal{P}(S)$. Regarding statement c), we notice that the elements of S (in case that S is nonempty) are not subsets of S and that S is therefore not a subset of the power set of S. Thus statement c) is not true for all sets S (it is true for $S = \emptyset$ but not for all S). Finally, statement d) is true because, according to b), S is an element of $\mathcal{P}(S)$, and the set containing S as its only element is therefore a subset of $\mathcal{P}(S)$.

Exercises

3.3.12. Prove Proposition 3.3.3a,b,d,e.

3.3.13. Find the power set of $S = \{0, 1, 2\}$.

3.3.14. Which of the statements listed below are true?

a) $\emptyset \subset \{\emptyset\}$

b) $\mathcal{P}(\emptyset) = \emptyset$

c) $\emptyset \subset \mathcal{P}(\emptyset)$

d) $\{\emptyset\} \subset \mathcal{P}(\emptyset)$

e) $\{\emptyset\} \in \mathcal{P}(\emptyset)$

3.3.15. Which of the following statements are necessarily true for all sets S?

    a) $\emptyset \subset \mathcal{P}(S)$

    b) $\{\emptyset\} \subset \mathcal{P}(S)$

    c) $\{\emptyset\} \in \mathcal{P}(\mathcal{P}(S))$

    d) $\emptyset \subset \mathcal{P}(\mathcal{P}(S))$

    e) $\emptyset \in \mathcal{P}(\mathcal{P}(S))$

    f) $\{S\} \in \mathcal{P}(\mathcal{P}(S))$

    g) $S \in \mathcal{P}(\mathcal{P}(S))$

    h) $\mathcal{P}(S) \subset \mathcal{P}(\mathcal{P}(S))$

    i) $\{\emptyset\} \subset \mathcal{P}(\mathcal{P}(S))$

    j) $\{\{\emptyset\}\} \subset \mathcal{P}(\mathcal{P}(S))$

3.4 Operations with Sets

3.4.1 Definition. Let A and B be sets.

    a) The *complement* of A in a universe $\mathcal{U}$ that contains A as a subset is
$$A^c := \{x \in \mathcal{U} \mid x \notin A\}.$$

    b) The *intersection* of A and B is
$$A \cap B := \{x \mid x \in A \wedge x \in B\}.$$

    c) The *union* of A and B is
$$A \cup B := \{x \mid x \in A \vee x \in B\}.$$

    d) The *set difference* of A and B is
$$A \smallsetminus B := \{x \mid x \in A \wedge x \notin B\}.$$

    e) The *symmetric difference* of A and B is
$$A \triangle B := (A \smallsetminus B) \cup (B \smallsetminus A).$$

3.4.2 Example. If $\mathcal{U} = \{0, 1, 2, 3, 4, 5\}$, $A = \{1, 2, 3\}$, and $B = \{3, 4\}$, then
$$A^c = \{0, 4, 5\},$$
$$A \cap B = \{3\},$$
$$A \cup B = \{1, 2, 3, 4\},$$
$$A \smallsetminus B = \{1, 2\},$$
$$A \triangle B = \{1, 2, 4\}.$$

3.4.3 Example. If $\mathcal{U} = \mathbb{R}$, $A = [1, 3]$, and $B = (2, 4]$, then

$$A^c = \{x \in \mathbb{R} \mid x < 1 \vee x > 3\} = (-\infty, 1) \cup (3, \infty),$$
$$A \cap B = (2, 3],$$
$$A \cup B = [1, 4],$$
$$A \smallsetminus B = [1, 2],$$
$$A \triangle B = [1, 2] \cup (3, 4].$$

3.4.4 Proposition. *The statements below are true for all sets A, B, and C.*

a) $A \cap B = B \cap A$,

b) $A \cup B = B \cup A$,

c) $A \cap (B \cap C) = (A \cap B) \cap C$,

d) $A \cup (B \cup C) = (A \cup B) \cup C$,

e) $A \cap (B \cup C) = (A \cap B) \cup (A \cap C)$,

f) $A \cup (B \cap C) = (A \cup B) \cap (A \cup C)$,

g) $A \subset A \cup B$,

h) $A \cap B \subset A$,

i) $A \cap \emptyset = \emptyset$,

j) $A \cup \emptyset = A$,

k) $A \smallsetminus \emptyset = A$,

l) $(A^c)^c = A$,

m) $(A \cap B)^c = A^c \cup B^c$,

n) $(A \cup B)^c = A^c \cap B^c$,

o) $A \smallsetminus B = A \cap B^c$,

p) $A \triangle B = (A \cup B) \smallsetminus (A \cap B)$.

q) $A \triangle B = A \cup B$ *if and only if* $A \cap B = \emptyset$.

r) $A \cup B = (A \cap B) \cup (A \triangle B)$.

Proof. We will pove i), k), p), and q) and leave the proofs of the remaining properties as exercises to the reader (see Exercise 3.4.5).
i) Let x be any given object (in any given universe). Then, according to Definition 3.4.1b, it is the case that

$$x \in A \cap \emptyset \Leftrightarrow x \in A \wedge x \in \emptyset$$
$$\Leftrightarrow x \in A \wedge F$$
$$\Leftrightarrow F$$
$$\Leftrightarrow x \in \emptyset,$$

and therefore, $A \cap \emptyset = \emptyset$, as desired.

k) Using Definition 3.4.1d and assuming that x is any given object (in any given universe), we may infer that

$$x \in A \setminus \emptyset \Leftrightarrow x \in A \wedge x \notin \emptyset$$
$$\Leftrightarrow x \in A \wedge T$$
$$\Leftrightarrow x \in A,$$

as desired.

p) Let x be any given object (in any given universe). Then, according to Definition 3.4.1e, we have

$$x \in A \triangle B \Leftrightarrow x \in (A \setminus B) \cup (B \setminus A)$$
$$\Leftrightarrow x \in A \setminus B \vee x \in B \setminus A$$
$$\Leftrightarrow (x \in A \wedge x \notin B) \vee (x \in B \wedge x \notin A)$$
$$\Leftrightarrow (x \in A \vee x \in B) \wedge (x \in A \vee x \notin A)$$
$$\wedge (x \notin B \vee x \in B) \wedge (x \notin B \vee x \notin A)$$
$$\Leftrightarrow (x \in A \vee x \in B) \wedge T \wedge T \wedge (x \notin B \vee x \notin A)$$
$$\Leftrightarrow (x \in A \vee x \in B) \wedge \neg(x \in B \wedge x \in A)$$
$$\Leftrightarrow (x \in A \cup B) \wedge \neg(x \in A \cap B)$$
$$\Leftrightarrow x \in (A \cup B) \setminus (A \cap B),$$

as desired.

q) Straight forward version: using p), it follows that

$$A \triangle B = A \cup B \Leftrightarrow \forall_x (x \in A \triangle B \Leftrightarrow x \in A \cup B)$$
$$\Leftrightarrow \forall_x ((x \in A \cup B \wedge x \notin A \cap B) \Leftrightarrow x \in A \cup B)$$
$$\Leftrightarrow \forall_x (((x \in A \vee x \in B) \wedge (x \notin A \vee x \notin B)) \Leftrightarrow (x \in A \vee x \in B))$$

and

$$A \cap B = \emptyset \Leftrightarrow \forall_x x \notin A \cap B$$
$$\Leftrightarrow \forall_x \neg(x \in A \wedge x \in B)$$
$$\Leftrightarrow \forall_x (x \notin A \vee x \notin B).$$

Setting $P :\Leftrightarrow x \in A$ and $Q :\Leftrightarrow x \in B$, it is thus sufficient to verify that the schema

$$(((P \vee Q) \wedge (\bar{P} \vee \bar{Q})) \Leftrightarrow (P \vee Q)) \Leftrightarrow (\bar{P} \vee \bar{Q})$$

is a tautology. Assuming P to be true, the schema reduces to

$$\bar{Q} \Leftrightarrow \bar{Q},$$

which is true. If P is false, then the schema assumes the form

$$(Q \Leftrightarrow Q) \Leftrightarrow T$$

which is true as well.

q) Less straight forward version:

"$\Leftarrow$" (Note: this arrow indicates, by convention, that we are going to prove the right-to-left implication in the material biconditional that we are trying to establish—in this case $(A \triangle B = A \cup B) \Leftrightarrow (A \cap B = \emptyset)$. That is to say, we are here again making use of the fact that $(P \Leftrightarrow Q) \Leftrightarrow ((P \Rightarrow Q) \wedge (Q \Rightarrow P))$ is a tautology and break up the proof of the biconditional in question into two separate proofs of a right-to-left and a left-to-right implication.) If $A \cap B = \emptyset$, then p) and k) in conjunction imply that $A \triangle B = A \cup B$, as desired.

"$\Rightarrow$" Again using p) yields

$$A \triangle B = A \cup B \Leftrightarrow \forall_x (x \in A \triangle B \Leftrightarrow x \in A \cup B)$$
$$\Leftrightarrow \forall_x ((x \in A \cup B \wedge x \notin A \cap B) \Leftrightarrow x \in A \cup B),$$

and setting $P(x) :\Leftrightarrow x \in A \cup B$ and $Q(x) :\Leftrightarrow x \in A \cap B$ and observing that $((R \wedge S) \vee (\bar{R} \wedge \bar{S})) \Leftrightarrow (R \Leftrightarrow S)$ is a tautology, we find that

$$A \triangle B = A \cup B \Leftrightarrow \forall_x ((P(x) \wedge \bar{Q}(x)) \Leftrightarrow P(x))$$
$$\Leftrightarrow \forall_x ((P(x) \wedge \bar{Q}(x)) \vee ((\bar{P}(x) \vee Q(x)) \wedge \bar{P}(x)))$$
$$\Leftrightarrow \forall_x ((P(x) \wedge \bar{Q}(x)) \vee \bar{P}(x)) \Leftrightarrow \forall_x (\bar{Q}(x) \vee \bar{P}(x)).$$

Thus, it remains to be shown that

$$\forall_x (\bar{Q}(x) \vee \bar{P}(x)) \Rightarrow \forall_x \bar{Q}(x).$$

At this point we quickly observe that the schema $(\bar{Q}(x) \vee \bar{P}(x)) \Rightarrow \bar{Q}(x)$ is not a tautology (it is false if $Q(x)$ is true and $P(x)$ is false), and we may therefore be tempted to immediately re-substitute the defining statements for $P(x)$ and $Q(x)$, but here we need to be careful, because $\forall_x (U(x) \Rightarrow V(x))$ is in general not equivalent to $\forall_x U(x) \Rightarrow \forall_x V(x)$. It certainly is true that $\forall_x (U(x) \Rightarrow V(x))$ implies $\forall_x U(x) \Rightarrow \forall_x V(x)$, but the reverse conclusion is in general false (just take some statement $U(x)$ that is not true for all x in a given universe, and set $V(x) :\Leftrightarrow \bar{U}(x)$). In other words, the weaker statement $\forall_x (\bar{Q}(x) \vee \bar{P}(x)) \Rightarrow \forall_x \bar{Q}(x)$ may still be valid even if the stronger statement $\forall_x ((\bar{Q}(x) \vee \bar{P}(x)) \Rightarrow \bar{Q}(x))$ is not. However, since

$$(\forall_x (\bar{Q}(x) \vee \bar{P}(x)) \Rightarrow \forall_x \bar{Q}(x)) \Leftrightarrow (\exists_x Q(x) \Rightarrow \exists_x (Q(x) \wedge P(x)))$$

it is fairly obvious that even the weaker statement cannot be inferred without any reference to the definitions of $P(x)$ and $Q(x)$. In spite of this observation, though, it is still possible to complete the proof of q) without going all the way down to the same level of resolution as in the first version above if we take into account that the definitions of $P(x)$ and $Q(x)$ imply that $\forall_x (Q(x) \Rightarrow P(x))$ is true (because $A \cap B \subset A \cup B$). Given this observation, the existence of an x for which $Q(x)$ is true clearly implies the existence of an x for which both $Q(x)$ and $P(x)$ are true (and the same argument can also be used to establish the validity of the stronger statement $\forall_x ((\bar{Q}(x) \vee \bar{P}(x)) \Rightarrow \bar{Q}(x))$ which is equivalent to $\forall_x (Q(x) \Rightarrow (P(x) \wedge Q(x)))$). $\qquad \square$

Exercises

3.4.5. Prove the remaining statements in Proposition 3.4.4.

3.4.6. Prove that the following statements are true for all sets A, B, and C:

  a) $(A \triangle B) \triangle C = A \triangle (B \triangle C)$,

  b) $A \triangle A = \emptyset$,

  c) $A \triangle \emptyset = A$.

  d) $A \triangle B = B \triangle A$.

Note: the validity of a) is the reason why it is permissible to write $A \triangle B \triangle C$ instead of $(A \triangle B) \triangle C$ or $A \triangle (B \triangle C)$.

3.4.7. Given two sets A and B, prove or disprove (by means of a counterexample) each of the following statements:

  a) $\mathcal{P}(A \cup B) \subset \mathcal{P}(A) \cup \mathcal{P}(B)$,

  b) $\mathcal{P}(A \cup B) \supset \mathcal{P}(A) \cup \mathcal{P}(B)$,

  c) $\mathcal{P}(A \cap B) \subset \mathcal{P}(A) \cap \mathcal{P}(B)$,

  d) $\mathcal{P}(A \cap B) \supset \mathcal{P}(A) \cap \mathcal{P}(B)$,

  e) $\mathcal{P}(A \triangle B) \subset \mathcal{P}(A) \triangle \mathcal{P}(B)$,

  f) $\mathcal{P}(A \triangle B) \supset \mathcal{P}(A) \triangle \mathcal{P}(B)$.

3.5   Ordered Pairs and Relations

In order to prepare our study of relations, we need to introduce first the notion of an *ordered pair*. This notion, most likely is well familiar to the reader from analytic geometry where the Cartesian xy-coordinate plane is understood to be the set of all ordered pairs (x, y) of real numbers x and y. However, given our present ambition to demonstrate convincingly that mathematics, at bottom, is the study of sets and nothing but sets, we somehow need to find a way to understand an ordered pair to also be a set. Naturally, we cannot define (x, y) to be $\{x, y\}$ because if $x \neq y$, then $(x, y) \neq (y, x)$ but $\{x, y\} = \{y, x\}$. So the essential property of orderedness would here be simply missing. As it turns out, a definition that adequately reflects this property is the following:

3.5.1 Definition. The *ordered pair* (x, y) of two objects or entities x and y is the set

$$\{\{x\}, \{x, y\}\}.$$

It is understandable if the reader finds this definition at first encounter weird and counterintuitive, but there really is no denying the fact that the spirit of abstract mathematics is beautifully captured by it. What is an ordered pair (x, y)? It is an entity that depends on x and y and is fully characterized by the property of orderedness: two pairs (x_1, y_1) and (x_2, y_2) are equal if and only if $x_1 = x_2$ and $y_1 = y_2$. If the set $\{\{x\}, \{x, y\}\}$ satisfies this property, then that is all we need to know to justify its being called an ordered pair. Mathematics is built up from sets, and ordered pairs are no exception.

3.5.2 Theorem. *Two ordered pairs (x_1, y_1) and (x_2, y_2) are equal if and only if $x_1 = x_2$ and $y_1 = y_2$.*

Proof. '$\Rightarrow$' Assume that $(x_1, y_1) = (x_2, y_2)$, i.e.,

$$\{\{x_1\}, \{x_1, y_1\}\} = \{\{x_2\}, \{x_2, y_2\}\}.$$

Then $\{x_1\} = \{x_2\}$ or $\{x_1\} = \{x_2, y_2\}$ and thus we need to consider two cases.
Case 1: $\{x_1\} = \{x_2, y_2\}$.
In this case we have $x_2, y_2 \in \{x_1\}$ and therefore $x_1 = x_2 = y_2$. Consequently,

$$\{\{x_1\}, \{x_1, y_1\}\} = \{\{y_2\}, \{y_2, y_2\}\} = \{\{y_2\}, \{y_2\}\} = \{\{y_2\}\},$$

and, by imlication, $\{x_1, y_1\} = \{y_2\}$. Thus $x_1 = y_1 = x_2 = y_2$.
Case 2: $\{x_1\} = \{x_2\}$.
In this case we have $x_1 = x_2$, and it remains to be shown that $y_1 = y_2$. Since $\{x_2, y_2\} \in \{\{x_1\}, \{x_1, y_1\}\}$ and since $\{x_2, y_2\} = \{x_1\}$ implies as we saw in Case 1 that $x_1 = x_2 = y_1 = y_2$, we may assume that $\{x_2, y_2\} = \{x_1, y_1\}$. Then $y_2 = y_1$ or $y_2 = x_1$. In the former case we are done, and in the latter case it follows as in Case 1 that $x_1 = y_1 = x_2 = y_2$.
'$\Leftarrow$' If $x_1 = x_2$ and $y_1 = y_2$, then $\{x_1\} = \{x_2\}$ and $\{x_1, y_1\} = \{x_2, y_2\}$, and therefore, $(x_1, y_1) = \{\{x_1\}, \{x_1, y_1\}\} = \{\{x_2\}, \{x_2, y_2\}\} = (x_2, y_2)$. $\qquad\square$

Remark. In the light of the definition of an ordered pair just given, it is tempting to think that an ordered triple might be properly defined as

$$(x, y, z) := \{\{x\}, \{x, y\}, \{x, y, z\}\}.$$

However, this is not the case because this definition does not guarantee that $(x_1, y_1, z_1) = (x_2, y_2, z_2)$ if and only if $x_1 = x_2 \wedge y_1 = y_2 \wedge z_1 = z_2$. A counterexample here is given by the triples $(0, 1, 1)$ and $(0, 1, 0)$ which should, of course, be different as triples but are in fact equal according to the proposed definition above:

$$(0,1,1) = \{\{0\}, \{0,1\}, \{0,1,1\}\} = \{\{0\}, \{0,1\}\}$$
$$= \{\{0\}, \{0,1\}, \{0,1,0\}\} = (0,1,0).$$

A definition of a triple that circumvents this difficulty (see Exercise 3.5.17) is

$$(x,y,z) := (x,(y,z)),$$

where

$$(x,(y,z)) = \{\{x\}, \{x,(y,z)\}\} = \{\{x\}, \{x, \{\{y\}, \{y,z\}\}\}\}.$$

3.5.3 Definition. The *Cartesian product* of two nonempty sets A and B is the set

$$A \times B := \{(x,y) \mid x \in A \wedge y \in B\}.$$

3.5.4 Example. The Cartesian product of $\mathbb{R}$ with itself is $\mathbb{R} \times \mathbb{R}$ and is commonly denoted by $\mathbb{R}^2$.

3.5.5 Example. The Cartesian product of $\{1,2\}$ and $\{a,b\}$ is

$$\{(1,a), (1,b), (2,a), (2,b)\}.$$

3.5.6 Definition. A *relation* r between two nonempty sets A and B is a subset of $A \times B$, i.e., $r \subset A \times B$. Furthermore, if r is a relation between A and B, then

$$D(r) := \{x \in A \mid \exists_{y \in B}(x,y) \in r\}$$

is the *domain* of r, and

$$R(r) := \{y \in B \mid \exists_{x \in A}(x,y) \in r\}$$

is the *range* of r.

Naturally, we may wonder at this point exactly how a subset of a Cartesian product $A \times B$ deserves to be called a 'relation'. What exactly relates here to what? When we think of relations in more down-to-earth intuitive terms there may come to mind, for example, a less-than or greater-than relation between two numbers, but not a set of ordered pairs. However, we need to remember that all of mathematics—supposedly—arises from sets and a less-than or greater-than relation between real numbers must therefore also be construed to be a set. It must be a set that adequately represents the information content of the relation. But when we further inquire as to the nature and extent of this information, we readily realize that the less-than relation on $\mathbb{R}$, for instance, is fully described by all the possible number comparisons: we take a pair of real numbers (x, y) and decide whether x is less than y or not. Seen from this angle, the less-than relation on $\mathbb{R}$ is completely captured by the set

$$r := \{(x, y) \in \mathbb{R}^2 \mid x < y\}, \tag{3.2}$$

and it is therefore this set r that is said to be the less-than relation. Moreover, this less-than relation on $\mathbb{R}$ is what mathematicians call an *order relation*, and more specifically, a *linear order relation*. The essential properties that a relation must possess in order to be so called are listed in the following definition:

3.5.7 Definition. Given a nonempty set S, a relation $r \subset S \times S$ is said to be a *partial order* on S if

a) r is *antisymmetric* in the sense that

$$\forall_{x,y \in S}((x, y) \in r \wedge (y, x) \in r) \Rightarrow x = y, \text{ and}$$

b) r is *transitive* in the sense that

$$\forall_{x,y,z \in S}((x, y) \in r \wedge (y, z) \in r) \Rightarrow (x, z) \in r.$$

Furthermore, a partial order r is said to be a *linear order* or *total order* if

$$\forall_{x,y \in S}((x, y) \in r \vee (y, x) \in r \vee x = y).$$

3.5.8 Example. The less-than relation defined in (3.2) is a linear order on $\mathbb{R}$ because it satisfies all the properties that are stated in Definition 3.5.7: it is antisymmetric because the antecedent $(x, y) \in r \wedge (y, x) \in r$ is equivalent to the assertion $x < y \wedge y < x$ which, by being always false, renders true the material conditional $((x, y) \in r \wedge (y, x) \in r) \Rightarrow x = y$ for all $x, y \in \mathbb{R}$. It is transitive because the antecedent $(x, y) \in r \wedge (y, z) \in r$ is equivalent to $x < y \wedge y < z$ and therefore implies that $x < z$, or equivalently, $(x, z) \in r$. Finally, it is linear because for all real numbers x and y it is the case that either $x < y$ or $y < x$ or $x = y$.

3.5.9 Example. Let us consider the relation

$$r := \{((x_1, y_1), (x_2, y_2)) \in \mathbb{R}^2 \times \mathbb{R}^2 \mid x_1^2 + y_1^2 \leq x_2^2 + y_2^2\}.$$

This relation is transitive because if $x_1^2 + y_1^2 \leq x_2^2 + y_2^2$ and $x_2^2 + y_2^2 \leq x_3^2 + y_3^2$, then certainly $x_1^2 + y_1^2 \leq x_3^2 + y_3^2$. In other words, the assumption that both $((x_1, y_1), (x_2, y_2))$ and $((x_2, y_2), (x_3, y_3))$ are in r always implies that $((x_1, y_1), (x_3, y_3))$ is in r as well. However, r is not a partial order on $\mathbb{R}^2$ because it is not antisymmetric: for $x_1 := 1$, $y_1 := 0$, $x_2 := 0$, and $y_2 := 1$ it is the case that $x_1^2 + y_1^2 = x_2^2 + y_2^2 = 1$, and therefore, it also is the case that both $((x_1, y_1), (x_2, y_2))$ and $((x_2, y_2), (x_1, y_1))$ are contained in r, but it is evidently not the case that $(x_1, y_1) = (x_2, y_2)$.

3.5.10 Example. The set

$$r := \{((x_1, y_1), (x_2, y_2)) \in \mathbb{R}^2 \times \mathbb{R}^2 \mid x_1^2 + y_1^2 < x_2^2 + y_2^2\}.$$

is a partial order on $\mathbb{R}^2$ because r is transitive and antisymmetric. This is so because if $x_1^2 + y_1^2 < x_2^2 + y_2^2$ and $x_2^2 + y_2^2 < x_3^2 + y_3^2$, then $x_1^2 + y_1^2 < x_3^2 + y_3^2$ and because the statement

$$(x_1^2 + y_1^2 < x_2^2 + y_2^2 \wedge x_2^2 + y_2^2 < x_1^2 + y_1^2) \Rightarrow (x_1, y_1) = (x_2, y_2)$$

is true for all $(x_1, y_1), (x_2, y_2) \in \mathbb{R}^2$ as the antecedent

$$x_1^2 + y_1^2 < x_2^2 + y_2^2 \wedge x_2^2 + y_2^2 < x_1^2 + y_1^2$$

is invariably false. However, r is not a linear order because if $x_1 := 1$, $y_1 := 0$, $x_2 := 0$, and $y_2 := 1$, then the statement

$$((x_1, y_1), (x_2, y_2)) \in r \vee ((x_2, y_2), (x_1, y_1)) \in r \vee (x_1, y_1) = (x_2, y_2)$$

is false (because $x_1^2 + y_1^2 = x_2^2 + y_2^2 = 1$ and $(x_1, y_1) \neq (x_2, y_2)$).

3.5.11 Example. An example of a linear order on $\mathbb{R}^2$ is the dictionary order

$$r := \{((x_1, y_1), (x_2, y_2)) \in \mathbb{R}^2 \times \mathbb{R}^2 \mid x_1 < x_2 \vee (x_1 = x_2 \wedge y_1 \leq y_2)\}.$$

In order to prove that r is antisymmetric, we observe that

$$((x_1, y_1), (x_2, y_2)) \in r \wedge ((x_2, y_2), (x_1, y_1)) \in r$$
$$\Leftrightarrow (x_1 < x_2 \vee (x_1 = x_2 \wedge y_1 \leq y_2)) \wedge (x_2 < x_1 \vee (x_2 = x_1 \wedge y_2 \leq y_1))$$
$$\Leftrightarrow (x_1 < x_2 \wedge x_2 < x_1) \vee (x_1 < x_2 \wedge (x_2 = x_1 \wedge y_2 \leq y_1))$$
$$\vee ((x_1 = x_2 \wedge y_1 \leq y_2) \wedge x_2 < x_1)$$
$$\vee ((x_1 = x_2 \wedge y_1 \leq y_2) \wedge (x_2 = x_1 \wedge y_2 \leq y_1))$$
$$\Leftrightarrow F \vee F \vee F \vee (x_1 = x_2 \wedge y_1 \leq y_2 \wedge y_2 \leq y_1)$$
$$\Leftrightarrow x_1 = x_2 \wedge y_1 = y_2.$$

Thus, the assumption $((x_1, y_1), (x_2, y_2)) \in r \wedge ((x_2, y_2), (x_1, y_1)) \in r$ implies that $(x_1, y_1) = (x_2, y_2)$, as desired. Similarly, in order to prove that r is transitive, we observe that

$$((x_1, y_1), (x_2, y_2)) \in r \wedge ((x_2, y_2), (x_3, y_3)) \in r$$
$$\Leftrightarrow (x_1 < x_2 \vee (x_1 = x_2 \wedge y_1 \leq y_2)) \wedge (x_2 < x_3 \vee (x_2 = x_3 \wedge y_2 \leq y_3))$$
$$\Leftrightarrow (x_1 < x_2 \wedge x_2 < x_3) \vee (x_1 < x_2 \wedge (x_2 = x_3 \wedge y_2 \leq y_3))$$
$$\vee ((x_1 = x_2 \wedge y_1 \leq y_2) \wedge x_2 < x_3)$$
$$\vee ((x_1 = x_2 \wedge y_1 \leq y_2) \wedge (x_2 = x_3 \wedge y_2 \leq y_3))$$
$$\Rightarrow x_1 < x_3 \vee (x_1 < x_3 \wedge y_2 \leq y_3) \vee (x_1 < x_3 \wedge y_1 \leq y_2)$$
$$\vee (x_1 = x_3 \wedge y_1 \leq y_3).$$

Since $P \vee (P \wedge Q)$ is equivalent to P, we may further conclude that

$$x_1 < x_3 \vee (x_1 < x_3 \wedge y_2 \leq y_3) \vee (x_1 < x_3 \wedge y_1 \leq y_2) \vee (x_1 = x_3 \wedge y_1 \leq y_3)$$
$$\Leftrightarrow x_1 < x_3 \vee (x_1 < x_3 \wedge y_1 \leq y_2) \vee (x_1 = x_3 \wedge y_1 \leq y_3)$$
$$\Leftrightarrow x_1 < x_3 \vee (x_1 = x_3 \wedge y_1 \leq y_3).$$

So the assumption $((x_1, y_1), (x_2, y_2)) \in r \wedge ((x_2, y_2), (x_3, y_3)) \in r$ implies that $((x_1, y_1), (x_3, y_3)) \in r$, as it should. Finally, r is linear because

$$((x_1, y_1), (x_2, y_2)) \in r \vee ((x_2, y_2), (x_1, y_1)) \in r \vee (x_1, y_1) = (x_2, y_2)$$
$$\Leftrightarrow (x_1 < x_2 \vee (x_1 = x_2 \wedge y_1 \leq y_2)) \vee (x_2 < x_1 \vee (x_2 = x_1 \wedge y_2 \leq y_1))$$
$$\vee (x_1 = x_2 \wedge y_1 = y_2)$$
$$\Leftrightarrow x_1 < x_2 \vee x_2 < x_1 \vee (x_1 = x_2 \wedge y_1 \leq y_2) \vee (x_2 = x_1 \wedge y_2 \leq y_1)$$
$$\vee (x_1 = x_2 \wedge y_1 = y_2),$$

and because the last of these statements is true (see Exercise 3.5.19).

3.5.12 Example. If S is a set, then $r := \{(A, B) \in \mathcal{P}(S) \times \mathcal{P}(S) \mid A \subset B\}$ is a partial order on $\mathcal{P}(S)$. This is so because for all $A, B, C \subset S$ it is the case that

$$(A \subset B \wedge B \subset A) \Rightarrow A = B \quad \text{(by Proposition 3.3.3g)},$$

and

$$(A \subset B \wedge B \subset C) \Rightarrow A \subset C \quad \text{(by Proposition 3.3.3f)}.$$

Apart from order relations we also encounter very frequently another type of relation, the so-called *equivalence relation*. A prototypical example of an equivalence relation is the relation of equality on any given nonempty set S:

$$r := \{(x, y) \in S \times S \mid x = y\}.$$

This relation is, as we say, *reflexive* because $x = x$ for all $x \in S$, it is *symmetric* because if $x = y$ then $y = x$, and it is *transitive* because the assumption that $x = y$ and $y = z$ implies that $x = z$. Inspired by this (admittedly extremely simple) example, we introduce the following definition:

3.5.13 Definition. Let S be a nonempty set. A relation $r \subset S \times S$ is said to be an *equivalence relation* on S if

a) r is *reflexive* in the sense that

$$\forall_{x \in S} (x, x) \in r,$$

b) r is *symmetric* in the sense that

$$\forall_{x, y \in S} (x, y) \in r \Rightarrow (y, x) \in r, \quad \text{and}$$

c) r is *transitive* in the sense that

$$\forall_{x,y,z \in S}((x,y) \in r \wedge (y,z) \in r) \Rightarrow (x,z) \in r.$$

3.5.14 Example. An example of an equivalence relation on $\mathbb{Z}$ is the relation of equality modulo n: given an integer $n \in \mathbb{N}$, we say that two numbers $x, y \in \mathbb{Z}$ are *equal modulo* n if x and y produce the same remainder when divided by n. Since the latter statement is evidently equivalent to the assertion that the difference $x - y$ is divisible by n, we proceed to claim that

$$r_n := \{(x,y) \in \mathbb{Z} \times \mathbb{Z} \mid n|x - y\}$$

is an equivalence relation on $\mathbb{Z}$. In order to prove this claim we need to show that r_n is reflexive, symmetric, and transitive. Clearly, r_n is reflexive because the fact that n divides $0 = x - x$ implies that $(x,x) \in r_n$ for all $x \in \mathbb{Z}$. Furthermore, r is symmetric because if n divides $x - y$, then n also divides $-(x - y) = y - x$. In other words, the assumption that (x,y) is contained in r_n always implies that (y,x) is contained in r_n as well. Finally, to show that r_n is transitive, we assume that $n|x - y$ and $n|y - z$ for some $x, y, z \in \mathbb{Z}$. Then there are integers $k, l \in \mathbb{Z}$ such that $nk = x - y$ and $nl = y - z$. Hence $n|x - z$, as desired, because

$$x - z = x - y + y - z = nk + nl = n(k + l).$$

3.5.15 Example. Another standard example of an equivalence relation is the relation of equality of fractions: two number pairs $(m,n), (p,q) \in S := \mathbb{Z} \times \mathbb{N}$ are said to be equivalent if $m/n = p/q$, or equivalently, if $mq = np$. Thus we wish show that

$$r := \{((m,n),(p,q)) \in S \times S \mid mq = np\}$$

is an equivalence relation on S. To do so, we observe to begin with that r is reflexive because $nm = mn$ for all $(m,n) \in S$. Moreover, r is symmetric because if $mq = np$ then, trivially, $pn = qm$. Finally, to prove that r is transitive, we assume that $mq = np$ and $pt = qs$ for some integers $m, p, s \in \mathbb{Z}$ and $n, q, t \in \mathbb{N}$. This yields $mtq = npt = nqs$ and, by implication, $0 = q(mt - ns)$. Since q is in $\mathbb{N}$ and therefore different from zero, we may infer that $mt = ns$, as desired. Note: to arrive at this latter conclusion we assume that it is known that for all $m, n \in \mathbb{Z}$ the assumption $mn = 0$ implies that $m = 0 \vee n = 0$ (see Theorem 6.1.2g, p.160).

3.5.16 Proposition. *If $r_1 \subset S_1 \times S_1$ and $r_2 \subset S_2 \times S_2$ are equivalence relations on two nonempty sets S_1 and S_2, respectively, then*

$$r := \{((x_1,x_2),(y_1,y_2)) \in (S_1 \times S_2) \times (S_1 \times S_2) \mid (x_1,y_1) \in r_1 \wedge (x_2,y_2) \in r_2\}$$

is an equivalence relation on $S_1 \times S_2$ (see Exercise 3.5.24).

Exercises

3.5.17. Prove that $(x_1, (y_1, z_1)) = (x_2, (y_2, z_2))$ if and only if $x_1 = x_2 \land y_1 = y_2 \land z_1 = z_2$.

3.5.18. Prove that the following sets are linear orders on $\mathbb{R}$:

  a) $\{(x, y) \in \mathbb{R}^2 \mid x \leq y\}$,
  b) $\{(x, y) \in \mathbb{R}^2 \mid x \geq y\}$,
  c) $\{(x, y) \in \mathbb{R}^2 \mid x > y\}$.

3.5.19. Explain why the statement $x_1 < x_2 \lor x_2 < x_1 \lor (x_1 = x_2 \land y_1 \leq y_2) \lor (x_2 = x_1 \land y_2 \leq y_1) \lor (x_1 = x_2 \land y_1 = y_2)$ is true for all $x_1, x_2, y_1, y_2 \in \mathbb{R}$.

3.5.20. Prove that $r := \{(n, m) \in \mathbb{N} \times \mathbb{N} \mid n \mid m\}$ is a partial order on $\mathbb{N}$ that is not linear.

3.5.21. What is the maximal number of elements that S may contain, if r, as defined in Example 3.5.12, is to be a linear order on $\mathcal{P}(S)$.

3.5.22. Let S be a nonempty set. Prove that $r := \emptyset$ and $s := \{(x, y) \in S \times S \mid x = y\}$ are both partial orders on S. Are r and/or s linear orders? Explain your answer.

3.5.23. How many antisymmetric relations on $S := \{1, 2, \ldots, n\}$ satisfy the linearity condition $\forall_{x,y \in S}((x, y) \in r \lor (y, x) \in r \lor x = y)$? *Hint:* you may want to use the fact that $\sum_{k=1}^{n} k = n(n + 1)/2$ for all $n \in \mathbb{N}$.

3.5.24. Prove Proposition 3.5.16.

3.5.25. Prove that the domain of an equivalence relation on S is always S itself.

3.5.26. Assume that T and S are nonempty sets such that $T \subset S$. Prove that

$$r := \{(A, B) \in \mathcal{P}(S) \times \mathcal{P}(S) \mid A \triangle B \in \mathcal{P}(T)\}$$

is an equivalence relation on $\mathcal{P}(S)$.

3.5.27. Assume that $r \subset S \times S$ is a relation on a nonempty set S which is transitive but not reflexive. Prove that r is not symmetric whenever $D(r) = S$.

3.5.28. Let S be a nonempty set. Prove that $r := S \times S$ and

$$s := \{(x, x) \mid x \in S\}$$

are equivalence relations on S.

3.5.29. Show that for any equivalence relation r the following statement is true: if $(x, y), (u, v) \in r$, then $(x, u) \in r$ if and only if $(y, v) \in r$.

3.5.30. Show that $r := \{(x, y) \in \mathbb{R} \setminus \{0\} \times \mathbb{R} \setminus \{0\} \mid x/y \in \mathbb{Q}\}$ is an equivalence relation on $\mathbb{R} \setminus \{0\}$.

3.5.31. If $r \subset S \times S$ is a partial order on a nonempty set S and $M \subset S$, then $x \in M$ is said to be *extreme* (i.e., *minimal* or *maximal*) in M if $(\forall_{z \in M}(x, z) \in r \Rightarrow x = z) \vee (\forall_{z \in M}(z, x) \in r \Rightarrow x = z)$. Prove that the following statement is true whenever r is linear: if $x, y \in M$ are either both minimal or both maximal in M, then $x = y$. (So in setting $P(x, z) :\Leftrightarrow ((x, z) \in r \Rightarrow x = z)$, the assumption that x and y are here supposed to satisfy is either $\forall_{z \in M} P(x, z) \wedge \forall_{z \in M} P(y, z)$ or $\forall_{z \in M} P(z, x) \wedge \forall_{z \in M} P(z, y)$.)

3.6   Partitions and Equivalence Classes

Looking back at Example 3.5.15, we notice that the nature of the eqivalence relation

$$r := \{((m, n), (p, q)) \in S \times S \mid mq = np\} \tag{3.3}$$

is to identify—or declare to be equivalent—any two pairs $(m, n), (p, q) \in S = \mathbb{Z} \times \mathbb{N}$ that produce the same fractional value. Put differently, we may say that for each fractional value, such as for instance the value $2/3$, there is an infinite set of pairs (m, n) that produce this particular value whenever the first component m is divided by the second component n. Moreover, these sets of pairs are pairwise disjoint and cover all of S because each pair produces exactly one fractional value m/n. To generalize this observation, we introduce the following definition:

3.6.1 Definition. Assume that r is an equivalence relation on a nonempty set S and let $x \in S$. Then

$$[x]_r := \{y \in S \mid (x, y) \in r\}$$

is said to be the *equivalence class* of x relative to r. Note: whenever the context is unambiguous, we will write $[x]$ instead of $[x]_r$.

3.6.2 Lemma. *If r is an equivalence relation on a nonempty set S, then for all $x, y \in S$ it is the case that*

  a) $x \in [x]$, *and*

  b) $[x] = [y]$ *if and only if $y \in [x]$.*

Proof. Statement a) is a trivial consequence of the fact that r is reflexive, and in order to prove b), we use a), with y in place of x, to infer that $y \in [x]$ whenever $[x] = [y]$. To prove the reverse implication in b), we assume that

$y \in [x]$ for some $x, y \in S$. Then $(x, y) \in r$ (by Definition 3.6.1), and therefore, $(y, x) \in r$ and $x \in [y]$ because r is symmetric. Consequently, it is sufficient to show that the assumption $y \in [x]$ implies that $[y] \subset [x]$, because if this implication is valid for all $x, y \in S$, then the fact that $x \in [y]$ implies that the reverse inclusion $[x] \subset [y]$ is satisfied as well and that $[x] = [y]$, as desired. So let $z \in [y]$, then $(y, z) \in r$, and since $(x, y) \in r$, it follows that $(x, z) \in r$ because r is transitive. Hence $z \in [x]$, and this shows that indeed $[y] \subset [x]$. $\square$

3.6.3 Example. For the equivalence relation r, defined in (3.3), we have

$$[(m, n)]_r = \{(p, q) \in \mathbb{Z} \times \mathbb{N} \mid mq = np\}.$$

That is to say, $[(m, n)]_r$ consists of all pairs $(p, q) \in \mathbb{Z} \times \mathbb{N}$ for which $m/n = p/q$. In this sense we may say that $[(m, n)]_r$ not only represents the fractional value m/n but is in fact identical with it. Indeed, in Chapter 6 we will adopt this point of view and proceed to *define* the set $\mathbb{Q}$ of rational numbers m/n to be equal to the set of all the equivalence classes $[(m, n)]_r$. Furthermore, since each fractional value is represented by—or equal to—exactly one equivalence class $[(m, n)]_r$, it follows that the entire set $\mathbb{Z} \times \mathbb{N}$ is a disjoint union of the equivalence classes $[(m, n)]_r$ or that, equivalently, the set of all the equivalence classes $[(m, n)]_r$ forms a *partition* of $\mathbb{Z} \times \mathbb{N}$. Hence we introduce the following definition:

3.6.4 Definition. Let S be a nonempty set. Then $\pi \subset \mathcal{P}(S)$ is said to be a *partition* of S if

    a) $\forall_{x \in S} \exists_{A \in \pi} \ x \in A$,

    b) $\forall_{A, B \in \pi} \ A \cap B \neq \emptyset \Rightarrow A = B$, and

    c) $\forall_{A \in \pi} A \neq \emptyset$.

3.6.5 Theorem. *If r is an equivalence relation on a nonempty set S, then*

$$\pi := \{[x] \mid x \in S\}$$

is a partition of S.

Proof. Properties a) and c) of Definition 3.6.4 are implied by Lemma 3.6.2a. Turning to b), we assume that we are given two equivalence classes $[x]$ and $[y]$ such that $[x] \cap [y] \neq \emptyset$. Given this assumption, we may pick an element $z \in S$ such that $z \in [x]$ and $z \in [y]$. Then $[z] = [x]$ and $[z] = [y]$ (by Lemma 3.6.2b), and therefore, $[x] = [y]$. Thus $\pi = \{[x] \mid x \in S\}$ is a partition. $\square$

3.6.6 Theorem. *If π is a partition of a nonempty set S, then*

$$r := \{(x, y) \in S \times S \mid \exists_{A \in \pi} \ x \in A \wedge y \in A\}$$

is an equivalence relation on S the equivalence classes of which are precisely the elements of π.

Proof. In order to demonstrate that r is an equivalence relation, we need to show that r is reflexive, symmetric, and transitive. To do so we assume first that x is an arbitrary element of S. Since π is assumed to be a partition, it follows that there exists an $A \in \pi$ such that $x \in A$ (see Definition 3.6.4a). Thus $(x, x) \in r$, and therefore, r is reflexive. In order to prove that r is symmetric, we assume that $(x, y) \in r$ for some $x, y \in S$. Then, by the definition of r, there exists an $A \in \pi$ such that $x \in A \wedge y \in A$. Since the latter statement is obviously equivalent to $y \in A \wedge x \in A$, it follows that $(y, x) \in A$. Consequently, r is symmetric. Finally, in order to establish transitivity, we assume that $x, y, z \in S$ such that $(x, y) \in r$ and $(y, z) \in r$. Using again the definition of r, it follows that there are sets $A, B \in \pi$ such that $x, y \in A$ and $y, z \in B$. Thus y is an element of both A and B, and, by implication, $A \cap B \neq \emptyset$. Using Definition 3.6.4b, it follows that $A = B$, and therefore $x, z \in A$. Hence $(x, z) \in r$ as desired. Turning now to the second assertion concerning the identity of the equivalence classes of r and the elements in π, it is sufficient to show that

$$\forall_{A \in \pi} \forall_{x \in A} \, A = [x].$$

So let $A \in \pi$ and $x \in A$ (note: $A \neq \emptyset$ by Definition 3.6.4c). If $y \in A$, then $(x, y) \in r$ and therefore $y \in [x]$. Thus $A \subset [x]$. If, on the other hand, $y \in [x]$, then $(x, y) \in r$ and therefore $x, y \in B$ for some $B \in \pi$. Since x is contained in both A and B, we may infer that $A \cap B \neq \emptyset$ and therefore $A = B$ by Definition 3.6.4b. So y is in A and, by implication, $[x] \subset A$. Thus $[x] = A$, as desired.   □

Exercises

3.6.7. Show that

$$r := \{(x, y) \in [0, 1] \times [0, 1] \mid x = y\} \cup \{(1/4, 3/4), (3/4, 1/4)\}$$

is an equivalence relation and describe the partition of $[0, 1]$ that r induces.

3.6.8. Let $\pi := \{[0, 1/2] \cup [5/6, 1], (1/2, 2/3], (2/3, 5/6)\}$. Sketch the equivalence relation corresponding to π as a subset of $[0, 1] \times [0, 1]$.

3.6.9. What are the equivalence classes of the equivalence relation defined in Example 3.5.14?

3.6.10. What are the equivalence classes of the equivalence relation defined in Exercise 3.5.26?

3.6.11. Show that $r := \{(x, y) \in \mathbb{R}^2 \mid x - y \in \mathbb{Z}\}$ is an equivalence relation on $\mathbb{R}$ and describe the partition of $\mathbb{R}$ that r induces.

3.6.12. Show that $r := \{(x, y) \in \mathbb{R} \smallsetminus \{0\} \times \mathbb{R} \smallsetminus \{0\} \mid x/y \in \{1, -1\}\}$ is an equivalence relation on $\mathbb{R} \smallsetminus \{0\}$ and describe the partition of $\mathbb{R} \smallsetminus \{0\}$ that r induces.

3.7 Functions

Among the various types of relations that we encounter in mathematics it is the functional relation that is by far the most common and also the most important.

3.7.1 Definition. Given two nonempty sets A and B, a relation $f \subset A \times B$ is said to be a *function* or *map* from A to B if

a) $D(f) = A$ (i.e., $\forall_{x \in A} \exists_{y \in B} (x, y) \in f$), and

b) $\forall_{x \in A} \forall_{y_1, y_2 \in B} ((x, y_1) \in f \wedge (x, y_2) \in f) \Rightarrow y_1 = y_2$.

Remark. The purpose of property b) in Definition 3.7.1 is to guarantee that to every x in the domain of f (that is, in A) there corresponds *exactly one* $y \in B$ for which the statement $(x, y) \in f$ is valid. Due to the uniqueness and existence of this element y, it is customary to write $y = f(x)$. In other words, y is uniquely determined by x, and a function from A to B is therefore best regarded as a rule of correspondence that assigns to each input element $x \in A$ the output element $y = f(x) \in B$. A standard notation that expresses this unique functional dependence of the output on the input is

$$f : A \to B$$
$$x \mapsto f(x).$$

Furthermore, if f is a function from a subset A of $\mathbb{R}$ to $\mathbb{R}$, then the property of being a function is equivalent to the demand that f passes the *vertical line test* in the sense that every line parallel to the y-axis intersects f (as a subset of $\mathbb{R}^2$) at most once.

3.7.2 Example. Using the notation of the preceding remark, the familiar function $f(x) := x^2$—as a function from $\mathbb{R}$ to $\mathbb{R}$—can be written as

$$f : \mathbb{R} \to \mathbb{R}$$
$$x \mapsto x^2.$$

Alternatively, we also may write $f : \mathbb{R} \to \mathbb{R}$, $f(x) := x^2$, and in the spirit of Definition 3.7.1, this function is equal to the *set*

$$f = \{(x, x^2) \mid x \in \mathbb{R}\}.$$

Put differently, Definition 3.7.1 identifies a function with its *graph*—the set of all points $(x, f(x))$ for which $x \in A = D(f)$.

3.7.3 Example. A real or rational sequence $(a_n)_{n \in \mathbb{N}}$ is a function of the form

$$a : \mathbb{N} \to \mathbb{R}$$
$$n \mapsto a_n$$

or

$$a : \mathbb{N} \to \mathbb{Q}$$
$$n \mapsto a_n.$$

So for all $n \in \mathbb{N}$ the value of the function a at n is a_n, that is, $a(n) = a_n$.

3.7.4 Definition. If f is a function from A to B, and $M \subset A$, then

$$f(M) := \{ y \in B \mid \exists_{x \in M} \, f(x) = y \}$$

is said to be the *image* of M under f.

3.7.5 Definition. If f is a function from A to B, and $N \subset B$, then

$$f^{-1}(N) := \{ x \in A \mid f(x) \in N \}$$

is said to be the *pre-image* of N under f.

Remark. If f is a function from A to B, then $f(A)$ is the range of f and the pre-image of the range is the domain, that is, $f(A) = R(f)$ and $f^{-1}(R(f)) = A$. Furthermore, it is important to understand that both '$f(M)$' and '$f^{-1}(N)$' are notations for sets. They are not values of a function but rather subsets of the range and the domain of f, respectively. In particular, for readers that are familiar with the concept of an inverse function from elementary calculus or pre-calculus courses, we wish to point out that the use of the symbol 'f^{-1}' does not in any way indicate that f is an invertible function and that f^{-1} is its inverse. Invertibility is not a property that Definition 3.7.5 makes any reference to, and the pre-image $f^{-1}(N)$ is therefore well defined for all functions $f \subset A \times B$ and all sets $N \subset B$.

3.7.6 Example. Let $f : \mathbb{R} \to \mathbb{R}$ be defined by the equation $f(x) := x^2 + 1$ and let $M := [-1, 2)$ (see Figure 3.1 on the left). Then $f(M) = [1, 5)$. Furthermore, for $N := (2, 5]$ we have $f^{-1}(N) = [-2, -1) \cup (1, 2]$ (see Figure 3.1 on the right).

3.7.7 Example. Let c be a real number and let $f : \mathbb{R} \to \mathbb{R}$ be the constant function whose value is c, that is, $f(x) := c$ for all $x \in \mathbb{R}$. Then for all sets $M, N \subset \mathbb{R}$ it is the case that

$$f(M) = \begin{cases} \{c\} & \text{if } M \neq \emptyset, \\ \emptyset & \text{if } M = \emptyset \end{cases}$$

and

$$f^{-1}(N) = \begin{cases} \mathbb{R} & \text{if } c \in N, \\ \emptyset & \text{if } c \notin N. \end{cases}$$

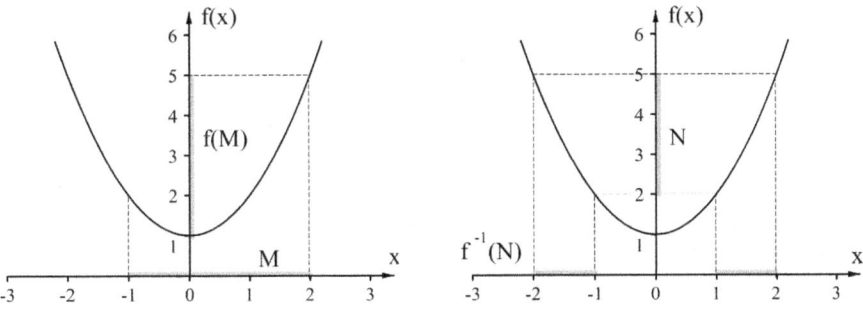

Figure 3.1: $f(M) = [1, 5)$ and $f^{-1}(N) = [-2, -1) \cup (1, 2]$.

3.7.8 Example. If $f : \mathbb{R}^2 \to \mathbb{R}$ (see Example 3.5.4) is defined by the equation $f(x, y) := x^2 + y^2$, then

$$f([-1, 3] \times [-2, 4]) = [0, 3^2 + 4^2] = [0, 25],$$
$$f([1, 3] \times [2, 4]) = [1^2 + 2^2, 3^2 + 4^2] = [5, 25],$$

and

$$f^{-1}([-1, 1]) = f^{-1}([0, 1]) = \{(x, y) \in \mathbb{R}^2 \mid x^2 + y^2 \le 1\},$$

where the latter set $\{(x, y) \in \mathbb{R}^2 \mid x^2 + y^2 \le 1\}$ is the unit disc in $\mathbb{R}^2$, that is, the region that is enclosed by the circle of radius one that is centered at the origin. Similarly, we also find that

$$f^{-1}((0, 1)) = \{(x, y) \in \mathbb{R}^2 \mid 0 < x^2 + y^2 < 1\}$$
$$= \{(x, y) \in \mathbb{R}^2 \mid x^2 + y^2 < 1\} \setminus \{(0, 0)\}.$$

In words: $f^{-1}((0, 1))$ is the open unit disc (without boundary) from which the center point $(0, 0)$ has been removed.

Exercises

3.7.9. Which of the following relations are functions from $\mathbb{R}$ to $\mathbb{R}$? Explain your answer.

   a) $\{(x, y) \in \mathbb{R}^2 \mid x^2 + y^2 = 1\}$
   b) $\{(x, y) \in \mathbb{R}^2 \mid x^2 + y = 1\}$
   c) $\{(x, y) \in \mathbb{R}^2 \mid x + y^2 = 1\}$
   d) $\{(x, y) \in \mathbb{R}^2 \mid x + y = 1\}$

3.7.10. Find the images $f(M)$ and the pre-images $f^{-1}(N)$ for the functions and sets listed below.

a) $f : \mathbb{R} \to \mathbb{R}$, $f(x) := x^3$, $M := [0, 1]$, $N := [-8, 1)$.

b) $f : \mathbb{N} \to \mathbb{N}$, $f(x) := x^2$, $M := \{1, 2, 3\}$, $N := \{n \in \mathbb{N} \mid \exists_{k \in \mathbb{N}} \, n = 2k\}$.

c) $f : \mathbb{R} \to \mathbb{R}$, $f(x) := |x|$, $M := [0, 1]$, $N := [1, 3]$.

d) $f : [-1, \infty) \to \mathbb{R}$, $f(x) := |x + 1|$, $M := [-2, 1]$, $N := [1, 3]$.

3.7.11. Given two nonempty sets A and B, explain why all relations $r \subset A \times B$ satisfy the condition

$$\forall_{x \in A} \exists_{y_1, y_2 \in B} \big((x, y_1) \in r \wedge (x, y_2) \in r \big) \Rightarrow y_1 = y_2.$$

3.7.12. Which of the relations $r \subset \mathbb{R}^2$ listed below satisfy the condition

$$\exists_{x \in D(r)} \forall_{y_1, y_2 \in \mathbb{R}} \big((x, y_1) \in r \wedge (x, y_2) \in r \big) \Rightarrow y_1 = y_2 ?$$

a) $r = [0, 1] \times [0, 1]$

b) $r = \{(x, y) \in \mathbb{R}^2 \mid x^2 + y^2 = 1\}$

c) $r = \{(x, y) \in \mathbb{R}^2 \mid y = x^2\}$

d) $r = \{(x, y) \in \mathbb{R}^2 \mid y = x^2\} \cup \{(x, y) \in \mathbb{R}^2 \mid y = x^2 + 1\}$

3.8   Injections, Surjections, and Bijections

3.8.1 Definition. A function f from A to B is said to be...

a) *injective* (or an *injection*) if

$$\forall_{y \in B} \forall_{x_1, x_2 \in A} \big((x_1, y) \in f \wedge (x_2, y) \in f \big) \Rightarrow x_1 = x_2$$

or equivalently, if

$$\forall_{x_1, x_2 \in A} \, f(x_1) = f(x_2) \Rightarrow x_1 = x_2.$$

b) *surjective* (or a *surjection*) if $R(f) = B$, that is, if

$$\forall_{y \in B} \exists_{x \in A} \, (x, y) \in f,$$

or equivalently, if

$$\forall_{y \in B} \, f^{-1}(\{y\}) \neq \emptyset.$$

c) *bijective* (or a *bijection*) if f is injective and surjective.

3.8.2 Example. The function $f : \mathbb{R} \setminus \{0\} \to \mathbb{R}$, $f(x) := 1/x$ is injective, because if $1/x_1 = f(x_1) = f(x_2) = 1/x_2$, then $x_1 = x_2$. However, f is not surjective (and therefore also not bijective) because there is no $x \in \mathbb{R} \setminus \{0\} = D(f)$ such that $f(x) = 0$ (or equivalently, because $f^{-1}(\{0\}) = \emptyset$).

Furthermore, since $y = 0$ is the only value in $B = \mathbb{R}$ for which there is no corresponding input value $x \in \mathbb{R} \setminus \{0\} = D(f)$ with $f(x) = y$ (this is so because if $y \neq 0$ and $x := 1/y$, then $f(x) = 1/(1/y) = y$), it follows that the function $g : \mathbb{R} \setminus \{0\} \to \mathbb{R} \setminus \{0\}$, $g(x) := 1/x$ is surjective, and since this function is also injective (just as f), we may infer that g is in fact bijective.

Remark. In generalizing the observations made in the preceding example, we notice that for a given function $f : A \to B$, the function $g : A \to R(f)$, $g(x) := f(x)$ is surjective. Moreover, if f is a function from a subset A of $\mathbb{R}$ to $\mathbb{R}$, then f is injective if and only if f passes the *horizontal line test* in the sense that every line parallel to the x-axis intersects the graph of f (as a subset of $\mathbb{R}^2$) at most once.

3.8.3 Example. The function $f : \mathbb{R} \to \mathbb{R}$, $f(x) := x^2$ is not injective because for $x_1 := 1$ and $x_2 := -1$ it is the case that $f(x_1) = f(x_2)$ and $x_1 \neq x_2$. Furthermore, f is also not surjective because $x^2 \geq 0$ for all $x \in \mathbb{R}$ and therefore $f^{-1}(\{y\}) = \emptyset$ for all $y < 0$. By contrast, the function $g : [0, \infty) \to [0, \infty)$, $g(x) := x^2$ is both injective and surjective and therefore bijective. It is injective, because if $x_1, x_2 \geq 0$ and $x_1^2 = g(x_1) = g(x_2) = x_2^2$, then $x_1 = \sqrt{x_1^2} = \sqrt{x_2^2} = x_2$, and it is surjective because if $y \in [0, \infty)$, then $g(x) = y$ for $x := \sqrt{y}$.

3.8.4 Definition. If S is a nonempty set, then the *identity map* on S is

$$\mathrm{id}_S : S \to S$$

$$x \mapsto x$$

Note: if the context makes it clear what set S we are dealing with, we will simply write id instead of id_S.

3.8.5 Example. If S is a nonempty set, then id_S is bijective (see Exercise 3.8.26).

3.8.6 Theorem. *Given two nonempty sets A and B, a function f from A to B is injective if and only if*

$$f^{-1} := \{(y, x) \in f(A) \times A \mid (x, y) \in f\}$$

is a function from $f(A)$ to A. In that case f is also said to be invertible and f^{-1} is said to be the inverse function of f. So the words 'injective' and 'invertible' can be used interchangeably.

Proof. "$\Rightarrow$" Assuming that $f : A \to B$ is injective, we need to show that

a) $\forall_{y \in f(A)} \exists_{x \in A} (y, x) \in f^{-1}$, and

b) $\forall_{y \in f(A)} \forall_{x_1, x_2 \in A} ((y, x_1) \in f^{-1} \wedge (y, x_2) \in f^{-1}) \Rightarrow x_1 = x_2$.

To prove a), we pick an arbitrary element $y \in f(A)$. Then, by the definition of $f(A)$, there exists an $x \in A$ such that $f(x) = y$ or, equivalently, $(x, y) \in f$. Using the definition of f^{-1}, it follows that $(y, x) \in f^{-1}$, as desired. Turning to b), we observe that, given the definition of f^{-1}, the statement

$$\forall_{y \in f(A)} \forall_{x_1, x_2 \in A} \left((y, x_1) \in f^{-1} \wedge (y, x_2) \in f^{-1} \right) \Rightarrow x_1 = x_2 \qquad (3.4)$$

is equivalent to

$$\forall_{y \in f(A)} \forall_{x_1, x_2 \in A} \left((x_1, y) \in f \wedge (x_2, y) \in f \right) \Rightarrow x_1 = x_2. \qquad (3.5)$$

Since the latter statement is true by the assumed injectivity of f, it follows that the former statement is true as well.
"$\Leftarrow$" If f^{-1} is a function from $f(A)$ to A, then f is injective by the above observed equivalence of (3.4) and (3.5). $\qquad \square$

3.8.7 Theorem. *If g is a function from A to B and f is a function from B to C, then*

$$f \circ g := \{ (x, z) \in A \times C \mid \exists_{y \in B} (x, y) \in g \wedge (y, z) \in f \}$$

is a function from A to C—the so-called composition of f and g.

Proof. We need to show that

a) $\forall_{x \in A} \exists_{z \in C} (x, z) \in f \circ g$, and

b) $\forall_{x \in A} \forall_{z_1, z_2 \in C} \left((x, z_1) \in f \circ g \wedge (x, z_2) \in f \circ g \right) \Rightarrow z_1 = z_2$.

To prove a) we pick an $x \in A$ and argue as follows: since g is a function defined on A, there exists a $y \in B$ such that $(x, y) \in g$, and since f is a function defined on B, there exists a $z \in C$ such that $(y, z) \in f$. Thus $(x, z) \in f \circ g$, as desired. To establish b), we assume that $x \in A$ and $z_1, z_2 \in C$ are given such that $(x, z_1), (x, z_2) \in f \circ g$. Then, given the definition of $f \circ g$, it follows that there exist $y_1, y_2 \in B$ such that $(x, y_1), (x, y_2) \in g$ and $(y_1, z_1), (y_2, z_2) \in f$. Since g is a function, it follows that $y_1 = y_2$. Setting $y := y_1$, it further follows that $(y, z_1), (y, z_2) \in f$, and therefore $z_1 = z_2$, as desired (because f is a function). $\qquad \square$

Remark. If $g : A \to B$ and $f : B \to C$ are given functions (as in the preceding theorem), and if $x \in A$, $y \in B$, and $z \in C$ such that $(x, y) \in g$ and $(y, z) \in f$, then $g(x) = y$, $f(y) = z$, and $(x, z) \in f \circ g$. Hence $f(g(x)) = f(y) = z = (f \circ g)(x)$. In other words, the defining equation for $f \circ g$ in standard function notation is

$$(f \circ g)(x) := f(g(x))$$

for any $x \in A$.

3.8.8 Corollary. *If* $f : A \to B$ *is injective, then*

$$f \circ f^{-1} = \mathrm{id}_{f(A)}$$

and

$$f^{-1} \circ f = \mathrm{id}_A .$$

Proof. In order to prove the first equation, we need to show that $f(f^{-1}(y)) = y$ for all $y \in f(A)$. So let y be in $f(A)$. Then, by the definition of $f(A)$, there exists an $x \in A$ such that $f(x) = y$, or equivalently, $(x, y) \in f$. Thus $(y, x) \in f^{-1}$ (by the definition of f^{-1}) and therefore $f(f^{-1}(y)) = f(x) = y$, as desired. The proof of the second equation above will be left as an exercise to the reader (see Exercise 3.8.27). □

Remark. If $f : A \to B$ is injective and $N \subset f(A)$, then the expression $f^{-1}(N)$ has a dual meaning. On the one hand it denotes the pre-image of N under f and on the other hand the image of N under f^{-1}. However, there is really no problem arising as the former interpretation is equivalent to the latter. For to say that $x \in A$ is contained in the pre-image of N under f is equivalent to the assertion that there exists a $y \in N$ such that $f(x) = y$, and this in turn is equivalent (by Corollary 3.8.8) to the statement that $x = f^{-1}(y)$ is contained in the image of N under f^{-1}.

3.8.9 Theorem. *A function* $f : A \to B$ *is injective if and only if there exists a function* $g : B \to A$ *such that* $g \circ f = \mathrm{id}_A$.

Proof. "$\Rightarrow$" If f is injective, then $f^{-1} : f(A) \to A$ is well defined and $f^{-1} \circ f = \mathrm{id}_A$. To proceed we consider two cases:
Case 1: $f(A) = B$.
Then $g := f^{-1}$ is a function from B to A such that $g \circ f = \mathrm{id}_A$ (see Corollary 3.8.8).
Case 2: $f(A) \neq B$.
In this case we pick an arbitrary $a \in A$ and define

$$g(y) := \begin{cases} f^{-1}(y) & \text{if } y \in f(A), \\ a & \text{if } y \in B \setminus f(A). \end{cases}$$

Then $(g \circ f)(x) = g(f(x)) = f^{-1}(f(x)) = x$ for all $x \in A$, and therefore $g \circ f = \mathrm{id}_A$, as desired.
"$\Leftarrow$" Given a function $g : B \to A$ such that $g \circ f = \mathrm{id}_A$, we need to show that

$$\forall_{x_1, x_2 \in A} \; f(x_1) = f(x_2) \Rightarrow x_1 = x_2.$$

So let us assume that $f(x_1) = f(x_2)$ for some $x_1, x_2 \in A$. Then $x_1 = g(f(x_1)) = g(f(x_2)) = x_2$, as desired. □

In order to characterize the property of sujectivity in a manner similar to the characterization of injectivity in Therorem 3.8.9, we need to introduce one of the fundamental axioms of set theory, the so-called *Axiom of Choice.*

The Axiom of Choice: *if I is a nonempty set and $\{M_i \mid i \in I\}$ is a collection of nonempty subsets of a given set M (i.e., $M_i \subset M$ for all $i \in I$), then there exists a function $g : I \to M$ such that $g(i) \in M_i$ for all $i \in I$.*

Intuitively speaking, this axiom merely asserts that the assumed nonemptiness of the sets M_i implies that we can in fact pick an element from each of them and call it $g(i)$. The truth of this assertion may strike us as utterly obvious, but the problem here is that there cannot be specified a general procedure or method by means of which the elements $g(i)$ are effectively chosen. That is to say, the axiom of choice is an entirely non-constructive existence statement, the truth of which has to be accepted by faith and cannot be proven.

3.8.10 Lemma. *If $f : A \to B$ is a surjection, then there exists a function $g : B \to A$ such that $g(y) \in f^{-1}(\{y\})$ for all $y \in B$.*

Proof. Applying the Axiom of Choice with B in place of I and $\{f^{-1}(\{y\}) \mid y \in B\}$ in place of $\{M_i \mid i \in I\}$, the existence of g, as stated, directly follows from the fact that the assumed surjectivity of f implies that $f^{-1}(\{y\}) \neq \emptyset$ for all $y \in B$. $\qquad\square$

Remark. For later reference, we wish to point out that the function g in Lemma 3.8.10 is injective. This is so because if $y_1 \neq y_2$, then $f^{-1}(\{y_1\}) \cap f^{-1}(\{y_2\}) = \emptyset$ and therefore, the conditions $g(y_1) \in f^{-1}(\{y_1\})$ and $g(y_2) \in f^{-1}(\{y_2\})$ imply that $g(y_1) \neq g(y_2)$.

3.8.11 Theorem. *A function $f : A \to B$ is surjective if and only if there exists a function $g : B \to A$ such that $f \circ g = \mathrm{id}_B$.*

Proof. "$\Rightarrow$" If f is surjective, then there exists a function $g : B \to A$ as described in Lemma 3.8.10. Since $g(y) \in f^{-1}(\{y\})$ for all $y \in B$, it follows that $(f \circ g)(y) = f(g(y)) = y = \mathrm{id}_B(y)$ for all $y \in B$. Hence $f \circ g = \mathrm{id}_B$. "$\Leftarrow$" If $g : B \to A$ satisfies the equation $f \circ g = \mathrm{id}_B$ and if $y \in B$, then, for $x := g(y)$ we have $f(x) = f(g(y)) = \mathrm{id}_B(y) = y$. Thus, f is surjective. $\qquad\square$

3.8.12 Theorem. *If g is a function from A to B and f is a function from B to C, then*

a) $f \circ g$ *is injective whenever f and g are both injective,*

b) $f \circ g$ *is surjective whenever f and g are both surjective,*

c) $f \circ g$ *is bijective whenever f and g are both bijective.*

Note: further information pertaining to this theorem is provided in Exercise 3.8.40.

Proof. Property c) is a trivial consequence of a) and b), the proof of b) is left as an exercise to the reader (see Exercise 3.8.28), and in order to prove a), we assume that f and g are injective and that $(f \circ g)(x_1) = (f \circ g)(x_2)$ for some $x_1, x_2 \in A$. Then $f(g(x_1)) = f(g(x_2))$, and therefore, the assumed injectivity of f implies that $g(x_1) = g(x_2)$. Since g is injective as well, it follows that $x_1 = x_2$, as desired. $\qquad\square$

3.8.13 Proposition. *Let $f : A \to B$ be a given function and $M_1, M_2 \subset A$.*

  a) *If $M_1 \subset M_2$, then $f(M_1) \subset f(M_2)$.*

  b) *If f is injective and $f(M_1) \subset f(M_2)$, then $M_1 \subset M_2$.*

  c) *If $\forall_{M_1, M_2 \subset A} \, f(M_1) \subset f(M_2) \Rightarrow M_1 \subset M_2$, then f is injective.*

Proof. **a)** Assume that $M_1 \subset M_2$ and let $y \in f(M_1)$. Then, by Definition 3.7.4, there exists an $x \in M_1$ such that $f(x) = y$. Since $M_1 \subset M_2$, it follows that $x \in M_2$, and therefore, $y = f(x) \in f(M_2)$, as desired.
b) Let f be injective and assume that $f(M_1) \subset f(M_2)$ for some $M_1, M_2 \subset A$ and that $x \in M_1$. We need to show that $x \in M_2$. Since $x \in M_1$, it follows that $f(x) \in f(M_1)$, and therefore $f(x) \in f(M_2)$ because $f(M_1) \subset f(M_2)$. Consequently, there is a $z \in M_2$ such that $f(z) = f(x)$. But since f is injective, we may infer that $x = z \in M_2$, as desired.
c) Assume that $\forall_{M_1, M_2 \subset A} \, f(M_1) \subset f(M_2) \Rightarrow M_1 \subset M_2$ and that $f(x_1) = f(x_2)$ for some $x_1, x_2 \in A$. We need to show that $x_1 = x_2$. Setting $M_1 := \{x_1\}$ and $M_2 := \{x_2\}$, it follows that

$$f(M_1) = f(\{x_1\}) = \{f(x_1)\} = \{f(x_2)\} = f(\{x_2\}) = f(M_2),$$

and in particular, $f(M_1) \subset f(M_2)$. Hence $M_1 \subset M_2$, and, by implication, $x_1 \in \{x_2\}$. Thus, $x_1 = x_2$, as desired. $\qquad\square$

Remark. If f is not injective, then the inference $f(M_1) \subset f(M_2) \Rightarrow M_1 \subset M_2$ is in general not valid as the following counterexample demonstrates: for $f(x) := x^2$ (with $D(f) = \mathbb{R}$), $M_1 := \{1\}$, and $M_2 := \{-1\}$, it is the case that $f(M_1) = \{1\} = f(M_2)$, and therefore in particular, $f(M_1) \subset f(M_2)$, but the consequent $M_1 \subset M_2$ is false.

3.8.14 Proposition. *Let $f : A \to B$ be a given function and $N_1, N_2 \subset B$.*

  a) *If $N_1 \subset N_2$, then $f^{-1}(N_1) \subset f^{-1}(N_2)$.*

  b) *If f is surjective and $f^{-1}(N_1) \subset f^{-1}(N_2)$, then $N_1 \subset N_2$.*

  c) *If $\forall_{N_1, N_2 \subset B} \, f^{-1}(N_1) \subset f^{-1}(N_2) \Rightarrow N_1 \subset N_2$, then f is surjective.*

Proof. We will prove only b) and c) and leave the proof of a) as an exercise to the reader (see Exercise 3.8.29).
b) Assume that f is surjective and that $f^{-1}(N_1) \subset f^{-1}(N_2)$ for some $N_1, N_2 \subset$

B. If $y \in N_1$, then, by the assumed surjectivity of f, there exists an $x \in A$ such that $f(x) = y$. Thus $x \in f^{-1}(N_1)$, and therefore, $x \in f^{-1}(N_2)$ because $f^{-1}(N_1) \subset f^{-1}(N_2)$. Consequently, $y = f(x) \in N_2$, as desired.

c) Since $R(f) \subset B$, we only need to show that the assumption

$$\forall_{N_1, N_2 \subset B} \, f^{-1}(N_1) \subset f^{-1}(N_2) \Rightarrow N_1 \subset N_2$$

implies that $B \subset R(f)$. Setting $N_1 := B$ and $N_2 := R(f)$, it follows that $f^{-1}(N_1) = A = f^{-1}(N_2)$, and therefore, in particular, $f^{-1}(N_1) \subset f^{-1}(N_2)$. Consequently, the assumption above allows us to infer that $B = N_1 \subset N_2 = R(f)$, as desired.   □

In order to explore, in the final part of this section, the properties of images and pre-images of arbitrary, potentially infinite unions and intersections of sets under functions in general and injections in particular, we introduce the following definition:

3.8.15 Definition. Assume that I is a nonempty set and that $\{S_i\}_{i \in I}$ is a corresponding collection of sets. Then

a) $\bigcup_{i \in I} S_i := \{x \mid \exists_{i \in I} \, x \in S_i\}$ and

b) $\bigcap_{i \in I} S_i := \{x \mid \forall_{i \in I} \, x \in S_i\}$.

Remark. If the sets S_i are subsets of some larger set S (i.e., $S_i \subset S$ for all $i \in I$), then the 'collection' $\{S_i\}_{i \in I}$, as a mathematical object, is properly understood to be the function from I to $\mathcal{P}(S)$ that maps i into S_i.

Remark. The restriction to nonempty index sets I in Definition 3.8.1 is necessary in order to guarantee that the relation

$$\bigcap_{i \in I} S_i \subset \bigcup_{i \in I} S_i$$

is always valid. For if I is nonempty, then $\forall_{i \in I} \, x \in S_i$ implies $\exists_{i \in I} \, x \in S_i$, and therefore, the inclusion holds as stated. By contrast, if $I = \emptyset$, then $\forall_{i \in I} \, x \in S_i$ is true for all x (because $\forall_{i \in I} \, x \in S_i \Leftrightarrow \forall_i (i \in I \Rightarrow x \in S_i)$ and because the antecedent $i \in I$ is false if $I = \emptyset$) whereas $\exists_{i \in I} \, x \in S_i$ is false for all x (because $\exists_{i \in I} \, x \in S_i \Leftrightarrow \exists_i (i \in I \wedge x \in S_i)$ and because $i \in I$ is false if $I = \emptyset$). In other words, if $I = \emptyset$, then $\bigcap_{i \in I} S_i$ equals the universe $\mathcal{U}$ from which the x are taken whereas $\bigcup_{i \in I} S_i$ equals the empty set. Thus the inclusion above does not hold if $I = \emptyset$ unless the universe $\mathcal{U}$ is empty as well.

3.8.16 Example. We wish to identify the subsets of $\mathbb{R}$ that are described by the following intersections and unions:

a) $\bigcap_{n \in \mathbb{N}} (0, 1/n)$,

b) $\bigcap_{n\in\mathbb{N}}[0, 1/n)$,

c) $\bigcup_{n\in\mathbb{N}}[0, 1/n)$,

d) $\bigcup_{n\in\mathbb{N}}(0, n)$.

Since $\lim_{n\to\infty} 1/n = 0$, there cannot exist a real number x that is greater than zero and less than $1/n$ for all $n \in \mathbb{N}$. Consequently, the intersection in a) is equal to the empty set. By contrast, the intersection in b) is nonempty and is in fact equal to $\{0\}$ because $x = 0$ is the unique real number that is contained in all the intervals $[0, 1/n)$. Furthermore, since $[0, 1/(n+1)) \subset [0, 1/n)$ for all $n \in \mathbb{N}$, it follows that the union in c) is equal to the first interval in this union, that is, it is equal to $[0, 1/1) = [0.1)$. Finally, since the upper bounds of the intervals $(0, n)$ in d) increase to ∞, we may infer that the union in d) is equal to $(0, \infty)$.

3.8.17 Example. Assume that $\{R_i\}_{i\in I}$ and $\{S_i\}_{i\in I}$ are given collections of sets. We wish to show that

$$\bigcup_{i\in I}(R_i \cap S_i) \subset \bigcup_{i\in I} R_i \cap \bigcup_{i\in I} S_i.$$

To do so, we use Proposition 1.7.7d to infer that

$$x \in \bigcup_{i\in I}(R_i \cap S_i) \Leftrightarrow \exists_{i\in I}\, x \in R_i \cap S_i \Leftrightarrow \exists_{i\in I}\,(x \in R_i \wedge x \in S_i)$$

$$\Rightarrow (\exists_{i\in I}\, x \in R_i) \wedge (\exists_{i\in I}\, x \in S_i) \Leftrightarrow x \in \bigcup_{i\in I} R_i \wedge x \in \bigcup_{i\in I} S_i$$

$$\Leftrightarrow x \in \bigcup_{i\in I} R_i \cap \bigcup_{i\in I} S_i,$$

as desired. Furthermore, the reverse inclusion

$$\bigcup_{i\in I}(R_i \cap S_i) \supset \bigcup_{i\in I} R_i \cap \bigcup_{i\in I} S_i.$$

is in general not valid because if $I = \{1, 2\}$, $R_1 = \{0\}$, $R_2 = \{1\}$, $S_1 = \{1\}$, and $S_2 = \{0\}$, then

$$\bigcup_{i\in I}(R_i \cap S_i) = \emptyset \cup \emptyset = \emptyset \not\supset \{0, 1\} = \{0, 1\} \cap \{0, 1\} = \bigcup_{i\in I} R_i \cap \bigcup_{i\in I} S_i.$$

3.8.18 Proposition. Let $\{R_i\}_{i\in I}$ and $\{S_i\}_{i\in I}$ be collections of subsets of a given set S. Then it is the case that

a) $S \setminus \bigcup_{i\in I} S_i = \bigcap_{i\in I} S \setminus S_i$,

b) $S \setminus \bigcap_{i\in I} S_i = \bigcup_{i\in I} S \setminus S_i$,

c) $\left(\bigcup_{i\in I} R_i\right) \smallsetminus \bigcap_{i\in I} S_i \supset \bigcup_{i\in I}(R_i \smallsetminus S_i),$

d) $\left(\bigcap_{i\in I} R_i\right) \smallsetminus \bigcup_{i\in I} S_i = \bigcap_{i\in I}(R_i \smallsetminus S_i).$

Proof. We will prove only d) and leave the proofs of the remaining statements as exercises to the reader (see Exercise 3.8.33). Using Proposition 1.7.7a, the proof of d) can be stated as follows:

$$x \in \left(\bigcap_{i\in I} R_i\right) \smallsetminus \bigcup_{i\in I} S_i \Leftrightarrow x \in \left(\bigcap_{i\in I} R_i\right) \wedge x \notin \bigcup_{i\in I} S_i$$

$$\Leftrightarrow \forall_{i\in I}\, x \in R_i \wedge \neg \exists_{i\in I}\, x \in S_i$$

$$\Leftrightarrow \forall_{i\in I}\, x \in R_i \wedge \forall_{i\in I}\, x \notin S_i$$

$$\Leftrightarrow \forall_{i\in I}\, (x \in R_i \wedge x \notin S_i)$$

$$\Leftrightarrow x \in \bigcap_{i\in I} R_i \smallsetminus S_i.$$

$\square$

3.8.19 Theorem. *Assume that A and B are nonempty sets, that f is a function from A to B, and that $\{M_i\}_{i\in I}$ and $\{N_i\}_{i\in I}$ are collections of subsets of A and B, respectively (i.e., $M_i \subset A$ and $N_i \subset B$ for all $i \in I$). Then*

a) $f\left(\bigcup_{i\in I} M_i\right) = \bigcup_{i\in I} f(M_i),$

b) $f\left(\bigcap_{i\in I} M_i\right) \subset \bigcap_{i\in I} f(M_i),$

c) $f^{-1}\left(\bigcup_{i\in I} N_i\right) = \bigcup_{i\in I} f^{-1}(N_i),$

d) $f^{-1}\left(\bigcap_{i\in I} N_i\right) = \bigcap_{i\in I} f^{-1}(N_i),$

Proof. **a)** Since

$$y \in f\left(\bigcup_{i\in I} M_i\right) \Leftrightarrow \exists_x\, x \in \bigcup_{i\in I} M_i \wedge f(x) = y$$

$$\Leftrightarrow \exists_x \exists_{i\in I}\, (x \in M_i \wedge f(x) = y)$$

$$\Leftrightarrow \exists_{i\in I} \exists_x\, (x \in M_i \wedge f(x) = y)$$

$$\Leftrightarrow \exists_{i\in I}\, y \in f(M_i)$$

$$\Leftrightarrow y \in \bigcup_{i\in I} f(M_i),$$

we may infer that $f\left(\bigcup_{i\in I} M_i\right) = \bigcup_{i\in I} f(M_i)$, as desired.

b)

$$y \in f\left(\bigcap_{i\in I} M_i\right) \Leftrightarrow \exists_x\, x \in \bigcap_{i\in I} M_i \wedge f(x) = y$$

$$\Leftrightarrow \exists_x \forall_{i \in I} \left(x \in M_i \land f(x) = y\right)$$
$$\Rightarrow \forall_{i \in I} \exists_x \left(x \in M_i \land f(x) = y\right)$$
$$\Leftrightarrow \forall_{i \in I} \, y \in f(M_i)$$
$$\Leftrightarrow y \in \bigcap_{i \in I} f(M_i).$$

Hence $f\left(\bigcap_{i \in I} M_i\right) \subset \bigcap_{i \in I} f(M_i)$. For the proofs of c) and d) see Exercise 3.8.36. $\qquad\square$

3.8.20 Theorem. *If f is injective, then $f\left(\bigcap_{i \in I} M_i\right) = \bigcap_{i \in I} f(M_i)$*

Proof. Given the proof of b) in Theorem 3.8.19, it is sufficient to demonstrate that the assumption of injectivity on f implies that

$$\exists_x \forall_{i \in I} \left(x \in M_i \land f(x) = y\right) \Leftarrow \forall_{i \in I} \exists_x \left(x \in M_i \land f(x) = y\right). \qquad (3.6)$$

To clarify the logical structure of this statement, we notice that

$$\forall_{i \in I} \exists_x \left(x \in M_i \land f(x) = y\right) \Leftrightarrow \forall_{i \in I} \exists_{x_i} \left(x_i \in M_i \land f(x_i) = y\right)$$

Assuming the statement on the right-hand side of this material biconditional to be true, the injectivity of f implies that $x_i = x_j$ for all $i, j \in I$. Thus, in setting $x := x_i$ for some $i \in I$, it follows that

$$\forall_{i \in I} \left(x \in M_i \land f(x) = y\right).$$

In other words, there exists an x such that

$$\forall_{i \in I} \left(x \in M_i \land f(x) = y\right),$$

and this proves the validity of the material conditional (3.6). $\qquad\square$

3.8.21 Theorem. *A map $f : A \to B$ is injective if and only if $f(M_1 \cap M_2) = f(M_1) \cap f(M_2)$ for all $M_1, M_2 \subset A$.*

Proof. Given the statement of Theorem 3.8.20, we only need to prove that

$$\left(\forall_{M_1, M_2 \subset A} \, f(M_1 \cap M_2) = f(M_1) \cap f(M_2)\right) \Rightarrow f \text{ is injective.}$$

So let us assume that the statement on the left of this material conditional is valid, and let x_1 and x_2 be elements in A such that $f(x_1) = f(x_2)$. To establish the injectivity of f, we need to show that $x_1 = x_2$. Setting $M_1 := \{x_1\}$ and $M_2 := \{x_2\}$, it follows that

$$\begin{aligned}
f(M_1 \cap M_2) = f(M_1) \cap f(M_2) &= \{f(x_1)\} \cap \{f(x_2)\} \\
&= \{f(x_1)\} \cap \{f(x_1)\} \quad (\text{because } f(x_1) = f(x_2)) \\
&= \{f(x_1)\} \neq \emptyset.
\end{aligned}$$

Consequently, $M_1 \cap M_2$ cannot be empty, and therefore, it must be the case that $x_1 = x_2$, as desired. $\qquad\square$

3.8.22 Proposition. $f(M_1 \smallsetminus M_2) \supset f(M_1) \smallsetminus f(M_2)$ for all $M_1, M_2 \subset A$.

Proof.

$$y \in f(M_1) \smallsetminus f(M_2) \Leftrightarrow \exists_x(x \in M_1 \wedge f(x) = y) \wedge \forall_x(x \in M_2 \Rightarrow f(x) \neq y)$$
$$\Rightarrow \exists_x(x \in M_1 \wedge x \notin M_2 \wedge f(x) = y)$$
$$\Leftrightarrow y \in f(M_1 \smallsetminus M_2).$$

Hence $f(M_1) \smallsetminus f(M_2) \subset f(M_1 \smallsetminus M_2)$. $\square$

3.8.23 Proposition. *A map* $f : A \to B$ *is injective if and only if*

$$f(A \smallsetminus M) = f(A) \smallsetminus f(M)$$

for all $M \subset A$.

Proof. "$\Rightarrow$" Assume that f is injective and let M be an arbitrary subset of A. In light of Proposition 3.8.22, we only need to show that $f(A \smallsetminus M) \subset f(A) \smallsetminus f(M)$. So let y be an element of $f(A \smallsetminus M)$. Then there exists an $x \in A \smallsetminus M$ such that $f(x) = y$. Consequently, the injectivity of f implies that $f(z) \neq y$ for all $z \in M$. Thus, $y \notin f(M)$, and by implication, $y \in f(A) \smallsetminus f(M)$. "$\Leftarrow$" We give a proof by contraposition. If x_1 and x_2 are distinct elements of A such that $f(x_1) = f(x_2)$ (i.e., if f is not injective), then $f(A \smallsetminus \{x_2\}) = f(A) \neq f(A) \smallsetminus f(\{x_2\})$. In other words, in setting $M := \{x_2\}$, we have found a set that satisfies the conditions $M \subset A$ and

$$f(A \smallsetminus M) \neq f(A) \smallsetminus f(M),$$

as desired. $\square$

3.8.24 Proposition. *A map* $f : A \to B$ *is injective if and only*

$$f(M_1 \smallsetminus M_2) = f(M_1) \smallsetminus f(M_2)$$

for all $M_1, M_2 \subset A$.

Proof. "$\Rightarrow$" Assume that f is injective and let M_1 and M_2 be arbitrary subsets of A. Then

$$f(M_1 \smallsetminus M_2) = f(M_1 \cap (A \smallsetminus M_2))$$
$$= f(M_1) \cap f(A \smallsetminus M_2) \quad \text{(by Theorem 3.8.21)}$$
$$= f(M_1) \cap (f(A) \smallsetminus f(M_2)) \quad \text{(by Proposition 3.8.23)}$$
$$= f(M_1) \smallsetminus f(M_2).$$

"$\Leftarrow$" Assume that $f(M_1 \smallsetminus M_2) = f(M_1) \smallsetminus f(M_2)$ for all $M_1, M_2 \subset A$. Then, in particular, $f(A \smallsetminus M) = f(A) \smallsetminus f(M)$ for all $M \subset A$, and therefore Proposition 3.8.23 implies that f is injective. $\square$

Exercises

3.8.25. Determine whether the functions listed in Exercise 3.7.10 are injective and/or surjective and briefly explain your answer.

3.8.26. Prove the statement of Example 3.8.5.

3.8.27. Prove the second equation in Corollary 3.8.8.

3.8.28. Prove Theorem 3.8.12b.

3.8.29. Prove Proposition 3.8.14a.

3.8.30. Show that the conclusion in Proposition 3.8.14b does not necessarily hold if f is not surjective.

3.8.31. Identify the subsets of $\mathbb{R}$ that are described by the following intersections and unions:

a) $\bigcap_{n \in \mathbb{N}} [-1/n, 1 - 1/n)$,
b) $\bigcap_{n \in \mathbb{N}} (-1/n, 1 + 1/n]$,
c) $\bigcup_{n \in \mathbb{N}} [-1/n, 1 - 1/n)$,
d) $\bigcup_{n \in \mathbb{N}} (1 - 1/n, 1 + 1/n]$.

3.8.32. Assume that $\{R_i\}_{i \in I}$ and $\{S_i\}_{i \in I}$ are given collections of sets. Show that

$$\bigcap_{i \in I} (R_i \cup S_i) \supset \bigcap_{i \in I} R_i \cup \bigcap_{i \in I} S_i.$$

Is the reverse inclusion valid as well? Explain your answer.

3.8.33. Prove Proposition 3.8.18a,b,c.

3.8.34. Show that the reverse inclusion in Proposition 3.8.18c is in general not valid.

3.8.35. ★ Assume that for every $r \in \mathbb{Q}$ you are given an open interval $I_r \subset \mathbb{R}$ containing r (i.e., $r \in I_r$ for all $r \in \mathbb{Q}$). Is it necessarily the case that $\bigcup_{r \in \mathbb{Q}} I_r = \mathbb{R}$? Give a proof or find a counterexample.

3.8.36. Prove Theorem 3.8.19c,d.

3.8.37. Show that the reverse inclusion relation in Theorem 3.8.19b is in general not satisfied. *Hint:* find a function f and sets M_1 and M_2 such that $f(M_1) \cap f(M_2)$ is not a subset of $f(M_1 \cap M_2)$.

3.8.38. Find a function f and sets M_1 and M_2 such that

$$f(M_1 \smallsetminus M_2) \not\subset f(M_1) \smallsetminus f(M_2).$$

3.8.39. Let $f : A \rightarrow B$ be a function.

a) Show that $f(M_1) \triangle f(M_2) \subset f(M_1 \triangle M_2)$ for all $M_1, M_2 \subset A$.

b) Show that in general

$$f(M_1) \triangle f(M_2) \triangle f(M_3) \not\subset f(M_1 \triangle M_2 \triangle M_3).$$

c) Show that f is injective if and only if $f(M_1 \triangle M_2) = f(M_1) \triangle f(M_2)$ for all $M_1, M_2 \subset A$.

3.8.40. Assume that g is a function from A to B and that f is a function from B to C. Prove the following statements:

a) If $f \circ g$ is injective, then so is g, but f is not necessarily injective.

b) If $f \circ g$ is surjective, then so is f, but g is not necessarily surjective.

c) If $f \circ g$ is bijective, then neither f nor g need to be bijective.

Chapter 4

Mathematical Induction

4.1 The Peano Axioms

In order to rigorously establish the notion of a positive integer, we postulate the existence of a set $\mathbb{N}$, an element $1 \in \mathbb{N}$ (i.e., $\mathbb{N}$ is nonempty), and a map $\varphi : \mathbb{N} \to \mathbb{N}$ such that

a) φ is injective,

b) $\varphi(n) \neq 1$ for all $n \in \mathbb{N}$, and

c) for all $A \subset \mathbb{N}$ it is the case that $(1 \in A \wedge \varphi(A) \subset A) \Rightarrow A = \mathbb{N}$.

These postulates are commonly referred to as the *Peano Axioms*, in honor of the Italian mathematician *Giuseppe Peano* who was one of the early pioneers in the fields of mathematical logic and set theory.

At first sight, it may not be directly obvious what in the world these axioms have to do with the concept of a positive counting number, but upon closer inspection we notice for instance that a) and b) in conjunction imply that the set $\mathbb{N}$ here in question must be infinite in size. This is so, because if $\mathbb{N}$ were finite, then the assumed injectivity of φ would imply that φ is surjective as well (as explained in the remark below), and condition b) would therefore be violated.

Remark. The fact that a function f from a finite set S into itself is surjective if it is injective will be rigorously established in Proposition 5.1.9. But intuitively speaking, the truth of this fact is quite readily apparent. For if a map f from a finite set S into itself is injective then the output elements $f(x)$ must be pairwise distinct (i.e., no two of them can be the same) and therefore equal in number to the input elements x. In other words, the number of ouput elements must be equal to the number of elements in S, and, by implication, the range of f must be equal to S.

To be sure, in order for the notion of infinity to have any meaning at all, there must be a set that is infinite, and it is precisely by means of the Peano Axioms that the existence of such a set is being asserted. That is to say, we cannot prove that $\mathbb{N}$, as described by the Peano Axioms, is infinite because the very notion of infinity is implicitly established by these very axioms. All the same, however, it is helpful here to realize that our intuitive understanding of the infinite extension of the set of counting numbers is formally coherently expressed in the conjunction of a) and b).

Furthermore—and again on a purely intuitive level—the map φ is properly thought of as the 'successor map' that assigns to a given element $n \in \mathbb{N}$ the sum $n+1$, that is, $\varphi(n) = n+1$ for all $n \in \mathbb{N}$. Naturally, the notion of a sum has not been defined as yet, and our use of the symbol '+' is therefore strictly speaking meaningless. But the purpose of the Peano Axioms just cannot be made clear and understood apart from elementary arithmetic. For it is precisely the rules and operations of arithmetic to which these axioms are meant to give rise.

In this context it is helpful as well to take a closer look at the so-called *Principle of Mathematical Induction*, stated in c), which in essence asserts that any subset A of $\mathbb{N}$ must actually be equal to $\mathbb{N}$ if it contains 1 and if it also contains the successor $\varphi(n) = n + 1$ of each of its elements n. In other words, a set $A \subset \mathbb{N}$ is here asserted to be necessarily equal to $\mathbb{N}$ if

$$1 \in \mathbb{N} \wedge \forall_{n \in A} \, \varphi(n) = n + 1 \in A.$$

Intuitively, the validity of this assertion is easily grasped. For if 1 is in A, then the statement $\forall_{n \in A} \, \varphi(n) = n + 1 \in A$ implies that $1 + 1 = 2$ is in A as well, and so is, by implication, $2 + 1 = 3$. Continuing in this manner, we find that in fact all numbers $n \in \mathbb{N}$ are in A and that therefore $A = \mathbb{N}$.

For later reference and in order to give the reader an initial impression of the critical role played by the Principle of Mathematical Induction in deriving the properties of $\mathbb{N}$, we will now establish, by means of this principle, the following two propositions:

4.1.1 Proposition. $\varphi(\mathbb{N}) = \mathbb{N} \smallsetminus \{1\}$.

Proof. Let $A := \{1\} \cup \varphi(\mathbb{N})$. Then trivially $1 \in A$ and if $n \in A$, then $\varphi(n) \in \varphi(A) \subset \varphi(\mathbb{N}) \subset A$. Hence $\varphi(A) \subset A$, and therefore, $A = \mathbb{N}$. Furthermore, since $1 \notin \varphi(\mathbb{N})$, it follows that $\varphi(\mathbb{N}) = A \smallsetminus \{1\} = \mathbb{N} \smallsetminus \{1\}$, as desired. □

4.1.2 Proposition. $\varphi(n) \neq n$ for all $n \in \mathbb{N}$.

Proof. Let $A := \{n \in \mathbb{N} \mid \varphi(n) \neq n\}$. Then $1 \in A$ because $\varphi(n) \neq 1$ for all $n \in \mathbb{N}$, and if $n \in A$, then $\varphi(n) \neq n$, and therefore, $\varphi(\varphi(n)) \neq \varphi(n)$ because φ is injective. Hence $\varphi(n) \in A$ and, by implication, $\varphi(A) \subset A$ and $A = \mathbb{N}$, as desired. □

4.2 Fuzzy Addition

Having introduced the Peano Axioms in the previous section, we will now turn our attention to the problem of defining the operation of addition on $\mathbb{N}$ on the basis of these axioms. Given this objective, it is helpful to observe to begin with that in adding up two numbers $m, n \in \mathbb{N}$, we produce a third number, called the 'sum', that we denote by $m + n$. So what we here are dealing with, more formally, is a function that assigns to any pair of integers (m, n) the single output integer $m + n$. In denoting this function by p (for 'plus'), we may construe addition on $\mathbb{N}$ to be a function of the following form:

$$p : \mathbb{N} \times \mathbb{N} \to \mathbb{N}$$
$$(m, n) \to p(m, n),$$

where, *intuitively speaking*, the value $p(m, n)$ is equal to the sum $m + n$, that is,

$$p(m, n) = m + n.$$

Naturally, the observation that addition is a certain type of function (or set) is not in itself sufficient to guarantee that addition has the computational properties that we expect it to have. Thus we need to dig a little deeper and somehow link the function p to the 'successor function' φ that the Peano Axioms provide us with. Clearly, if $\varphi(n)$ is supposed to represent the sum $n+1$ (as explained above), then it ought to be the case that

$$p(m, 1) = m + 1 = \varphi(m)$$

for all $m \in \mathbb{N}$, and it also ought to be true that

$$\varphi(p(m, n)) = p(m, n) + 1 = (m + n) + 1 = m + (n + 1) = p(m, n + 1)$$
$$= p(m, \varphi(n))$$

for all $m, n \in \mathbb{N}$. In other words, using quantification notation, it ought to be the case that

$$\forall_{m \in \mathbb{N}}\, p(m, 1) = \varphi(m), \tag{4.1}$$

and

$$\forall_{m,n \in \mathbb{N}}\, \varphi(p(m, n)) = p(m, \varphi(n)). \tag{4.2}$$

As it turns out and as we shall see, these two properties, taken in conjunction, are sufficient to uniquely characterize the map p as the familiar operation of addition on $\mathbb{N}$. The uniqueness part of this claim is the content of the following theorem:

4.2.1 Theorem. *If $p, q : \mathbb{N} \times \mathbb{N} \to \mathbb{N}$ are functions that both satisfy the statements (4.1) and (4.2), then $p = q$.*

Proof. Setting

$$A := \{n \in \mathbb{N} \mid p(m, n) = q(m, n) \text{ for all } m \in \mathbb{N}\},$$

it follows that 1 is in A because p and q are both assumed to satisfy (4.1), and if n is in A, that is, if $p(m, n) = q(m, n)$ for all $m \in \mathbb{N}$, then $\varphi(n)$ is in A as well because (4.2) implies that

$$p(m, \varphi(n)) = \varphi(p(m, n)) = \varphi(q(m, n)) = q(m, \varphi(n))$$

for all $m \in \mathbb{N}$. Thus $1 \in A \land \varphi(A) \subset A$, and therefore, $A = \mathbb{N}$, as desired. ☐

Having established the uniqueness of p in the preceding theorem, we now need to turn to the problem of proving its existence. Given this objective, the first step we must take is to *define* $p(m, 1)$ to be equal to $\varphi(m)$ for all $m \in \mathbb{N}$. This guarantees that (4.1) is satisfied *by definition*. But what about property (4.2)? Here the common and convenient but also fatally flawed approach is to invoke the Principle of Mathematical Induction in the following manner: setting

$$A := \{n \in \mathbb{N} \mid p(m, n) \text{ has meaning for all } m \in \mathbb{N}\}, \qquad (4.3)$$

we observe that 1 is in A because, *by definition*, the 'meaning' of $p(m, 1)$ for all $m \in \mathbb{N}$ precisely is that $p(m, 1) = \varphi(m)$, and then, in order to show that $\varphi(A) \subset A$, we further observe that, for a given $n \in \mathbb{N}$, the assumption that $p(m, n)$ has 'meaning' for all $m \in \mathbb{N}$, allows us to assign a 'meaning' as well to $p(m, \varphi(n))$ because, in reference to (4.2), we may *define* $p(m, \varphi(n))$ to be equal to $\varphi(p(m, n))$ for all $m \in \mathbb{N}$. Consequently, the assumption that n is in A leads to the conclusion that $\varphi(n)$ is in A as well and that therefore $\varphi(A)$ is a subset of A, as desired. Having thus established the validity of the conjunction $1 \in A \land \varphi(A) \subset A$, the Principle of Mathematical Induction, as stated in the Peano Axioms in property c), allows us to infer that $A = \mathbb{N}$. Hence the expression $p(m, n)$ has 'meaning' for all m and n in $\mathbb{N}$, and the way in which this meaning is assigned guarantees that (4.1) and (4.2) are satisfied.

Unfortunately, the problem with this approach is that it leaves us pondering the critical question as to what exactly it means to say, in the definition of A in (4.3), that $p(m, n)$ has 'meaning' for all $m \in \mathbb{N}$. Does it mean merely that for a given $n \in A$ there exists a map $q_n : \mathbb{N} \times \{n\} \to \mathbb{N}$ that assigns to each pair (m, n) some 'value' in $\mathbb{N}$ which we then choose to denote by '$p(m, n)$'? Surely not for the existence of such a map q_n is utterly trivial: simply set $q_n(m, n) := q_1(m, 1) = \varphi(m)$ (in accordance with (4.1)). The natural objection that this latter definition makes no reference to the fact that $q_{\varphi(n)}(m, \varphi(n))$ ought to be equal to $\varphi(q_n(m, n))$ by (4.2) is certainly valid but not in itself very helpful because it still leaves us confused concerning the meaning of 'meaning' in (4.3).

Evidently, the 'meaning' referred to in (4.3) is meant to be consistent with equations (4.1) and (4.2). But (4.2) presupposes the term $p(m, n)$ in the argument of φ to be well defined, and the attempt to make sense of the 'meaning' in (4.3) by way of the requirement that (4.2) be satisfied is therefore viciously circular. Moreover, the complete inadequacy of (4.3) becomes apparent as well when we ask ourselves how the statement '$m + n$ has meaning for all $m \in \mathbb{N}$' ought to be quantified. If we cannot express this statement in a purely symbolic form, using nothing but sets, elements of sets, and formal logic quantification notation, then our overarching claim that mathematics, at bottom, is the study of sets and nothing but sets is either untenable or, alternatively, (4.3) has to be discarded. As it turns out, there really is no way to coherently quantify the statement here in question, nor is it doubtful in the least that the assertion that mathematics and set theory are at bottom congruent is perfectly cogent and valid. By implication, the conclusion here must be that (4.3) is indeed completely inadequate and has to be discarded.

That said, there arises the obvious question as to how we ought to proceed. One option is to follow tradition and to simply cast aside the logical problems that the defining equation (4.3) is clearly beset with. After all, what is so bad about a little imprecision? Can we not just relax our standards a bit and happily accept as meaningful a defining equation that really makes no sense at all? There is no question that we can, but there is room for doubt whether we would serve the reader well in doing so. For if indeed the purpose of this text is to familiarize the reader with rigorous mathematical thought, then rigorous mathematical thought is arguably the kind of thought that we ought to practice in composing this text.

This stance, of course, may readily seem inconsistent in light of the fact that we were willing to forego the demand of absolute rational rigor not long ago in our discussion of the notion of a set. For while we did explain quite carefully why Cantor's definition was flawed, we never ventured to remove that flaw by offering an alternative definition or axiom. All we did was to refer the reader to some relevant literature sources. So why not be content to do the same with respect to the notion of addition? The proper answer is, so we believe, that the former problem is simply too involved and too advanced to be satisfactorily treated in an introductory text on abstract mathematics whereas the latter is not. The foundations of set theory are very elusive a topic that ultimately confronts us with the limits of human rationality as such and that perhaps can never be treated with final, incontrovertible clarity. By contrast, the problem of establishing the operation of addition—on the basis of the notion of a set—is fairly straightforward and well within the scope of this present exposition.

Consequently, what we will do is to settle for a compromise: we will offer the reader a dual treatment by pursuing the standard flawed approach in the current section *and* the correct non-standard one in the section that follows. Concerning the first track we wish to add for clarity that we will derive the properties of $\mathbb{N}$ from (4.1) and (4.2) in a fully rigorous manner despite the fact that the derivation of (4.1) and (4.2) from (4.3) was fatally flawed.

So let us now assume that the defining equation (4.3) is actually meaningful and that our derivation of (4.1) and (4.2) by way of the Principle of Mathematical Induction was fully coherent and valid. Given this assumption, we wish to show that the addition map p satisfies familiar computational properties such as associativity and commutativity and that it can be used to define a linear order on $\mathbb{N}$.

4.2.2 Theorem. *The addition map $p(m, n) = m + n$ satisfies the following properties:*

a) *p is associative in the sense that $p(k, p(m, n)) = p(p(k, m), n)$ or, equivalently, $k + (m + n) = (k + m) + n$ for all $k, m, n \in \mathbb{N}$.*

b) *p is commutative in the sense that $p(m, n) = p(n, m)$ or, equivalently, $m + n = n + m$ for all $m, n \in \mathbb{N}$.*

Proof. **a)** Let

$$A := \{n \in \mathbb{N} \mid p(k, p(m, n)) = p(p(k, m), n) \text{ for all } k, m \in \mathbb{N}\}.$$

Then $1 \in A$ because

$$p(k, p(m, 1)) = p(k, \varphi(m)) = \varphi(p(k, m)) = p(p(k, m), 1)$$

for all $k, m \in \mathbb{N}$. Moreover, if $n \in A$, then

$$p(k, p(m, \varphi(n))) = p(k, \varphi(p(m, n))) = \varphi(p(k, p(m, n))) = \varphi(p(p(k, m), n))$$
$$= p(p(k, m), \varphi(n)),$$

and therefore, $\varphi(n) \in A$. Thus $A = \mathbb{N}$, as desired.

b) Here we define to begin with

$$A := \{m \in \mathbb{N} \mid p(m, 1) = p(1, m)\}.$$

Then, trivially, $1 \in A$, and if $m \in A$ then

$$p(1, \varphi(m)) = \varphi(p(1, m)) = \varphi(p(m, 1)) = \varphi(\varphi(m)) = p(\varphi(m), 1).$$

Hence $\varphi(m) \in A$, and therefore, $A = \mathbb{N}$. To proceed, we set

$$B := \{n \in \mathbb{N} \mid p(m, n) = p(n, m) \text{ for all } m \in \mathbb{N}\}.$$

Since $A = \mathbb{N}$, we may infer that $1 \in B$. Finally, if $n \in B$, then we may use a) in conjunction with (4.1), (4.2), and the fact that $A = \mathbb{N}$ to conclude that

$$p(m, \varphi(n)) = \varphi(p(m, n)) = \varphi(p(n, m)) = p(p(n, m), 1) = p(1, p(n, m))$$
$$= p(p(1, n), m) = p(p(n, 1), m) = p(\varphi(n), m).$$

Thus $\varphi(n) \in B$, and therefore, $B = \mathbb{N}$, as desired. $\qquad\square$

The Lemmas 4.2.3 and 4.2.4 that follow will be helpful in establishing the familiar linear orders on $\mathbb{N}$—the strictly-less-than and less-than-or-equal-to relations.

4.2.3 Lemma. *If* $r \subset S \times S$ *is a linear order on a given set* S, *then*

$$s := \{(x, y) \in S \times S \mid (x, y) \in r \vee x = y\}$$

is a linear order on S *as well.*

Proof. We need to show that s is antisymmetric and transitive and that

$$(x, y) \in s \vee (y, x) \in s \vee x = y \tag{4.4}$$

for all $x, y \in S$. Since

$$(x, y) \in s \Leftrightarrow ((x, y) \in r \vee x = y),$$

it follows that (4.4) is equivalent to

$$(x, y) \in r \vee (y, x) \in r \vee x = y,$$

and this latter statement is true for all $x, y \in S$ because r is linear. In order to prove, by way of contradiction, that s is antisymmetric, we assume that $(x, y) \in s$, $(y, x) \in s$, and $x \neq y$ for some $x, y \in S$. Then the statement

$$((x, y) \in r \vee x = y) \wedge ((y, x) \in r \vee x = y)$$

is true, and since $x \neq y$, this statement is equivalent to

$$(x, y) \in r \wedge (y, x) \in r.$$

Using that fact that r is antisymmetric, it follows that $x = y$, in contradiction to the assumption $x \neq y$. Thus s is indeed antisymmetric. Finally, in order to show that s is transitive, we assume that $(x, y) \in s$ and $(y, z) \in s$ for some elements $x, y, z \in S$. Then it is the case that

$$((x, y) \in r \vee x = y) \wedge ((y, z) \in r \vee y = z).$$

If $x = y$ and $y = z$, then $x = z$, and therefore, $(x, z) \in s$, as desired. Hence we may assume that $x \neq y \vee y \neq z$. Considering w.l.o.g. (without loss of generality) the case $y \neq z$ (the case $x \neq y$ is completely analogous), we may infer that

$$((x, y) \in r \vee x = y) \wedge (y, z) \in r$$

or, equivalently, that

$$((x, y) \in r \wedge (y, z) \in r) \vee (x = y \wedge (y, z) \in r).$$

If $(x, y) \in r \wedge (y, z) \in r$, then $(x, z) \in r$ because r is transitive, and if $x = y \wedge (y, z) \in r$, then $(x, z) = (y, z) \in r$. So in either case we find that (x, z) is in r, and therefore it follows that (x, z) is in s. $\square$

4.2.4 Lemma. *For all* $n \in \mathbb{N}$ *the map*

$$u_n : \mathbb{N} \to \mathbb{N}$$
$$m \mapsto p(m, n) = m + n$$

is injective and satisfies the following property: $u_n(m) \neq m$ *for all* $m \in \mathbb{N}$.

Proof. In order to establish that u_n is injective for all $n \in \mathbb{N}$, we set

$$A := \{n \in \mathbb{N} \mid u_n \text{ is injective}\}.$$

Given this definition, it follows that 1 is in A because, according to (4.1), we have $u_1 = \varphi$, and therefore, u_1 is injective by the first of the Peano Axioms. Furthermore, if n is A, that is, if u_n is injective, then $u_{\varphi(n)}$ is injective as well, by Theorem 3.8.12a, because (4.2) implies that $u_{\varphi(n)} = \varphi \circ u_n$. Thus $A = \mathbb{N}$, as desired. Furthermore, in order to show that $u_n(m) \neq m$ for all $n, m \in \mathbb{N}$, we give a proof by contradiction: if $m = u_n(m) = m + n$ for some $m, n \in \mathbb{N}$, then

$$u_m(1) = 1 + m = m + 1 = (m + n) + 1 = m + (n + 1) = u_m(n + 1),$$

and therefore $1 = n + 1 = \varphi(n)$ because u_m is injective. But this is impossible by the second of the Peano Axioms. $\qquad\qquad\qquad\qquad\qquad\qquad\qquad\qquad$ □

4.2.5 Theorem. *The relation*

$$r := \{(m, n) \in \mathbb{N} \times \mathbb{N} \mid \exists_{k \in \mathbb{N}}\, p(m, k) = m + k = n\}$$

is a linear order on $\mathbb{N}$. *Note:* r *is the strictly-less-than relation on* $\mathbb{N}$, *that is,*

$$m < n :\Leftrightarrow (m, n) \in r$$

for all $m, n \in \mathbb{N}$. *In particular,* $m < n \wedge n < m$ *is false for all* $m, n \in \mathbb{N}$, *and* $m < n \Rightarrow m \neq n$ *is true for all* $m, n \in \mathbb{N}$.

Proof. Starting with the last statement above, we notice that the assumption $m < n$ implies that there is a $k \in \mathbb{N}$ such that $n = m + k$, and therefore Lemma 4.2.4 implies that $m \neq u_k(m) = n$, as desired. Next we observe that in order to prove that r is a linear order on $\mathbb{N}$, we need to show that r is antisymmetric and transitive and that

$$(m, n) \in r \vee (n, m) \in r \vee m = n \qquad\qquad\qquad (4.5)$$

for all $m, n \in \mathbb{N}$. To establish antisymmetry, we need to show that

$$\forall_{m,n \in \mathbb{N}} ((m, n) \in r \wedge (n, m) \in r) \Rightarrow m = n.$$

To do so, it is sufficient to prove that the antecedent

$$(m, n) \in r \wedge (n, m) \in r$$

is false for all $m, n \in \mathbb{N}$. In order to establish this latter claim by way of contradiction, we assume that $(m, n) \in r$ and $(n, m) \in r$ for some $m, n \in \mathbb{N}$. Then there are elements $k, l \in \mathbb{N}$ such that $p(m, k) = m + k = n$ and $p(n, l) = n + l = m$. Hence

$$n = (n + l) + k = n + (l + k) = u_{l+k}(n)$$

in contradiction to the fact that $u_n(m) \neq m$ for all $m, n \in \mathbb{N}$ (as stated in Lemma 4.2.4). To prove that r is transitive, we assume that $(k, m) \in r$ and $(m, n) \in r$ for some $k, m, n \in \mathbb{N}$. Then there are elements $i, j \in \mathbb{N}$ such that $k + i = m$ and $m + j = n$. Thus

$$k + (i + j) = (k + i) + j = m + j = n,$$

and therefore, $(k, n) \in r$, as desired. Finally, in order to show that (4.5) is valid, we set

$$A := \{n \in \mathbb{N} \mid \forall_{m \in \mathbb{N}} (m, n) \in r \vee (n, m) \in r \vee m = n\}.$$

Since $\varphi(\mathbb{N}) = \mathbb{N} \setminus \{1\}$ (by Proposition 4.1.1), it follows that for every $m \neq 1$ there exists a $k \in \mathbb{N}$ such that $m = \varphi(k) = k + 1 = 1 + k$. Thus, for all $m \in \mathbb{N}$ it is the case that

$$(1, m) \in r \vee m = 1, \tag{4.6}$$

and this shows that 1 is in A. Furthermore, if n is in A, then

$$\forall_{m \in \mathbb{N}} (m, n) \in r \vee (n, m) \in r \vee m = n,$$

and we need to show that

$$(m, \varphi(n)) \in r \vee (\varphi(n), m) \in r \vee m = \varphi(n) \tag{4.7}$$

for all $m \in \mathbb{N}$. So let m be in $\mathbb{N}$. If $m = n$, then $m + 1 = n + 1 = \varphi(n)$, and therefore, (4.7) is satisfied because $(m, \varphi(n)) \in r$. If $(m, n) \in r$, then $m + k = n$ for some $k \in \mathbb{N}$, and, by implication,

$$m + (k + 1) = (m + k) + 1 = n + 1 = \varphi(n).$$

Consequently, we find again that $(m, \varphi(n)) \in r$ and that therefore (4.7) is satisfied. Moreover, if $(n, m) \in r$, then $n + k = m$ for some $k \in \mathbb{N}$. If $k = 1$, then $\varphi(n) = n + 1 = m$, and once again we find that (4.7) is satisfied. Finally, if $k \neq 1$, then $(1, k) \in r$ (by (4.6)), and therefore, $k = 1 + i$ for some $i \in \mathbb{N}$. Hence

$$m = n + (1 + i) = (n + 1) + i = \varphi(n) + i,$$

and therefore $(\varphi(n), m) \in r$, as desired. $\qquad \square$

4.2.6 Corollary. *The less-than-or-equal relation*

$$m \leq n :\Leftrightarrow (m < n \vee m = n)$$

is a linear order on $\mathbb{N}$.

Proof. The statement of this corollary is a direct consequence of Theorem 4.2.5 and Lemma 4.2.3. □

Having defined the strictly-less-than relation by means of addition, we can now also define subtraction:

4.2.7 Theorem. *There exists a unique map*

$$s : \{(m,n) \in \mathbb{N} \times \mathbb{N} \mid m < n\} \to \mathbb{N}$$
$$(m,n) \mapsto s(m,n) = n - m$$

such that $p(s(m,n), m) = (n - m) + m = n$ *for all* $m, n \in \mathbb{N}$ *with* $m < n$.

Proof. If $m < n$, then there is a $k_{m,n} \in \mathbb{N}$ such that $u_m(k_{m,n}) = k_{m,n} + m = n$. Since u_m is injective (by Lemma 4.2.4), it follows that the map $s(m,n) := k_{m,n}$ is well defined and is uniquely determined by the requirement that $s(m,n) + m$ be equal to n. □

Our final objective in this section is to prove the so-called *Well-Ordering Principle* which asserts that every subset of $\mathbb{N}$ has a minimal element. To do so, we need to use the following lemma:

4.2.8 Lemma. $m < n \Rightarrow m + 1 \leq n$ *for all* $m, n \in \mathbb{N}$.

Proof. If $m < n$, then $m + k = n$ for some $k \in \mathbb{N}$. If $k = 1$, then $m + 1 = n$, and therefore, $m + 1 \leq n$, as desired. If $k \neq 1$, then $1 < k$ (by (4.6)), and therefore, $k = 1 + l$ for some $l \in \mathbb{N}$. Hence $m + 1 \leq n$ (and, in fact, $m + 1 < n$) because $n = m + (1 + l) = (m + 1) + l$. □

4.2.9 Theorem. *(The Well-Ordering Principle) Every nonempty subset of* $\mathbb{N}$ *has a unique minimal element.*

Proof. In order to establish the existence of a minimal element, we need to prove the following statement:

$$\forall_{C \subseteq \mathbb{N}} (C \neq \emptyset \Rightarrow \exists_{n \in C} \forall_{m \in C}\ n \leq m)$$

To give a proof by contraposition, we will assume that a given set $C \subset \mathbb{N}$ satisfies the assumption

$$\forall_{n \in C} \exists_{m \in C}\ n \nleq m \tag{4.8}$$

and then demonstrate that $C = \emptyset$. To do so, we define

$$A := \{n \in \mathbb{N} \mid \forall_{m \in \mathbb{N}} \; m \le n \Rightarrow m \notin C\}.$$

To prove that $1 \in A$, we observe that $m \le 1$ implies that $m = 1$ (by antisymmetry because $1 \le m$ by (4.6)). So we only need to prove that $1 \notin C$. Since $1 \le m$ for all $m \in \mathbb{N}$ (again by (4.6)), assumption (4.8) implies that $1 \notin C$, as desired. Next we assume that n is an arbitrary element in A. In order to show that $\varphi(n)$ is in A as well, we will prove first that

$$\forall_{m \in C} \; \varphi(n) \le m. \tag{4.9}$$

So let $m \in C$. Then $n < m$, because if $m \le n$ (which is the only alternative by linearity), then the assumption $n \in A$ would imply that $m \notin C$. Consequently, Lemma 4.2.8 implies that $\varphi(n) = n + 1 \le m$, as desired. Combining (4.8) and (4.9), it follows that

$$\varphi(n) \notin C. \tag{4.10}$$

In order to prove that $\varphi(n)$ is in A, we will now assume that $m \le \varphi(n)$ and then show that $m \notin C$. If $m \le n$, then $m \notin C$, as desired (because $n \in A$), and if $n < m$ (which again is the only alternative by linearity), then $\varphi(n) = n+1 \le m$ (by Lemma 4.2.8). Hence $\varphi(n) = m$ (by antisymmetry), and therefore, (4.10) implies $m \notin C$. Thus $\varphi(n) \in A$ and $A = \mathbb{N}$, as desired. Given this result, it follows that $n \notin C$ for all $n \in \mathbb{N}$ (because $n \le n$ for all $n \in \mathbb{N}$), and therefore, $C = \emptyset$. Finally, in order establish the uniqueness of minimal elements (see also Exercise 3.5.31), we assume that n_1 and n_2 are minimal elements of a given set $C \subset \mathbb{N}$. Then it is the case that $n_1 \le n_2$ because n_1 is minimal and $n_2 \in C$, and it also is the case that $n_2 \le n_1$ because n_2 is minimal and $n_1 \in C$. Hence $n_1 = n_2$ by antisymmetry, as desired. $\square$

To conclude this section, we list some additional elementary properties of addition and subtraction, most of which involve the order on $\mathbb{N}$.

4.2.10 Proposition. *The following statements are true for all* $k, m, n \in \mathbb{N}$:

a) $1 \le n$,

b) $m < n \Rightarrow \forall_{l \in \mathbb{N}} \; m + l < n + l$,

c) $m \le n \Rightarrow \forall_{l \in \mathbb{N}} \; m + l \le n + l$,

d) $(\exists_{l \in \mathbb{N}} \; m + l < n + l) \Rightarrow m < n$,

e) $(\exists_{l \in \mathbb{N}} \; m + l \le n + l) \Rightarrow m \le n$,

f) $m \not< n \Leftrightarrow n \le m$,

g) $(n \le m \le n + 1) \Rightarrow (m = n \lor m = n + 1)$.

h) $(n + m) - m = n$,

i) $n - m < n$ whenever $m < n$,

j) $(n - m) - k = n - (m + k)$ whenever $m + k < n$,

k) $n - m = n - k \Leftrightarrow m = k$ whenever $m, k < n$.

l) $m - k < n - k$ whenever $k < m < n$.

Proof. We will prove only a), b), d), g), and h) and leave the proofs of the remaining properties as exercises to the reader (see Exercise 4.2.11).

a) According to (4.6), it is the case that $1 < n \vee 1 = n$, and therefore, $1 \leq n$ for all $n \in \mathbb{N}$.

b) If $m < n$, then $m + k = n$ for some $k \in \mathbb{N}$. Thus $n + l = (m + k) + l = m + (k + l) = m + (l + k) = (m + l) + k$, and therefore, $m + l < n + l$ for all $l \in \mathbb{N}$.

d) If $m + l < n + l$ for some $l \in \mathbb{N}$, then there is a $k \in \mathbb{N}$ such that $(m + k) + l = (m + l) + k = n + l$. Hence $u_l(m + k) = u_l(n)$, and since u_l is injective, we may infer that $m + k = n$. Thus $m < n$, as desired.

g) Let $n \leq m \leq n + 1$. Since $n \leq m \Leftrightarrow (n = m \vee n < m)$, we may assume that $n < m$ because if $n = m$, then the consequent $m = n \vee m = n + 1$ is true. Given this assumption, Lemma 4.2.8 implies that $n + 1 \leq m$, and therefore, the antisymmetry of the less-than-or equal-to relation allows us to conclude that $m = n + 1$. So again we find that the consequent $m = n \vee m = n + 1$ is true.

h) Since $m < n + m$ (because $m + n = n + m$), the difference $(n + m) - m$ is well defined and $u_m((n + m) - m) = ((n + m) - m) + m = n + m = u_m(n)$. Consequently, the injectivity of u_m implies that $(n + m) - m = n$, as desired. $\square$

Exercises

4.2.11. Prove Proposition 4.2.10c,e,f,i,j,k,l.

4.2.12. Prove the following statement: if $r \subset S \times S$ is a linear order on a given set S, then

$$s := \{(x, y) \in S \times S \mid (x, y) \in r \wedge x \neq y\}$$

is a linear order on S as well.

4.3 ★ Addition Done Right

In order to construct the addition function p in a fully coherent manner, we will define for each $n \in \mathbb{N}$ a partial addition function p_n whose domain is not $\mathbb{N} \times \{n\}$, as suggested on p.98, but rather $\mathbb{N} \times \{1, \ldots, n\}$. That is to say, for all $(k, m) \in \mathbb{N} \times \{1, \ldots, n\}$ the value $p_n(k, m)$ will be understood to be $k + m$. Furthermore and as we shall see, the functions p_n (construed as sets by being identified with their graphs) will be fully determined by the following conditions:

$$\forall_{m \in \mathbb{N}} \, p_1(m, 1) = \varphi(m),$$
$$\forall_{n \in \mathbb{N}} \, p_n \subset p_{\varphi(n)},$$
$$\forall_{m,n \in \mathbb{N}} \, p_{\varphi(n)}(m, \varphi(n)) = \varphi(p_n(m, n)).$$

Given the close relatedness of these conditions to the properties (4.1) and (4.2), the addition function p on $\mathbb{N} \times \mathbb{N}$ which maps (m, n) into $m+n$ for all $m, n \in \mathbb{N}$ will be found to be the union of the functions p_n, that is,

$$p = \bigcup_{n \in \mathbb{N}} p_n. \tag{4.11}$$

However, in order for this approach to work, we must antecedently define $\{1, \ldots, n\}$ to be the set of all $k \in \mathbb{N}$ that are less than or equal to n, and this in turn requires that we define a less-than-or-equal relation on $\mathbb{N}$ *entirely independently of the notion of addition*. To do so, we will establish in the Theorems 4.3.1 and 4.3.2 the critical fact that for each $n \in \mathbb{N}$ there exists a unique set $B_n \subset \mathbb{N}$ which, intuitively speaking, is equal to the set of all integers greater than or equal to n. Having constructed these sets B_n, we will then be able to show (in Theorem 4.3.4) that $m \le n :\Leftrightarrow B_m \supset B_n$ defines a linear order on $\mathbb{N}$. Note: the use of the familiar symbol '$\le$' indicates that the linear order here in question will be found to be the same as the one defined in Corollary 4.2.6.

4.3.1 Theorem. *For every $n \in \mathbb{N}$ there exists a set $B \subset \mathbb{N}$ such that $\varphi(B) = B \setminus \{n\}$ and $n \in B$.*

Proof. Let $A \subset \mathbb{N}$ be the set of all elements in $\mathbb{N}$ for which a B as stated exists, that is,

$$A := \{n \in \mathbb{N} \mid \exists_{B \subset \mathbb{N}} \, \varphi(B) = B \setminus \{n\} \wedge n \in B\}.$$

Then $1 \in A$, because for $B := \mathbb{N}$ we have $\varphi(B) = B \setminus \{1\}$ by Proposition 4.1.1 and trivially $1 \in B$. Now let $n \in A$ and let $C \subset \mathbb{N}$ such that $\varphi(C) = C \setminus \{n\}$ and $n \in C$. Setting $B := \varphi(C)$, it follows that $\varphi(n) \in B$, and therefore, the assumed injectivity of φ in conjunction with Proposition 3.8.24 allows us to infer that

$$\varphi(B) = \varphi(\varphi(C)) = \varphi(C \setminus \{n\}) = \varphi(C) \setminus \varphi(\{n\}) = B \setminus \{\varphi(n)\},$$

and therefore, $\varphi(n) \in A$ and $A = \mathbb{N}$, as desired. $\qquad\square$

4.3.2 Theorem. *Let $n \in \mathbb{N}$. If B_1 and B_2 are subsets of $\mathbb{N}$ such that $\varphi(B_i) = B_i \setminus \{n\}$ and $n \in B_i$ for $i \in \{1, 2\}$, then $B_1 = B_2$. Note: at this point in our development the symbol '2' is really just a symbol. For '2'—as a* *number—is defined to be equal to $1 + 1$, and since addition has so far not been properly constructed, the latter expression is strictly speaking still completely meaningless.*

Proof. Let A be the set of all $n \in \mathbb{N}$ for which the statement is true, that is,

$$A := \{n \in \mathbb{N} \mid \forall_{B_1, B_2 \subseteq \mathbb{N}} (\forall_{i \in \{1,2\}} \varphi(B_i) = B_i \smallsetminus \{n\} \wedge n \in B_i) \Rightarrow B_1 = B_2\}.$$

Then $1 \in A$ because if $B_1, B_2 \subseteq \mathbb{N}$ such that $1 \in B_i$ and $\varphi(B_i) = B_i \smallsetminus \{1\} \subset B_i$ for $i \in \{1, 2\}$, then $B_i = \mathbb{N}$ for $i \in \{1, 2\}$, and in particular, $B_1 = B_2$. Now let n be an arbitrary element of A and let B_1 and B_2 be subsets of $\mathbb{N}$ that satisfy the conditions $\varphi(n) \in B_i$ and $\varphi(B_i) = B_i \smallsetminus \{\varphi(n)\}$ for $i \in \{1, 2\}$. We need to show that $B_1 = B_2$. Setting $C_i := \varphi^{-1}(B_i)$, it follows that $\varphi(C_i) = B_i$ (because $1 \notin \varphi(\mathbb{N})$ implies that $1 \notin \varphi(B_i) = B_i \smallsetminus \{\varphi(n)\}$ which in turn implies that $1 \notin B_i$ and $B_i \subset \mathbb{N} \smallsetminus \{1\} = \varphi(\mathbb{N})$ and $\varphi(\varphi^{-1}(B_i)) = B_i$), and since φ is injective, we may apply Proposition 3.8.24 to infer that

$$\varphi(\varphi(C_i)) = \varphi(B_i) = B_i \smallsetminus \{\varphi(n)\} = \varphi(C_i) \smallsetminus \{\varphi(n)\} = \varphi(C_i \smallsetminus \{n\})$$

for $i \in \{1, 2\}$. Hence $\varphi(C_i) = C_i \smallsetminus \{n\}$ (because φ is injective) and trivially also $n \in C_i$ (because $\varphi(n) \in B_i$). Using the assumption $n \in A$, it follows that $C_1 = C_2$, and therefore, $B_1 = \varphi(C_1) = \varphi(C_2) = B_2$, as desired. $\square$

Given the statement of Theorem 4.3.1, the Axiom of Choice (as stated on p.86) implies that there exists a map $B : \mathbb{N} \to \mathcal{P}(\mathbb{N})$, $n \mapsto B_n$ such that $\varphi(B_n) = B_n \smallsetminus \{n\}$ and $n \in B_n$ for all $n \in \mathbb{N}$. Furthermore, Theorem 4.3.2 shows that the conditions $\varphi(B_n) = B_n \smallsetminus \{n\}$ and $n \in B_n$ determine the map B uniquely, and the proof of Theorem 4.3.1 therefore implies that

$$B_1 = \mathbb{N} \tag{4.12}$$

and also that

$$B_{\varphi(n)} = \varphi(B_n) \tag{4.13}$$

for all $n \in \mathbb{N}$ (because in that proof the set $B = B_{\varphi(n)}$ was defined to be equal to $\varphi(C) = \varphi(B_n)$). Thus the injectivity of φ implies (see Exercise 4.3.16) that

$$B_n = \varphi^{-1}(\varphi(B_n)) = \varphi^{-1}(B_{\varphi(n)}) \tag{4.14}$$

for all $n \in \mathbb{N}$.

4.3.3 Proposition. *For all $m, n \in \mathbb{N}$ it is the case that $B_n \subset B_m$ whenever $n \in B_m$.*

Proof. Let A be the set of all $n \in \mathbb{N}$ for which the implication holds as stated for all $m \in \mathbb{N}$, that is,

$$A := \{n \in \mathbb{N} \mid \forall_{m \in \mathbb{N}} \, n \in B_m \Rightarrow B_n \subset B_m\}.$$

Then we have $1 \in A$, because if $1 \in B_m$, then $B_m = \mathbb{N}$ (because $\varphi(B_n) = B_n \setminus \{n\} \subset B_n$) and therefore $B_1 \subset B_m$. Now let $n \in A$ and assume that $\varphi(n) \in B_m$. Since $B_1 = \mathbb{N}$ (by (4.12)), we may assume w.l.o.g. that $m \neq 1$, so that $\varphi^{-1}(m)$ is well defined (because $\varphi(\mathbb{N}) = \mathbb{N} \setminus \{1\}$). Furthermore, since $\varphi(n) \in B_m$, it follows that $n \in \varphi^{-1}(B_m) = B_{\varphi^{-1}(m)}$ (by (4.14)), and therefore, $B_n \subset B_{\varphi^{-1}(m)}$ by the inductive assumption. Hence $B_{\varphi(n)} = \varphi(B_n) \subset \varphi(B_{\varphi^{-1}(m)}) = B_m$ (by (4.13)), and, by implication, $\varphi(n) \in A$, as desired. □

4.3.4 Theorem. *The relation* $m \leq n :\Leftrightarrow B_m \supset B_n$ *is a linear order on* $\mathbb{N}$ *that satisfies the following, slightly stronger linearity condition:*

$$\forall_{n,m \in \mathbb{N}} \ n \leq m \lor m \leq n.$$

Proof. It is sufficient show that

$$n \leq m \land m \leq n \Rightarrow m = n, \tag{4.15}$$

$$(k \leq m \land m \leq n) \Rightarrow k \leq n, \tag{4.16}$$

$$m \leq n \lor n \leq m \tag{4.17}$$

for all $k, m, n \in \mathbb{N}$. Regarding (4.15), we observe that $n \leq m$ and $m \leq n$ implies that $B_m = B_n$. Consequently, we have $B_m \setminus \{m\} = \varphi(B_m) = \varphi(B_n) = B_n \setminus \{n\} = B_m \setminus \{n\}$, and this shows that $m = n$ because $m \in B_m$. Turning to (4.16), we assume that $k \leq m \land m \leq n$ for some $k, m, n \in \mathbb{N}$. Then $B_k \supset B_m$ and $B_m \supset B_n$, and therefore, $B_k \supset B_n$ and $k \leq n$, as desired. Finally, in order to establish (4.17), we set

$$A := \{n \in \mathbb{N} \mid \forall_{m \in \mathbb{N}} \ m \leq n \lor n \leq m\}.$$

Given this definition, it follows that $1 \in A$, because the fact that $B_m \subset \mathbb{N} = B_1$ for all $m \in \mathbb{N}$ implies that

$$1 \leq m \tag{4.18}$$

for all $m \in \mathbb{N}$. To show that $\varphi(n) \in A$ for all $n \in A$ (and to show that $A = \mathbb{N}$ thereby), we pick arbitrary elements $n \in A$ and $m \in \mathbb{N}$. Then $m \leq n$ or $n \leq m$ by the definition of A.

Case 1: If $m \leq n$, then $\varphi(n) \in \varphi(B_n) = B_n \setminus \{n\} \subset B_n \subset B_m$, and therefore, $m \leq \varphi(n)$ by Proposition 4.3.3.

Case 2: If $n \leq m$, then $B_{\varphi(n)} = \varphi(B_n) = B_n \setminus \{n\} \supset B_m \setminus \{n\}$ by (4.13). Consequently, $\varphi(n) \leq m$ whenever $n \notin B_m$. Moreover, if $n \in B_m$, then $B_n \subset B_m$ by Proposition 4.3.3, and therefore, $B_n = B_m$ (because $(n \leq m \land B_n \subset B_m) \Leftrightarrow (B_m \subset B_n \land B_n \subset B_m) \Leftrightarrow B_n = B_m$) and $B_{\varphi(n)} = \varphi(B_n) = \varphi(B_m) \subset B_m$. Hence $m \leq \varphi(n)$, as desired. □

4.3.5 Proposition. $B_n = \{m \in \mathbb{N} \mid n \leq m\}$ for all $n \in \mathbb{N}$.

Proof. If $m \in B_n$, then $B_m \subset B_n$ by Proposition 4.3.3, and therefore, $n \leq m$. Conversely, if $n \leq m$ for some $m \in \mathbb{N}$, then $B_m \subset B_n$, and, by implication, $m \in B_n$ because $m \in B_m$. $\qquad\qquad\qquad\qquad\qquad\qquad\qquad\qquad\qquad\qquad\qquad\square$

4.3.6 Proposition. $n \leq \varphi(n)$ *for all* $n \in \mathbb{N}$.

Proof. Since $B_{\varphi(n)} = \varphi(B_n) = B_n \smallsetminus \{n\} \subset B_n$, it follows that $n \leq \varphi(n)$, as desired. $\qquad\qquad\qquad\qquad\qquad\qquad\qquad\qquad\qquad\qquad\qquad\qquad\qquad\square$

4.3.7 Lemma. *Let* $n, m \in \mathbb{N}$. *If* $n \leq m \leq \varphi(n)$, *then* $n = m$ *or* $m = \varphi(n)$.

Proof. Let $A := \{n \in \mathbb{N} \mid \forall_{m \in \mathbb{N}} \ n \leq m \leq \varphi(n) \Rightarrow (n = m \vee m = \varphi(n))\}$. To prove that $1 \in A$ we assume $1 \leq m \leq \varphi(1)$. Then $\mathbb{N} \smallsetminus \{1\} = \varphi(B_1) = B_{\varphi(1)} \subset B_m \subset B_1 = \mathbb{N}$ and therefore $B_m = B_1$ or $B_m = B_{\varphi(1)}$. Hence $m = 1$ or $m = \varphi(1)$ (by antisymmetry). Now assume that $n \in A$ and $\varphi(n) \leq m \leq \varphi(\varphi(n))$. Then $m \neq 1$ (because otherwise $B_1 = \mathbb{N} \supset B_{\varphi(n)}$ would imply that $\varphi(n) = 1$ by antisymmetry), $\varphi(B_{\varphi(n)}) = B_{\varphi(\varphi(n))} \subset B_m = \varphi(B_{\varphi^{-1}(m)})$, and $B_m \subset B_{\varphi(n)} = \varphi(B_n)$. Hence $B_{\varphi(n)} \subset B_{\varphi^{-1}(m)} \subset B_n$ (because φ is injective), and since $n \in A$, it follows that $n = \varphi^{-1}(m)$ or $\varphi(n) = \varphi^{-1}(m)$. Consequently, $\varphi(n) = m$ or $\varphi(\varphi(n)) = m$, as desired. $\qquad\qquad\qquad\square$

4.3.8 Corollary. *Let* $n, m \in \mathbb{N}$. *If* $m \leq n$ *and* $m \neq n$, *then*

a) $\varphi(m) \leq n$, *and*

b) $n \neq 1$ *and* $m \leq \varphi^{-1}(n)$.

Proof. **a)** Assume that $m \leq n$ and $m \neq n$. Then, according to Theorem 4.3.4, it is the case that $n \leq \varphi(m)$ or $\varphi(m) \leq n$. Since the latter of these two inequalities represents the desired conclusion, we only need to consider the case where $n \leq \varphi(m)$. Then $m \leq n \leq \varphi(m)$, and therefore, $n = \varphi(m)$ (by Lemma 4.3.7 because $n \neq m$). Thus $\varphi(m) \leq n$ because trivially $B_{\varphi(m)} = B_n \supset B_n$.

b) Again we assume that $m \leq n$ and $m \neq n$. Since $1 \leq m$ (because $B_1 = \mathbb{N} \supset B_m$), it follows that $n \neq 1$, for otherwise the inequalities $n = 1 \leq m$ and $m \leq n$ would imply that $n = m$ (by antisymmetry). Thus $\varphi^{-1}(n)$ is well defined. Moreover, if it were the case that $m = \varphi^{-1}(n)$, then, trivially, $m \leq \varphi^{-1}(n)$, as desired. Consequently, we may assume that $m \neq \varphi^{-1}(n)$. Since $m \leq \varphi^{-1}(n)$ or $\varphi^{-1}(n) \leq m$ (by linearity), we only need to show that $\varphi^{-1}(n) \leq m \wedge m \neq \varphi^{-1}(n)$ is false. To do so, we assume that this statement is true. Then $n \leq m$ (by a)), and therefore, $n = m$ by antisymmetry. Since this latter conclusion contradicts the initial assumption $m \neq n$, the proof is complete. $\qquad\qquad\qquad\qquad\qquad\qquad\qquad\qquad\qquad\qquad\qquad\qquad\square$

In order to construct the addition function p on $\mathbb{N} \times \mathbb{N}$ from a sequence of partial addition functions p_n on $\{1, \ldots, n\} \times \mathbb{N}$, as indicated in equation (4.11), we need to show that the union of a nested sequence of functions (that is, a sequence of functions f_n that satisfy the condition $f_n \subset f_{\varphi(n)}$ for all $n \in \mathbb{N}$) is itself a function as well. To do so, we need to first re-establish the Well-Ordering Principle, relying solely on the results that we derived in the present section, and then use this principle to prove a preparatory proposition.

4.3.9 Theorem. *(The Well-Ordering Principle) Every nonempty subset of* $\mathbb{N}$ *has a unique minimal element.*

Proof. Proceeding as in the proof of Theorem 4.2.9, we need to show that

$$\forall_{C \subseteq \mathbb{N}} (C \neq \emptyset \Rightarrow \exists_{n \in C} \forall_{m \in C} \; n \leq m)$$

To give a proof by contraposition, we will assume that a given set $C \subset \mathbb{N}$ satisfies the assumption

$$\forall_{n \in C} \exists_{m \in C} \; n \not\leq m \qquad (4.19)$$

and then demonstrate that $C = \emptyset$. To do so, we define

$$A := \{ n \in \mathbb{N} \mid \forall_{m \in \mathbb{N}} \; m \leq n \Rightarrow m \notin C \}.$$

To prove that $1 \in A$, we observe that $m \leq 1$ implies that $m = 1$ (by antisymmetry because $1 \leq m$ by (4.18)). So we only need to prove that $1 \notin C$. Since $1 \leq m$ for all $m \in \mathbb{N}$ (again by (4.18)), assumption (4.19) implies that $1 \notin C$, as desired. Next we assume that $n \in A$. In order to show that $\varphi(n) \in A$, we will prove first that

$$\forall_{m \in C} \; \varphi(n) \leq m. \qquad (4.20)$$

So let $m \in C$. Then $n \leq m$ and $m \neq n$, because if $m \leq n$ (which is the only alternative by linearity), then the assumption $n \in A$ would imply that $m \notin C$. Hence $\varphi(n) \leq m$ by Corollary 4.3.8a, as desired. Combining (4.19) and (4.20), it follows that

$$\varphi(n) \notin C. \qquad (4.21)$$

In order to prove that $\varphi(n) \in A$, we need to show that $m \notin C$ whenever $m \leq \varphi(n)$. So let $m \leq \varphi(n)$. If $m \leq n$, then $m \notin C$, as desired (because $n \in A$), and if $m \geq n$, then $m = n$ or $m = \varphi(n)$ by Lemma 4.3.7. If $m = n$ then $m \notin C$ because $n \in A$, and if $m = \varphi(n)$, then (4.21) implies $m \notin C$. Thus $\varphi(n) \in A$ and $A = \mathbb{N}$, as desired. Given this result, it follows that $n \notin C$ for all $n \in \mathbb{N}$ (because $n \leq n$ for all $n \in \mathbb{N}$), and therefore, $C = \emptyset$. Finally, the uniqueness proof is identical to the uniqueness proof for Theorem 4.2.9. $\qquad \square$

4.3.10 Proposition. *If A is a nonempty subset of $\mathbb{N}$ and if n is the unique minimal element of A (which exists according to the Well-Ordering Principle), then $A = B_n$ whenever $\varphi(A) \subset A$.*

Proof. According to Proposition 4.3.5, we need to show that $A = \{m \in \mathbb{N} \mid n \leq m\}$. If $m \in A$, then $n \leq m$ because n is the minimal element of A, and therefore, $A \subset \{m \in \mathbb{N} \mid n \leq m\}$. In order to prove the reverse inclusion by way of contradiction, we assume that $C := \{m \in \mathbb{N} \mid n \leq m \wedge m \notin A\}$ is nonempty. Then C has a minimal element m_0, which, by the definition of C, is not in A and thus different from n (because n is in A). So it is the case that $n \leq m_0$ and $n \neq m_0$, and therefore, Corollary 4.3.8b implies that $n \leq \varphi^{-1}(m_0)$. If $\varphi^{-1}(m_0)$ were contained in A, then the assumption $\varphi(A) \subset A$ would allow us to infer that $m_0 = \varphi(\varphi^{-1}(m_0))$ is contained in A as well, but this is not the case. Consequently, $\varphi^{-1}(m_0)$ is not in A, and this shows that $\varphi^{-1}(m_0) \in C$. Hence the minimality of m_0 implies that $m_0 \leq \varphi^{-1}(m_0)$, and this in turn implies that $\varphi(m_0) \leq m_0$ (by Corollary 4.3.8a). Using antisymmetry in conjunction with Proposition 4.3.6, it follows that $\varphi(m_0) = m_0$, but this is impossible by Proposition 4.1.2. Thus C must be empty and $\{m \in \mathbb{N} \mid n \leq m\} \subset A$, as desired. $\square$

4.3.11 Corollary. *Let $A \subset \mathbb{N}$ and $n \in \mathbb{N}$. If $n \in A$ and $\varphi(A) \subset A$, then $B_n \subset A$.*

Proof. This is a trivial consequence of the Propositions 4.3.10 and 4.3.3. $\square$

4.3.12 Lemma. *If $\{S_n\}_{n \in \mathbb{N}}$ is a collection of sets such that $S_n \subset S_{\varphi(n)}$ for all $n \in \mathbb{N}$, then $n \leq m \Rightarrow S_n \subset S_m$ for all $n, m \in \mathbb{N}$.*

Proof. Let $n \in \mathbb{N}$ and $A := \{m \in \mathbb{N} \mid S_n \subset S_m\}$. We need to show that $\{m \in \mathbb{N} \mid n \leq m\} \subset A$. In light of Proposition 4.3.5 and Corollary 4.3.11, it is sufficient to prove that $n \in A$ and $\varphi(m) \in A$ for all $m \in A$. The first of these statements—$n \in A$—trivially follows from the fact that $S_n \subset S_n$. To prove the second we pick an arbitrary $m \in A$ and use the definition of A in conjunction with the assumption $\forall_{m \in \mathbb{N}} S_m \subset S_{\varphi(m)}$ to conclude that $S_n \subset S_m \subset S_{\varphi(m)}$. Hence $\varphi(m) \in A$, as desired. $\square$

4.3.13 Proposition. *Let $\{A_n\}_{n \in \mathbb{N}}$ be a collection of nonempty sets and let $\{f_n\}_{n \in \mathbb{N}}$ be a collection of functions such that for each $n \in \mathbb{N}$, f_n is a function from A_n to a given nonempty set B. If $f_n \subset f_{\varphi(n)}$ for all $n \in \mathbb{N}$, then $f := \bigcup_{n \in \mathbb{N}} f_n$ is a function from $A := \bigcup_{n \in \mathbb{N}} A_n$ to B.*

Proof. f is a relation between A and B because $f_n \subset A_n \times B \subset A \times B$ for all $n \in \mathbb{N}$. In order to prove that the domain of f is A, we pick an arbitrary $x \in A$. Then, by the definition of A, there exists an $n \in \mathbb{N}$ such that $x \in A_n$. Since the domain of f_n is A_n, there exists a $y \in B$ such that $(x, y) \in f_n \subset f$. Hence $x \in D(f)$, as desired. Finally, in order to show that f is a function, we assume that $(x, y) \in f$ and $(x, z) \in f$ for some $x \in A$ and some $y, z \in B$. Then, by the definition of f, there are elements $m, n \in \mathbb{N}$ such that $(x, y) \in f_m$ and $(x, z) \in f_n$. Since $m \leq n \vee n \leq m$ (by linearity), we may assume w.l.o.g.

that $n \leq m$. Applying now Lemma 4.3.12 to the collection $\{f_n\}_{n \in \mathbb{N}}$ in place of $\{S_n\}_{n \in \mathbb{N}}$, it follows that $f_n \subset f_m$. Consequently, *both* (x, y) and (x, z) are elements of f_m. Since f_m is a function, we may infer that $y = z$, as desired. $\square$

4.3.14 Theorem. *There exists a unique map* $p : \mathbb{N} \times \mathbb{N} \to \mathbb{N}$ *that satisfies the conditions* (4.1) *and* (4.2).

Proof. Since the uniqueness of p was rigorously established in Theorem 4.2.1, we only need to prove that p exists. To do so, we denote by A the set of all $n \in \mathbb{N}$ for which there exists a map $p_n : \mathbb{N} \times \mathbb{N} \setminus B_{\varphi(n)} \to \mathbb{N}$ such that

$$p_n(m, 1) = \varphi(m), \tag{4.22}$$
$$p_n(m, \varphi(k)) = \varphi(p_n(m, k)) \tag{4.23}$$

for all $m \in \mathbb{N}$ and all $k \in \mathbb{N} \setminus B_n$. To prove that 1 is in A, we observe that $\mathbb{N} \setminus B_{\varphi(1)} = \mathbb{N} \setminus \varphi(B_1) = \mathbb{N} \setminus \varphi(\mathbb{N}) = \{1\}$ and define $p_1 : \mathbb{N} \times \{1\} \to \mathbb{N}$ via the equation $p_1(m, 1) := \varphi(m)$. Given this definition, it is easy to see that p_1 satisfies (4.22) and (4.23). To prove that $\varphi(A) \subset A$, we pick an arbitrary $n \in \mathbb{N}$ and assume the existence of a map p_n as specified above. Since $B_{\varphi(\varphi(n))} = \varphi(B_{\varphi(n)}) = B_{\varphi(n)} \setminus \{\varphi(n)\}$, it follows that $\mathbb{N} \setminus B_{\varphi(\varphi(n))} = (\mathbb{N} \setminus B_{\varphi(n)}) \cup \{\varphi(n)\}$. Given this observation, we may define $p_{\varphi(n)} : \mathbb{N} \times \mathbb{N} \setminus B_{\varphi(\varphi(n))}$ via the equation

$$p_{\varphi(n)}(m, k) := \begin{cases} p_n(m, k) & \text{for } (m, k) \in \mathbb{N} \times \mathbb{N} \setminus B_{\varphi(n)}, \\ \varphi(p_n(m, n)) & \text{for } k = \varphi(n) \text{ and } m \in \mathbb{N}. \end{cases}$$

To verify that $p_{\varphi(n)}$ satisfies (4.22) and (4.23), it is sufficient to observe that, by assumption, p_n satisfies these conditions and that

$$p_{\varphi(n)}(m, \varphi(n)) = \varphi(p_n(m, n)) = \varphi(p_{\varphi(n)}(m, n)).$$

Thus $A = \mathbb{N}$. To proceed, we consider p_n to be a subset of $(\mathbb{N} \times \mathbb{N} \setminus B_{\varphi(n)}) \times \mathbb{N}$ and define

$$p := \bigcup_{n \in \mathbb{N}} p_n.$$

Given the definition of $p_{\varphi(n)}$ above, it is obvious that every element

$$((m, k), p_n(m, k)) \in p_n$$

is also an element of $p_{\varphi(n)}$. Thus $p_n \subset p_{\varphi(n)}$ for all $n \in \mathbb{N}$, and therefore Proposition 4.3.13 implies that p is a function from $\bigcup_{n \in \mathbb{N}} \mathbb{N} \times (\mathbb{N} \setminus B_{\varphi(n)})$ to $\mathbb{N}$. Using Proposition 3.8.18b,d, it follows that

$$\mathbb{N} \times \bigcup_{n \in \mathbb{N}} (\mathbb{N} \setminus B_{\varphi(n)}) = \mathbb{N} \times \left(\mathbb{N} \setminus \bigcap_{n \in \mathbb{N}} B_{\varphi(n)} \right) = \mathbb{N} \times \left(\mathbb{N} \setminus \bigcap_{n \in \mathbb{N}} \varphi(B_n) \right)$$

$$= \mathbb{N} \times \left(\mathbb{N} \setminus \bigcap_{n \in \mathbb{N}} (B_n \setminus \{n\}) \right)$$

$$= \mathbb{N} \times \left(\mathbb{N} \setminus \left(\left(\bigcap_{n \in \mathbb{N}} B_n \right) \setminus \bigcup_{n \in \mathbb{N}} \{n\} \right) \right)$$

$$= \mathbb{N} \times \left(\mathbb{N} \setminus \left(\left(\bigcap_{n \in \mathbb{N}} B_n \right) \setminus \mathbb{N} \right) \right) = \mathbb{N} \times (\mathbb{N} \setminus \emptyset)$$

$$= \mathbb{N} \times \mathbb{N}.$$

Hence the domain of p is $\mathbb{N} \times \mathbb{N}$, as desired. Furthermore, since $p_1 \subset p$, we may infer that $((m,1), \varphi(m)) = ((m,1), p_1(m,1)) \in p$, and therefore, $p(m,1) = \varphi(m)$ for all $m \in \mathbb{N}$. Finally, to show that $p(m, \varphi(n)) = \varphi(p(m,n))$ for all $m, n \in \mathbb{N}$, we pick arbitrary elements $m, n \in \mathbb{N}$ and use the fact that $p_n \subset p_{\varphi(n)} \subset p$ to conclude that

$$p(m, \varphi(n)) = p_{\varphi(n)}(m, \varphi(n)) = \varphi(p_n(m,n)) = \varphi(p(m,n)),$$

as desired. $\qquad\qquad\qquad\qquad\qquad\qquad\qquad\qquad\qquad\qquad\qquad\qquad\square$

In the light of the theorem just established we are now finally allowed to speak of the *addition* of two integers and to even use symbols like 1 or 2 with a certain measure of confidence. That is to say, for any integers $n, m \in \mathbb{N}$ we may write

$$\boxed{n + m = p(n, m)}$$

and even more daringly

$$\boxed{1 + 1 = 2}$$

because in setting $2 := \varphi(1)$, we find that

$$1 + 1 = p(1,1) = \varphi(1) = 2.$$

Having properly constructed the addition function p, we will turn our attention in the next section to the problem of defining multiplication. But before we can get to that, there is one more problem to be solved: we need to show that the less-than-or-equal-to relations defined in Corollary 4.2.6 and Theorem 4.3.4, respectively, are indeed identical. This is important in particular because it shows that the definition of subtraction in Theorem 4.2.7 is feasible as well in the non-standard case where p is constructed as in the proof of Theorem 4.3.14 and that Proposition 4.2.10 remains valid as well.

4.3.15 Theorem. *For all $m, n \in \mathbb{N}$ it is the case that*

$$B_m \supset B_n \Leftrightarrow (m < n \vee m = n),$$

where $m < n \Leftrightarrow \exists_{k \in \mathbb{N}} \, n = m + k$.

Proof. "⇒" Let $B_m \supset B_n$ and $m \neq n$ for some $m, n \in \mathbb{N}$. We need to show that $m < n$, that is, we need to show that $n = m + k$ for some $k \in \mathbb{N}$. Using Corollary 4.3.8a, it follows that $\varphi(m) \leq n$. Since

$$B_{\varphi(m)} = \{n \in \mathbb{N} \mid \varphi(m) \leq n\} \quad \text{(by Proposition 4.3.5)},$$

it is therefore sufficient to show that

$$B_{\varphi(m)} \subset A := \{n \in \mathbb{N} \mid \exists_{k \in \mathbb{N}}\, n = m + k\}.$$

Consequently, according to Corollary 4.3.11, it suffices to prove that $\varphi(m) \in A$ and $\varphi(n) \in A$ whenever $n \in A$. Since $\varphi(m) = p(m, 1) = m + 1$, we find that $\varphi(m) \in A$, as desired. Furthermore, if $n \in A$, that is, if $n = m + k$ for some $k \in \mathbb{N}$, then $\varphi(n) = n + 1 = (m + k) + 1 = m + (k + 1)$ (by Theorem 4.2.2a because Theorem 4.2.2 was proven using nothing but (4.1) and (4.2) and is therefore valid regardless of how the addition map p is constructed), and, by implication, $\varphi(n) \in A$.

"⇐" We wish to show that the set

$$A := \{k \in \mathbb{N} \mid \forall_{m,n \in \mathbb{N}}\, (n = m + k \Rightarrow B_m \supset B_n)\}$$

is equal to $\mathbb{N}$. To do so, we observe that 1 is in A because if $n = m + 1 = \varphi(m)$, then $B_n = B_{\varphi(m)} = \varphi(B_m) = B_m \smallsetminus \{m\} \subset B_m$. Moreover, if $k \in A$ and $n = m + \varphi(k)$ for some $m, n \in \mathbb{N}$, then $n = m + (k + 1) = m + (1 + k) = (m + 1) + k$, and therefore, $B_m \supset B_{\varphi(m)} = B_{m+1} \supset B_n$ because $k \in A$. Hence $\varphi(k) \in A$, as desired. Having thus established that $A = \mathbb{N}$, we proceed to assume that $m < n \lor m = n$ or, equivalently, that $n = m + k \lor m = n$ for some $k, m, n \in \mathbb{N}$. If $m = n$, then, trivially, $B_m = B_n \supset B_n$, and we are done. If $n = m + k$, then $B_m \supset B_n$ because $A = \mathbb{N}$, and again we are done. $\qquad\square$

Exercises

4.3.16. Assume that $f : A \to B$ is an injective function. Show that for all $C \subset A$ we have $f^{-1}(f(C)) = C$, and find a counterexample to this equation in a case where f is not injective.

4.3.17. Explain why for all $m, n \in \mathbb{N}$ it is the case that $m < n$ if and only if $B_m \supset B_n \land m \neq n$.

4.4 ★ Constructing Multiplication

The following theorem provides a construction of multiplication on $\mathbb{N}$ that is completely analogous to the construction of addition in Theorem 4.3.14. Readers who skipped the previous section and the proof of Theorem 4.3.14 thereby may skip this present section as well, as long as they are sure, that is, that they truly understood in elementary school why $1 \cdot 1 = 1$.

4.4.1 Theorem. *There exists a unique map* $\mu : \mathbb{N} \times \mathbb{N} \to \mathbb{N}$ *that satisfies the following conditions:*

a) $\mu(1, m) = m$ *and*

b) $\mu(\varphi(n), m) = \mu(n, m) + m$

for all $n, m \in \mathbb{N}$.

Proof. Uniqueness: assume that ν is another map that satisfies a) and b) above. Setting

$$A := \{n \in \mathbb{N} \mid \forall_{m \in \mathbb{N}} \, \mu(n, m) = \nu(n, m)\},$$

it follows that $1 \in A$ by property a). Moreover, in assuming that $n \in A$, we find that $\mu(n, m) = \nu(n, m)$ for all $m \in \mathbb{N}$, and b) therefore implies that $\mu(\varphi(n), m) = \mu(n, m) + m = \nu(n, m) + m = \nu(\varphi(n), m)$. Hence $A = \mathbb{N}$ and $\mu = \nu$.

Existence: in direct analogy to the proof of Theorem 4.3.14, we define A to be the set of all $n \in \mathbb{N}$ for which there exists a map $\mu_n : \mathbb{N} \smallsetminus B_{\varphi(n)} \times \mathbb{N} \to \mathbb{N}$ such that

$$\mu_n(1, m) = m, \tag{4.24}$$
$$\mu_n(\varphi(k), m) = \mu_n(k, m) + m \tag{4.25}$$

for all $m \in \mathbb{N}$ and all $k \in \mathbb{N} \smallsetminus B_n$. To prove that 1 is in A we observe, as the proof of Theorem 4.3.14, that $\mathbb{N} \smallsetminus B_{\varphi(1)} = \{1\}$ and define $\mu_1 : \{1\} \times \mathbb{N} \to \mathbb{N}$ via the equation $\mu_1(1, m) := m$. Given this definition, μ_1 evidently satisfies (4.24) and (4.25), as desired. In order to show that $\varphi(A) \subset A$ we pick an $n \in \mathbb{N}$ and assume the existence of a map μ_n as stated above. Taking again as our model the proof of Theorem 4.3.14, we define $\mu_{\varphi(n)} : \mathbb{N} \smallsetminus B_{\varphi(\varphi(n))} \times \mathbb{N}$ via the equation

$$\mu_{\varphi(n)}(k, m) := \begin{cases} \mu_n(k, m) & \text{for } (k, m) \in \mathbb{N} \smallsetminus B_{\varphi(n)} \times \mathbb{N}, \\ \mu_n(n, m) + m & \text{for } k = \varphi(n) \text{ and } m \in \mathbb{N}. \end{cases}$$

To verify that $\mu_{\varphi(n)}$ satisfies (4.24) and (4.25), it is sufficient to observe that, by assumption, μ_n satisfies these conditions and that

$$\mu_{\varphi(n)}(\varphi(n), m) = \mu_n(n, m) + m = \mu_{\varphi(n)}(n, m) + m.$$

Hence $A = \mathbb{N}$. Setting

$$\mu := \bigcup_{n \in \mathbb{N}} \mu_n,$$

we may invoke Proposition 4.3.13, as in the proof of Theorem 4.3.14, to infer that μ is a function from $\bigcup_{n \in \mathbb{N}} (\mathbb{N} \smallsetminus B_{\varphi(n)}) \times \mathbb{N} = \mathbb{N} \times \mathbb{N}$ to $\mathbb{N}$ (because evidently

$\mu_n \subset \mu_{\varphi(n)}$ for all $n \in \mathbb{N}$). Since $\mu_1 \subset \mu$, it follows that μ satisfies property a) above, because $\mu(1, m) = \mu_1(1, m) = m$ for all $m \in \mathbb{N}$. Finally, in order to show that μ satisfies property b) as well, we pick arbitrary elements $n, m \in \mathbb{N}$ and use the fact that $\mu_n \subset \mu_{\varphi(n)} \subset m$ to conclude that

$$\mu(\varphi(n), m) = \mu_{\varphi(n)}(\varphi(n), k) = \mu_n(n, m) + m = \mu(n, m) + m,$$

as desired. □

Given the construction of the multiplication function μ in the previous theorem, we naturally write

$$\boxed{\mu(n, m) = n \cdot m = nm}$$

for all $n, m \in \mathbb{N}$. To further justify this notation, we note that multiplication is associative and commutative and that multiplication and addition are distributive:

4.4.2 Theorem. *For all $k, m, n \in \mathbb{N}$ it is the case that $1 \cdot n = n$, $(nm)k = n(mk)$, $nm = mn$, and $n(m + k) = nm + nk$.*

Proof. See Exercise 4.4.3. □

Exercises

4.4.3. Prove Theorem 4.4.2.

4.4.4. Prove that $n \leq nm$ for all $n, m \in \mathbb{N}$.

4.4.5. Prove the following statement: $\forall_{m,n \in \mathbb{N}} \, nm = n \Leftrightarrow m = 1$.

4.5 Proofs by Induction

Assume that for each $n \in \mathbb{N}$ we are given a statement $P(n)$ and that we wish to show that $P(n)$ is true for all $n \in \mathbb{N}$. Setting

$$A := \{n \in \mathbb{N} \mid P(n)\},$$

it follows that we need to show that $A = \mathbb{N}$. In other words, according to the Principle of Mathematical Induction, as stated in the Peano Axioms, we need to show that $1 \in A$ (i.e., that $P(1)$ is true) and that $\varphi(A) \subset A$. Since

$$\varphi(A) \subset A \Leftrightarrow \forall_{n \in A} \, \varphi(n) \in A \Leftrightarrow \forall_{n \in \mathbb{N}} \, n \in A \Rightarrow n + 1 = \varphi(n) \in A$$
$$\Leftrightarrow \forall_{n \in \mathbb{N}} \, P(n) \Rightarrow P(n+1),$$

it follows that we need to show the following:

a) $P(1)$ and

b) $P(n) \Rightarrow P(n+1)$ for all $n \in \mathbb{N}$.

Remark. In the proofs of Theorems 4.2.9 and 4.3.9 we saw that the Principle of Mathematical Induction implies that every nonempty subset of $\mathbb{N}$ has a minimal element. For reasons of completeness, we wish here to add that the reverse conclusion is valid as well. More precisely, it is the case that the following material conditional is valid for any sequence of statements $(P(n))_{n\in\mathbb{N}}$:

$$
\begin{aligned}
&(\forall_{C\subseteq\mathbb{N}} \, C \neq \emptyset \Rightarrow \exists_{n\in C} \forall_{m\in C} \, n \leq m) \\
&\Rightarrow ((P(1) \wedge \forall_{n\in\mathbb{N}} P(n) \Rightarrow P(n+1)) \Rightarrow \forall_{n\in\mathbb{N}} P(n)).
\end{aligned}
\tag{4.26}
$$

To prove this assertion, we us assume that the antecedent

$$
\forall_{C\subseteq\mathbb{N}} \, C \neq \emptyset \Rightarrow \exists_{n\in C} \forall_{m\in C} \, n \leq m
\tag{4.27}
$$

is true and that we are given a sequence of statements $(P(n))_{n\in\mathbb{N}}$ such that

$$
P(1) \wedge \forall_{n\in\mathbb{N}} P(n) \Rightarrow P(n+1).
\tag{4.28}
$$

In order to demonstrate that $P(n)$ is true for all $n \in \mathbb{N}$, we further assume that there is an $n \in \mathbb{N}$ such that $P(n)$ is false. In other words, we assume that

$$
C := \{n \in \mathbb{N} \mid \neg P(n)\}
$$

is nonempty. Given this assumption, statement (4.27) allows us to infer the existence of a number $n_0 \in C$ such that $n_0 \leq m$ for all $m \in C$. Since $P(n_0)$ is false (because $n_0 \in C$) and since $P(1)$ is true (by (4.28)), it follows that $n_0 > 1$ and therefore $n_0 - 1 \in \mathbb{N}$. Furthermore, since $n_0 - 1 < n_0$ (see Proposition 4.2.10j), the minimality of n_0 in C implies that $n_0 - 1 \notin C$. Hence $P(n_0 - 1)$ is true, and therefore $P(n_0)$ is true as well (by (4.28)). Having thus arrived at a contradiction, the proof of (4.26) is complete.

4.5.1 Example. We wish to show that

$$
\sum_{k=1}^{n} k = \frac{n(n+1)}{2}
$$

for all $n \in \mathbb{N}$. In this case we have

$$
P(n) :\Leftrightarrow \sum_{k=1}^{n} k = \frac{n(n+1)}{2},
$$

and therefore, we need to prove that

$$
\sum_{k=1}^{1} k = \frac{1(1+1)}{2}
\tag{4.29}
$$

and that

$$\sum_{k=1}^{n+1} k = \frac{(n+1)((n+1)+1)}{2} \quad \text{whenever} \quad \sum_{k=1}^{n} k = \frac{n(n+1)}{2}. \qquad (4.30)$$

To establish (4.29), we only need to observe that both $\sum_{k=1}^{1} k$ and $1(1+1)/2$ are equal to 1, and to establish (4.30), we assume that

$$\sum_{k=1}^{n} k = \frac{n(n+1)}{2}$$

for some $n \in \mathbb{N}$. Given this assumption, it follows that

$$\sum_{k=1}^{n+1} k = \sum_{k=1}^{n} k + n + 1 = \frac{n(n+1)}{2} + n + 1 = \frac{n(n+1) + 2(n+1)}{2}$$
$$= \frac{(n+1)((n+1)+1)}{2},$$

as desired.

4.5.2 Example. We wish to show that $1 + nx \le (1+x)^n$ for all $x \in [-1, \infty)$ and all $n \in \mathbb{N}$. In this case we have

$$P(n) \Leftrightarrow \forall_{x \in [-1,\infty)} 1 + nx \le (1+x)^n,$$

and in particular $P(1) \Leftrightarrow 1 + x \le 1 + x$. Thus $P(1)$ is true. Assuming now that $P(n)$ is true for some $n \in \mathbb{N}$, we need to show that $P(n+1)$ is true as well, that is, we need to show that

$$\forall_{x \in [-1,\infty)} 1 + (n+1)x \le (1+x)^{n+1}.$$

So let $x \in [-1, \infty)$. Then, by assumption, $1 + nx \le (1+x)^n$, and since $1 + x \ge 0$, it follows that

$$(1+x)^{n+1} = (1+x)^n (1+x) \ge (1+nx)(1+x) = 1 + (n+1)x + nx^2$$
$$\ge 1 + (n+1)x,$$

as desired.

4.5.3 Example. We wish to prove that

$$x^n - y^n = (x-y)(x^{n-1} + x^{n-2}y + \cdots + xy^{n-2} + y^{n-1})$$
$$= (x-y) \sum_{k=0}^{n-1} x^{n-1-k} y^k \qquad (4.31)$$

for all $n \in \mathbb{N}$ and all $x, y \in \mathbb{R}$. Since for $n = 1$ the sum on the right in (4.31) is equal to 1, the equation reduces in this case to $x - y = x - y$ which is evidently true. Assuming now that (4.31) is valid for a given n, we need to show that

$$x^{n+1} - y^{n+1} = (x - y) \sum_{k=0}^{n} x^{n-k} y^k.$$

To do so we transform the product on the right-hand side of this equation as follows:

$$(x - y) \sum_{k=0}^{n} x^{n-k} y^k = (x - y) \left(y^n + \sum_{k=0}^{n-1} x^{n-k} y^k \right)$$

$$= (x - y) \left(y^n + x \sum_{k=0}^{n-1} x^{n-1-k} y^k \right)$$

$$= (x - y) y^n + x(x - y) \sum_{k=0}^{n-1} x^{n-1-k} y^k$$

$$= (x - y) y^n + x(x^n - y^n)$$

$$\text{(by the inductive assumption)}$$

$$= x^{n+1} - y^{n+1},$$

as desired.

4.5.4 Example. We wish to show that

$$\sum_{k=1}^{n} \sqrt{k} \le \frac{2}{3} \left(\sqrt{n+1}^{\,3} - 1 \right)$$

for all $n \in \mathbb{N}$. For $n = 1$ the inequality above reduces to the true in equality $1 \le 2(2\sqrt{2} - 1)/3$. Assuming that the inequality above is satisfied for some $n \in \mathbb{N}$, it follows that

$$\sum_{k=1}^{n+1} \sqrt{k} = \sqrt{n+1} + \sum_{k=1}^{n} \sqrt{k} \le \sqrt{n+1} + \frac{2}{3} \left(\sqrt{n+1}^{\,3} - 1 \right).$$

Thus we only need to show that

$$\sqrt{n+1} + \frac{2}{3} \left(\sqrt{n+1}^{\,3} - 1 \right) \le \frac{2}{3} \left(\sqrt{n+2}^{\,3} - 1 \right),$$

and to do so, we equivalently transform this remaining in equality as follows:

$$\sqrt{n+1} + \frac{2}{3} \left(\sqrt{n+1}^{\,3} - 1 \right) \le \frac{2}{3} \left(\sqrt{n+2}^{\,3} - 1 \right)$$

$$\Leftrightarrow 3\sqrt{n+1} \le 2 \left(\sqrt{n+2}^{\,3} - \sqrt{n+1}^{\,3} \right)$$

$$\Leftrightarrow 3\sqrt{n+1} \le 2 \left(\sqrt{n+2} - \sqrt{n+1} \right) \left(2n + 3 + \sqrt{n+2}\sqrt{n+1} \right) \quad \text{(by (4.31))}$$

$$\Leftrightarrow 3\sqrt{n+1}\left(\sqrt{n+2}+\sqrt{n+1}\right) \le 2\left(2n+3+\sqrt{n+2}\sqrt{n+1}\right)$$
$$\Leftrightarrow \sqrt{n+1}\sqrt{n+2} \le n+3$$
$$\Leftrightarrow n^2 + 3n + 2 \le n^2 + 6n + 9$$
$$\Leftrightarrow 0 \le 3n + 7.$$

Since the last of these inequalities is evidently true for all $n \in \mathbb{N}$, the proof is complete.

4.5.5 Example. We wish to prove the *binomial formula:*

$$(a+b)^n = \sum_{k=0}^{n} \binom{n}{k} a^k b^{n-k} \tag{4.32}$$

for all $a, b \in \mathbb{R}$ and all $n \in \mathbb{N}$, where

$$\binom{n}{k} := \frac{n!}{k!(n-k)!}$$

and

$$n! = 1 \cdot 2 \cdots n = \prod_{k=1}^{n} k.$$

Since

$$\sum_{k=0}^{1} \binom{1}{k} a^k b^{1-k} = \binom{1}{0} a^0 b^1 + \binom{1}{1} a^1 b^0 = b + a = (a+b)^1,$$

it follows that the formula is valid for $n = 1$. Thus, we only need to show that it is valid for $n+1$ whenever it is valid for n. So let us assume that for a given positive integer n we have

$$(a+b)^n = \sum_{k=0}^{n} \binom{n}{k} a^k b^{n-k}.$$

Then

$$(a+b)^{n+1} = (a+b)(a+b)^n = (a+b) \sum_{k=0}^{n} \binom{n}{k} a^k b^{n-k}$$

$$= \sum_{k=0}^{n} \binom{n}{k} a^{k+1} b^{n-k} + \sum_{k=0}^{n} \binom{n}{k} a^k b^{n+1-k}$$

$$= \sum_{k=1}^{n+1} \binom{n}{k-1} a^k b^{n-(k-1)} + \sum_{k=0}^{n} \binom{n}{k} a^k b^{n+1-k}$$

$$= \binom{n}{n} a^{n+1} b^0 + \sum_{k=1}^{n} \left(\binom{n}{k-1} + \binom{n}{k}\right) a^k b^{n+1-k} + \binom{n}{0} a^0 b^{n+1}$$

$$= \binom{n+1}{n+1} a^{n+1} b^0 + \sum_{k=1}^{n} \left(\binom{n}{k-1} + \binom{n}{k} \right) a^k b^{n+1-k} + \binom{n+1}{0} a^0 b^{n+1},$$

because

$$\binom{n}{n} = \binom{n}{0} = \binom{n+1}{n+1} = \binom{n+1}{0} = 1.$$

Furthermore,

$$\binom{n}{k-1} + \binom{n}{k} = \frac{n!}{(k-1)!(n+1-k)!} + \frac{n!}{k!(n-k)!}$$

$$= \frac{n!(k+n+1-k)}{k!(n+1-k)!} = \frac{(n+1)!}{k!(n+1-k)!} = \binom{n+1}{k},$$

and therefore,

$$(a+b)^{n+1} = \binom{n+1}{n+1} a^{n+1} b^0 + \sum_{k=1}^{n} \binom{n+1}{k} a^k b^{n+1-k} + \binom{n+1}{0} a^0 b^{n+1}$$

$$= \sum_{k=0}^{n+1} \binom{n+1}{k} a^k b^{n+1-k},$$

as desired.

4.5.6 Example. We wish to show that

$$\frac{1}{n} \sum_{k=1}^{n} x_k \geq \sqrt[n]{\prod_{k=1}^{n} x_k} \tag{4.33}$$

for all $x_1, \ldots, x_n \in (0, \infty)$ (in words: the arithmetic mean is greater than or equal to the geometric mean). In applying the natural logarithm to both sides, inequality (4.33) is readily seen to be equivalent to

$$\ln \left(\frac{1}{n} \sum_{k=1}^{n} x_k \right) \geq \frac{1}{n} \sum_{k=1}^{n} \ln(x_k). \tag{4.34}$$

In order to prove this latter inequality, we need to establish first a couple of calculus lemmas:

4.5.7 Lemma. *If f is twice continuously differentiable on an open interval containing the closed interval $[a, b]$ and if $f''(x) \leq 0$ for all $x \in [a, b]$ and $f(a) = f(b) = 0$, then $f(x) \geq 0$ for all $x \in [a, b]$.*

Proof. To give a proof by contradiction, we assume that there is a value $x \in [a, b]$ such that $f(x) < 0$. Since $f(a) = 0$, it follows that $a < x$, and therefore, the mean value theorem allows us to infer the existence of a $t \in (a, x)$ such that

$$f'(t) = \frac{f(x) - f(a)}{x - a} = \frac{f(x)}{x - a} < 0.$$

Given the assumption that f'' is less than or equal to zero, it further follows that f' is decreasing, and therefore $f'(s) \leq f'(t) < 0$ for all $s \in [t, b]$ and in particular for all $s \in [x, b]$. Hence

$$f(b) = f(x) + \int_x^b f'(s) \, ds \leq f(x) < 0$$

in contradiction to the assumption $f(b) = 0$. □

4.5.8 Lemma. *If f is twice continuously differentiable on an open interval I and if $f''(x) \leq 0$ for all $x \in I$, then $f(\lambda c + \mu d) \geq \lambda f(c) + \mu f(d)$ for all $c, d \in I$ and all $\lambda, \mu \in [0, 1]$ that satisfy the equation $\lambda + \mu = 1$.*

Proof. Since the inequality $f(\lambda c + \mu d) \geq \lambda f(c) + \mu f(d)$ is trivially satisfied if $c = d$, we may assume that $c \neq d$. Setting

$$g(x) := f(x) - \frac{f(d) - f(c)}{d - c}(x - c) - f(c)$$

for all $x \in I$, we find by inspection that $g(c) = g(d) = 0$ and that $g''(x) = f''(x) \leq 0$ for all $x \in I$. Consequently, Lemma 4.5.7 implies that $g(x) \geq 0$ for all $x \in [c, d] \cup [d, c]$, and, in particular, it follows that $g(\lambda c + \mu d) \geq 0$ because $\lambda c + \mu d \in [c, d] \cup [d, c]$. Hence

$$0 \leq f(\lambda c + \mu d) - \frac{f(d) - f(c)}{d - c}(\lambda c + \mu d - c) - f(c)$$
$$= f(\lambda c + \mu d) - \frac{f(d) - f(c)}{d - c}(-\mu c + \mu d) - f(c)$$
$$= f(\lambda c + \mu d) - \mu(f(d) - f(c)) - f(c)$$
$$= f(\lambda c + \mu d) - (\lambda f(c) + \mu f(d))$$

or, equivalently,

$$f(\lambda c + \mu d) \geq \lambda f(c) + \mu f(d),$$

as desired. □

To proceed, we now will use Lemma 4.5.8 to prove by induction the following theorem:

4.5.9 Theorem. *Assume that f is twice continuously differentiable on an open interval I and that $f''(x) \leq 0$ for all $x \in I$. Then $\sum_{k=1}^{n} \lambda_k x_k \in I$ and*

$$f\left(\sum_{k=1}^{n} \lambda_k x_k\right) \geq \sum_{k=1}^{n} \lambda_k f(x_k)$$

for all $n \in \mathbb{N}$, for all $x_1, \ldots x_n \in I$, and for all $\lambda_1, \ldots, \lambda_n \in [0, 1]$ that satisfy the equation $\sum_{k=1}^{n} \lambda_k = 1$.

Proof. Since the case $n = 1$ is trivial, we only need to show that the assumed validity of the statement of the theorem for a given $n \in \mathbb{N}$ implies the validity of the same statement for $n + 1$. To do so, we assume that that we are given values $x_1, \ldots, x_{n+1} \in I$ and $\lambda_1, \ldots, \lambda_{n+1} \in [0, 1]$ such that $\sum_{k=1}^{n+1} \lambda_k = 1$. Since addition is commutative, we may assume w.l.o.g. that $x_1 \leq \cdots \leq x_{n+1}$. Then $x_1 = \sum_{k=1}^{n+1} \lambda_k x_1 \leq \sum_{k=1}^{n+1} \lambda_k x_k \leq \sum_{k=1}^{n+1} \lambda_k x_{n+1} = x_{n+1}$, and therefore, $\sum_{k=1}^{n+1} \lambda_k x_k \in I$ because $x_1, x_{n+1} \in I$ and because I is an interval. Furthermore, since the inequality

$$f\left(\sum_{k=1}^{n+1} \lambda_k x_k\right) \geq \sum_{k=1}^{n+1} \lambda_k f(x_k)$$

is evidently satisfied in the case where $\lambda_{n+1} = 1$ and $\sum_{k=1}^{n} \lambda_k = 0$, we may assume that $\sum_{k=1}^{n} \lambda_k > 0$. Using Lemma 4.5.8 with $\lambda := \sum_{k=1}^{n} \lambda_k$, $\mu := \lambda_{n+1}$, $c := \left(\sum_{k=1}^{n} \lambda_k\right)^{-1} \sum_{k=1}^{n} \lambda_k x_k$, and $d := x_{n+1}$, it follows that

$$f\left(\sum_{k=1}^{n+1} \lambda_k x_k\right) = f(\lambda c + \mu d) \geq \lambda f(c) + \mu f(d)$$

$$= \left(\sum_{k=1}^{n} \lambda_k\right) f\left(\sum_{k=1}^{n} \frac{\lambda_k x_k}{\sum_{k=1}^{n} \lambda_k}\right) + \lambda_{n+1} f(x_{n+1}),$$

and therefore, the inductive assumption implies that

$$f\left(\sum_{k=1}^{n+1} \lambda_k x_k\right) \geq \left(\sum_{k=1}^{n} \lambda_k\right) \sum_{k=1}^{n} \frac{\lambda_k f(x_k)}{\sum_{k=1}^{n} \lambda_k} + \lambda_{n+1} f(x_{n+1}) = \sum_{k=1}^{n+1} \lambda_k f(x_k),$$

as desired. $\qquad\square$

To conclude our proof of (4.33), we observe that the equivalent inequality (4.34) is a direct consequence of Theorem 4.5.9 with $\lambda_1 = \cdots = \lambda_n = 1/n$, $I = (0, \infty)$, and $f(x) = \ln(x)$. (Note: $d^2 \ln(x)/dx^2 = -1/x^2 < 0$.)

A slight modification of the induction pattern that we have been using so far is the following:

$$(P(1) \wedge \forall_{n \in \mathbb{N}} (\forall_{k \in \mathbb{N}} k \leq n \Rightarrow P(k)) \Rightarrow P(n+1)) \Rightarrow \forall_{n \in \mathbb{N}} P(n). \qquad (4.35)$$

In other words, what is here being asserted is the truth of $P(n)$ for all $n \in \mathbb{N}$ conditional upon the assumption that $P(1)$ is true and that $P(n+1)$ is true whenever $P(k)$ is true for all $k \leq n$. To see why this alternative principle is valid as well, we set

$$Q(n) :\Leftrightarrow \forall_{k \in \mathbb{N}}\, k \leq n \Rightarrow P(k)$$

for all $n \in \mathbb{N}$. Then the antecedent of statement (4.35) is equivalent to

$$Q(1) \wedge \forall_{n \in \mathbb{N}}\, Q(n) \Rightarrow P(n+1) \tag{4.36}$$

because $P(1) \Leftrightarrow (\forall_{k \in \mathbb{N}}\, k \leq 1 \Rightarrow P(k)) \Leftrightarrow Q(1)$. Furthermore, since the definition of $Q(n)$ implies that $(Q(n) \wedge P(n+1)) \Leftrightarrow Q(n+1)$, and since $(A \Rightarrow B) \Leftrightarrow (A \Rightarrow (A \wedge B))$ is a tautology, it follows that (4.36) is equivalent to

$$Q(1) \wedge \forall_{n \in \mathbb{N}}\, Q(n) \Rightarrow Q(n+1).$$

Using the ordinary Principle of Mathematical Induction, this latter statement implies that $Q(n)$ is true for all $n \in \mathbb{N}$ which in turn implies that $P(n)$ is true for all $n \in \mathbb{N}$ as well. Hence the principle stated in (4.35) is indeed valid.

4.5.10 Example. We wish to show that every positive integer greater than 1 can be written as a product of primes. Setting

$$P(n) :\Leftrightarrow n+1 \text{ is a product of primes}$$

for all $n \in \mathbb{N}$, it follows that $P(1)$ is true, because $1 + 1 = 2$ is prime. To proceed, we assume that $P(k)$ is true for all $k \leq n$. In order to prove that $P(n+1)$ is true as well, we observe that $P(n+1)$ is trivially true if $n+2$ is itself a prime number. Thus we may assume that $n+2$ is not prime. Then there are integers $k, j \in \{2, \ldots, n+1\}$ such that $n+2 = kj$. According to our inductive assumption $P(k-1)$ and $P(j-1)$ are true. Hence both k and j are products of primes and by implication, so is $n+2 = kj$.

Exercises

4.5.11. Prove that $\sum_{k=1}^{n} k^2 = n(n+1)(2n+1)/6$ for all $n \in \mathbb{N}$.

4.5.12. Find a formula for $\sum_{k=1}^{n} k^3$ and prove it by means of mathematical induction.

4.5.13. Prove that $\sum_{k=0}^{n} x^k = (1 - x^{n+1})/(1-x)$ for all $n \in \mathbb{N}$ and all $x \in \mathbb{R} \setminus \{1\}$.

4.5.14. Prove that $n! \geq 2^{n-1}$ for all $n \in \mathbb{N}$.

4.5.15. Show that $\sum_{k=1}^{n} k^{-2} \leq 2 - n^{-1}$ for all $n \in \mathbb{N}$.

4.5.16. Let n be a positive integer greater than one and assume that $x_1, \ldots, x_n$ are positive real numbers such that $x_1 \geq \cdots \geq x_n$. Show that $x_1/x_n + \sum_{k=1}^{n-1} x_{k+1}/x_k \geq n$.

4.5.17. Show that $\sum_{i=0}^{k} \binom{n}{i}(-1)^i = (-1)^k \binom{n-1}{k}$ for all $n \in \mathbb{N}$ and all $k \in \{0, \ldots, n-1\}$.

4.5.18. Prove that $e^{\sum_{k=1}^{n} x_k/n} \leq \sum_{k=1}^{n} e^{x_k}/n$ for all $n \in \mathbb{N}$ and all $x_1, \ldots, x_n \in \mathbb{R}$.

4.5.19. Prove that $\left| \sum_{k=1}^{n} x_k \right| \leq \sum_{k=1}^{n} |x_k|$ for all $n \in \mathbb{N}$ and all $x_1, \ldots, x_n \in \mathbb{R}$.

4.5.20. A function $f : \mathbb{R} \to \mathbb{R}$ is said to be *affine* if there are constants $m, b \in \mathbb{R}$ such that $f(x) = mx + b$ for all $x \in \mathbb{R}$. Show that a function $f : \mathbb{R} \to \mathbb{R}$ is affine if and only if

$$f(\lambda x + (1 - \lambda)y) = \lambda f(x) + (1 - \lambda)f(y)$$

for all $x, y \in \mathbb{R}$ and all $\lambda \in [0, 1]$.

4.5.21. ★ Show that f is not necessarily affine if we only require $f((x+y)/2) = (f(x) + f(y))/2$ for all $x, y \in \mathbb{R}$. *Hint.* You may use the fact that there exists an infinite set $B \subset \mathbb{R} \setminus \{0\}$ such that for every $x \in \mathbb{R} \setminus \{0\}$ there exists a unique element $n \in \mathbb{N}$, unique pairwise distinct values $\{b_1, \ldots, b_n\} \subset B$ (unique up to re-orderings), and unique coefficients $r_1, \ldots, r_n \in \mathbb{Q} \setminus \{0\}$ such that $x = r_1 b_1 + \cdots + r_n b_n$. Note: any such set B is said to be a Hamel basis of $\mathbb{R}$ over $\mathbb{Q}$.

4.5.22. *Assume that f is twice continuously differentiable on an open interval I and that $f''(x) \geq 0$ for all $x \in I$. Show that*

$$f\left(\sum_{k=1}^{n} \lambda_k x_k \right) \leq \sum_{k=1}^{n} \lambda_k f(x_k)$$

for all $n \in \mathbb{N}$, for all $x_1, \ldots x_n \in I$, and for all $\lambda_1, \ldots, \lambda_n \in [0, 1]$ that satisfy the equation $\sum_{k=1}^{n} \lambda_k = 1$.

4.5.23. Prove that for all integers $m, n \in \mathbb{N}$ it is the case that m^n is even if and only if m is even.

4.6 Inductive Definition

In order to illustrate the principle of *inductive definition*, let us suppose that $f : \mathbb{N} \to \mathbb{N}$ is a function that satisfies the following properties:

a) $f(1) = 1$,

b) $f(n+1) = (n+1)f(n)$ for all $n \in \mathbb{N}$.

Then

$$f(2) = f(1+1) = (1+1)f(1) = 2 \cdot 1 = 2!,$$
$$f(3) = f(2+1) = (2+1)f(2) = 3 \cdot 2! = 3!$$
$$f(4) = f(3+1) = (3+1)f(3) = 4 \cdot 3! = 4!$$
$$\vdots$$

and therefore, $f(n) = n!$ for all $n \in \mathbb{N}$. In other words, given the intial value of f at 1, the condition in b) above allows us to successively identify the value of f at n for any $n \in \mathbb{N}$. A proper generalization of this inductive method of definition is stated in Theorem 4.6.2, and for the proof of this theorem it is convenient to introduce the following definition:

4.6.1 Definition. If A, B, and C are sets such that $C \subset A$ and if $f : A \to B$ is a given function, then

$$f \upharpoonright_C : C \to B$$
$$x \mapsto f(x)$$

is said to be the *restriction* of f to C.

4.6.2 Theorem. *Let M be a nonempty set. If $g : \mathbb{N} \times M \to M$ is a given function and $z \in M$, then there exists a unique function $f : \mathbb{N} \to M$ such that*

a) $f(1) = z$ *and*

b) $f(n+1) = g(n, f(n))$ *for all $n \in \mathbb{N}$.*

Proof. Let A be the set of all values $n \in \mathbb{N}$ for which there exists a function $f_n : \{1, \ldots, n\} \to M$ such that

a) $f_n(1) = z$,

b) $\forall_{k \in \{1,\ldots,n\}} \; k < n \Rightarrow f_n(k+1) = g(k, f_n(k))$.

We wish to show that $A = \mathbb{N}$. For $n = 1$ we define $f_1 : \{1\} \to M$ via the equation $f_1(1) := z$. Since f_1 evidently satisfies a) and b) above, it follows that $1 \in A$. Proceeding by induction and assuming that $n \in A$ for some $n \in \mathbb{N}$, we need to show that $n + 1 \in A$. To do so, we observe that the assumption $n \in A$ implies that there exists a function $f_n : \{1, \ldots, n\} \to M$ that satisfies a) and b) above. To define $f_{n+1} : \{1, \ldots, n+1\} \to M$, we set

$f_{n+1}(k) := f_n(k)$ for all $k \in \{1, \ldots, n\}$ and $f_{n+1}(n+1) := g(n, f_n(n))$. Then $f_{n+1}(1) = f_n(1) = z$ (by a) above) and

$$f_{n+1}(k+1) = \begin{cases} f_n(k+1) = g(k, f_n(k)) = g(k, f_{n+1}(k)) & \text{if } k < n, \\ g(n, f_n(n)) = g(n, f_{n+1}(n)) & \text{if } k = n. \end{cases}$$

This shows that f_{n+1} satisfies a) and b) above, and that therefore $n+1 \in A$, as desired. Since the definition of f_{n+1} implies that $f_n = f_{n+1} \restriction \{1,\ldots,n\}$ and since this latter equation is satisfied if and only if $f_n \subset f_{n+1}$ (see Exercise 4.6.9), we may apply Proposition 4.3.13 to infer that

$$f := \bigcup_{n \in \mathbb{N}} f_n$$

is a well defined function from $\mathbb{N}$ to M. (Note: readers who skipped Section 4.3 are advised at this point to take a look at Proposition 4.3.13 and the prior Lemma 4.3.12 because, together, these two results are essentially self-contained and do not require familiarity with all the results in Section 4.3 that preceded them.) Since $(1, z) \in f_1 \subset f$, it follows that $f(1) = z$, and since $(n, f_{n+1}(n)), (n+1, f_{n+1}(n+1)) \in f_{n+1} \subset f$, it also follows that $f(n+1) = f_{n+1}(n+1) = g(n, f_{n+1}(n)) = g(n, f(n))$, as desired. In order to show that f is unique, we assume that we are given a second function $h : \mathbb{N} \to M$ such that $h(1) = z$ and $h(n+1) = g(n, h(n))$ for all $n \in \mathbb{N}$. Using mathematical induction and observing that $f(1) = z = h(1)$, we need to prove that $f(n+1) = h(n+1)$ whenever $f(n) = h(n)$: if $f(n) = h(n)$ for some $n \in \mathbb{N}$, then $f(n+1) = g(n, f(n)) = g(n, h(n)) = h(n+1)$. Thus $f(n) = h(n)$ for all $n \in \mathbb{N}$. □

Remark. In essence, the proof of Theorem 4.6.2 amounts to an inductive definition of a sequence of functions $(f_n)_{n \in \mathbb{N}}$ starting from f_1. Thus it may appear that we are facing here a case of circular reasoning as the principle of inductive definition is established by means of an inductive construction of functions. However, it is precisely to circumvent this fallacy that the set A is introduced as the set of all $n \in \mathbb{N}$ for which a function f_n satisfying the stated properties a) and b) exists. For in defining A in this manner a direct appeal to the principle of inductive definition is avoided and the charge of circular reasoning therefore refuted.

4.6.3 Example. Defining $g : \mathbb{N} \times \mathbb{N} \to \mathbb{N}$ via the equation $g(n, m) := (n+1)m$ and setting $z := 1$, the preceding theorem allows us to infer that there exists a unique function $f : \mathbb{N} \to \mathbb{N}$ such that $f(1) = 1$ and $f(n+1) = g(n, f(n)) = (n+1)f(n)$ for all $n \in \mathbb{N}$. We wish to show that $f(n) = n!$ for all $n \in \mathbb{N}$. Since $f(1) = 1 = 1!$, we only need to show that $\forall_{n \in \mathbb{N}} f(n) = n! \Rightarrow f(n+1) = (n+1)!$. So let n be in $\mathbb{N}$ such that $f(n) = n!$. Then $f(n+1) = (n+1)f(n) = (n+1)n! = (n+1)!$.

4.6.4 Example. Defining $g : \mathbb{N} \times \mathbb{R} \to \mathbb{R}$ via the equation $g(n, x) := \sqrt{2 + x}$ and setting $z := \sqrt{2}$, we may apply Theorem 4.6.2 to conclude that there exists a unique function $f : \mathbb{N} \to \mathbb{R}$ that satisfies the following properties:

a) $f(1) = \sqrt{2}$,

b) $f(n + 1) = g(n, f(n)) = \sqrt{2 + f(n)}$ for all $n \in \mathbb{N}$.

We wish to show that $f(n) < 2$ for all $n \in \mathbb{N}$. Since $f(1) = \sqrt{2} < 2$, we only need to prove that $f(n + 1) < 2$ whenever $f(n) < 2$. So let us assume that $f(n) < 2$ for some $n \in \mathbb{N}$. Then $f(n + 1) = \sqrt{2 + f(n)} < \sqrt{4} = 2$, as desired.

4.6.5 Example. For a given sequence of real numbers $(a_k)_{k \in \mathbb{N}}$, the sum $s_n = \sum_{k=1}^{n} a_k$ is inductively defined via the following equations:

a) $s_1 := a_1$,

b) $s_{n+1} := s_n + a_{n+1}$.

In this case the corresponding function $g : \mathbb{N} \times \mathbb{R} \to \mathbb{R}$ is $g(n, x) := x + a_{n+1}$. Similarly, the product $p_n := \prod_{k=1}^{n} a_k$ can be inductively defined via the equations

a) $p_1 := a_1$,

b) $p_{n+1} := p_n a_{n+1}$.

4.6.6 Example. If $a \in \mathbb{R} \setminus \{0\}$ and $g(n, x) := ax$ for all $(n, x) \in \mathbb{N} \times \mathbb{R}$, then, according to Theorem 4.6.2, there is a function $f : \mathbb{N} \to \mathbb{R}$ such that $f(1) = a$ and $f(n + 1) = g(n, f(n)) = af(n)$. We wish to show that $f(n) = a^n$ for all $n \in \mathbb{N}$. Since $f(1) = a = a^1$, we only need to show that $f(n + 1) = a^{n+1}$ whenever $f(n) = a^n$: if $f(n) = a^n$, then $f(n + 1) = af(n) = a \cdot a^n = a^{n+1}$, as desired.

To conclude this section, we wish to point out that Theorem 4.6.2 can be easily extended to higher-order inductive patterns. A theorem that establishes a second-order recursion is the following:

4.6.7 Theorem. *Let M be a nonempty set. If $g : \mathbb{N} \times M \times M \to M$ is a given function and $z_1, z_2 \in M$, then there exists a unique function $f : \mathbb{N} \to M$ such that*

a) $f(1) = z_1$, $f(2) = z_2$, *and*

b) $f(n + 2) = g(n, f(n), f(n + 1))$ *for all $n \in \mathbb{N}$.*

Proof. Let the function $G : \mathbb{N} \times M \times M \to M \times M$ be defined by the equation

$$G(n, x_1, x_2) := (x_2, g(n, x_1, x_2))$$

for all $(n, x_1, x_2) \in \mathbb{N} \times M \times M$. Then, according to Theorem 4.6.2, there exists a unique function $F : \mathbb{N} \to M \times M$ such that $F(1) = (z_1, z_2)$ and $F(n + 1) = G(n, F(n))$. Denoting the two component functions of F by F_1 and F_2, respectively (i.e., $F(n) = (F_1(n), F_2(n))$ for all $n \in \mathbb{N}$), it follows that

$$(F_1(n + 1), F_2(n + 1)) = F(n + 1) = G(n, F(n)) = G(n, F_1(n), F_2(n))$$
$$= (F_2(n), g(n, F_1(n), F_2(n))),$$

and therefore,

$$F_1(n + 1) = F_2(n),$$
$$F_2(n + 1) = g(n, F_1(n), F_2(n))$$

for all $n \in \mathbb{N}$. Setting $f(n) := F_1(n)$ for all $n \in \mathbb{N}$, we may infer that $f(1) = z_1$ because

$$(F_1(1), F_2(1)) = F(1) = (z_1, z_2),$$

$f(2) = z_2$ because

$$(F_1(2), F_2(2)) = (F_2(1), g(1, F_1(1), F_2(1))) = (z_2, g(1, F_1(1), F_2(1))),$$

and

$$f(n + 2) = F_1(n + 2) = F_2(n + 1) = g(n, F_1(n), F_2(n))$$
$$= g(n, F_1(n), F_1(n + 1)) = g(n, f(n), f(n + 1)).$$

Hence the function f exists as claimed. To prove that f is unique, we assume that $h : \mathbb{N} \to M$ is a function that satisfies the same recursion conditions as f (i.e., $h(1) = z_1$, $h(2) = z_2$, and $h(n + 2) = g(n, h(n), h(n + 1))$) and define $H : \mathbb{N} \to M \times M$ via the equation $H(n) := (h(n), h(n+1))$ for all $n \in \mathbb{N}$. Then $H(1) = (z_1, z_2)$ and

$$H(n + 1) = (h(n + 1), h(n + 2)) = (h(n + 1), g(n, h(n), h(n + 1)))$$
$$= G(n, h(n), h(n + 1)) = G(n, H(n))$$

for all $n \in \mathbb{N}$. Consequently, the uniqueness of F implies that $H = F$, and this in turn implies that $h = f$, as desired, because

$$(f(n), f(n + 1)) = (F_1(n), F_1(n + 1)) = (F_1(n), F_2(n)) = F(n) = H(n)$$
$$= (h(n), h(n + 1))$$

for all $n \in \mathbb{N}$. $\qquad\qquad\square$

4.6.8 Example. The sequence of Fibonacci numbers $(F_n)_{n \in \mathbb{N}}$, which is encountered in a wide variety of organic growth patterns, is inductively defined by means of the equations $F_1 = 1$, $F_2 = 1$, and $F_{n+2} = F_n + F_{n+1}$. The fact that these equations do indeed specify the sequence $(F_n)_{n\in\mathbb{N}}$ uniquely follows from the statement of Theorem 4.6.7 by setting $g(n, x_1, x_2) := x_1 + x_2$ for all $(n, x_1, x_2) \in \mathbb{N} \times \mathbb{N} \times \mathbb{N}$. For in doing so, we find, as desired, that $F_{n+2} = g(n, F_n, F_{n+1}) = F_n + F_{n+1}$ for all $n \in \mathbb{N}$. Furthermore, given the initial values $F_1 = 1$ and $F_2 = 1$, the recursion equation $F_{n+2} = F_n + F_{n+1}$ implies that

$$F_3 = 1 + 1 = 2, \ F_4 = 1 + 2 = 3, \ F_5 = 2 + 3 = 5, \ldots ad\ infinitum.$$

Exercises

4.6.9. Let $f : A \to B$ and $g : C \to D$ be given functions. Show that $f = g \restriction_A$ if and only if $f \subset g$.

4.6.10. For the function f given in Example 4.6.4 do the following:

  a) determine $f(2)$, $f(3)$, and $f(4)$,

  b) show that $f(n + 1) > f(n)$ for all $n \in \mathbb{N}$,

  c) show that $2 - f(n + 1) < (2 - f(n))/2$ for all $n \in \mathbb{N}$,

  d) use c) to prove that $\lim_{n\to\infty} f(n) = 2$.

4.6.11. Let $f(1) := 1$ and $f(n + 1) := (1 + 1/n)f(n)$ for all $n \in \mathbb{N}$ (i.e., $g(n, x) := (1 + 1/n)x$). Show that $f(n) = n$ for all $n \in \mathbb{N}$.

4.6.12. Let $f(1) := 1$ and $f(n + 1) := (1 + 1/n^2)f(n)$ for all $n \in \mathbb{N}$ (i.e., $g(n, x) := (1 + 1/n^2)x$). Write $f(n)$ as a product and then use elementary calculus to show that $f(n) < e^2$ for all $n \in \mathbb{N}$.

4.6.13. Show that there exists a sequence $(a_n)_{n\in\mathbb{N}}$ such that $a_1 = 2$ and $a_{n+1} = 2a_n + 3n$ for all $n \in \mathbb{N}$.

4.6.14. **a)** Show that there exists a sequence $(a_n)_{n\in\mathbb{N}}$ such that $a_1 = 1$ and $a_{n+1} = a_n/2 + 1$ for all $n \in \mathbb{N}$.

  b) Use mathematical induction to prove that $a_n = (2^n - 1)/2^{n-1}$ for all $n \in \mathbb{N}$.

  c) Use b) to show that $\lim_{n\to\infty} a_n = 2$.

4.6.15. **a)** Show that there exists a sequence $(a_n)_{n\in\mathbb{N}}$ such that $a_1 = 1$ and $a_{n+1} = a_n/3 + 1$ for all $n \in \mathbb{N}$.

  b) Use mathematical induction to prove that $a_n = (3^n - 1)/(2 \cdot 3^{n-1})$ for all $n \in \mathbb{N}$.

  c) Use b) to show that $\lim_{n\to\infty} a_n = 3/2$.

4.6.16. Use the uniqueness assertion of Theorem 4.6.7 to establish the fact that the Fibonacci sequence defined in Example 4.6.8 satisfies the equation

$$F_n = \frac{1}{\sqrt{5}} \left(\left(\frac{1 + \sqrt{5}}{2} \right)^n - \left(\frac{1 - \sqrt{5}}{2} \right)^n \right)$$

for all $n \in \mathbb{N}$.

4.6.17. Prove that there exists a sequence $(a_n)_{n \in \mathbb{N}}$ such that $a_1 = 2$, $a_2 = 3$, and $a_{n+2} - n^2 a_{n+1} + 2a_n = 0$ for all $n \in \mathbb{N}$.

Chapter 5

Cardinality

5.1 Finite Sets

The purpose of this chapter is to introduce and study the concepts and criteria that allow us to compare the 'cardinality' or 'quantity' of elements that two given sets contain. As a matter of course, in the case where the sets in question are finite, comparing their respective cardinalities is trivially a matter of counting their elements, but for infinite sets the problem is considerably more involved. For intuitively speaking, it is not at all obvious whether and in what sense the infinite quantity of elements in one set can be greater or smaller than the equally infinite quantity in another. In other words, the notion that infinity can actually be larger or less than infinity seems very weird at first encounter. However, before we can discuss in more detail this issue of distinguishing infinities, we need to take a closer look at quantities of elements in finite sets. Thus we introduce the following definition:

5.1.1 Definition. For every $n \in \mathbb{N}$ we set $M_n := \{k \in \mathbb{N} \mid k \leq n\} = \{1, \ldots, n\}$, and we say that a given set S is *finite* if $S = \emptyset$ or if there is an $n \in \mathbb{N}$ for which there exists a bijective map $f : M_n \to S$. (Note: this definition trivially implies that M_n is finite for all $n \in \mathbb{N}$, because for any $n \in N$ the identity map id_{M_n} is a bijection from M_n onto itself.)

In the light of this definition, it seems perfectly natural to think that the existence of a bijective map $f : M_n \to S$ should allow us to say that n is the number of elements of S or, equivalently, the *cardinality* of S. But in order for this to truly make sense, we need to show that S determines n uniquely, and to this end it is helpful to establish the following lemma:

5.1.2 Lemma. *Let $n, m \in \mathbb{N}$ and let $f : M_n \to M_m$ be a given function.*

   a) *If f is injective, then $n \leq m$,*

b) *if f is surjective, then $m \leq n$, and*

c) *if f is bijective, then $n = m$.*

Proof. **a)** To give a proof by induction, we set

$$P(n) :\Leftrightarrow \forall_{m \in \mathbb{N}} \; (\exists_{f:M_n \to M_m} \; f \text{ injective}) \Rightarrow n \leq m$$

and observe that $P(1)$ is true because $1 \leq m$ for all $m \in \mathbb{N}$. Assuming that $P(n)$ is true for some $n \in \mathbb{N}$, we need to show that $P(n+1)$ is true as well, and to do so, we pick an arbitrary $m \in \mathbb{N}$ and assume that there is an injective map $f : M_{n+1} \to M_m$. Since $n + 1 \geq 2$, the injectivity of f implies that $f(1)$ and $f(2)$ are two distinct elements of M_m. Hence $m > 1$ and $m - 1 \in \mathbb{N}$. Furthermore, if $f(k) < m$ for all $k \in M_{n+1}$, then $R(f) \subset M_{m-1}$, and therefore, $f \upharpoonright_{M_n}$ is an injective map from M_n to M_{m-1}. Consequently, the truth of $P(n)$ allows us to infer that $n \leq m - 1$ or, equivalently, that $n + 1 \leq m$, as desired. Alternatively, if there is a $k_0 \in M_{n+1}$ such that $f(k_0) = m$, then $f(k) < m$ for all $k \in M_{n+1} \setminus \{k_0\}$ because f is injective, and therefore, the equation

$$g(k) := \begin{cases} f(k) & \text{if } k < k_0, \\ f(k+1) & \text{if } k \geq k_0. \end{cases}$$

defines g as a function from M_n to M_{m-1}. Since f is injective, all output values of f are pairwise distinct, and by implication, all output values of g are pairwise distinct as well. Thus g is injective, and therefore, the assumed truth of $P(n)$ again allows us to infer that $n \leq m - 1$ and $n + 1 \leq m$. Consequently, $P(n+1)$ is true, as desired.

b) Setting

$$P(n) :\Leftrightarrow \forall_{m \in \mathbb{N}} \; (\exists_{f:M_n \to M_m} \; f \text{ surjective}) \Rightarrow m \leq n,$$

we notice that $P(1)$ is true, because if $f : M_1 \to M_m$ is surjective, then $\{f(1)\} = R(f) = M_m$ and $m = 1 \leq 1 = n$ because the only set M_m that contains only one element is M_1. In order to show that $P(n)$ implies $P(n+1)$, we assume that $P(n)$ is true for some $n \in \mathbb{N}$ and that we are given an $m \in \mathbb{N}$ for which there exists a surjective map $f : M_{n+1} \to M_m$. If $f \upharpoonright_{M_n}$ is surjective as well, then $P(n)$ implies that $m \leq n < n + 1$, as desired. Alternatively, if $f \upharpoonright_{M_n}$ is not surjective, then $m > 1$ and $f(M_n) = M_m \setminus \{f(n+1)\}$ because f is surjective. Furthermore, setting $A := f^{-1}(\{1, \ldots, f(n+1) - 1\})$, $B := f^{-1}(\{f(n+1) + 1, \ldots, m\})$, and

$$g(k) := \begin{cases} f(k) & \text{if } k \in A, \\ f(k) - 1 & \text{if } k \in B, \end{cases}$$

it follows that g is a surjective map from $M_n = A \cup B$ to M_{m-1} because

$$g(M_n) = g(A) \cup g(B) = f(A) \cup \{j - 1 \mid j \in f(B)\}$$
$$= \{1, \ldots, f(n+1) - 1\} \cup \{f(n+1), \ldots, m - 1\} = M_{m-1}.$$

Hence $P(n)$ implies that $m - 1 \le n$, and therefore, $m \le n + 1$, as desired.

c) This is a trivial consequence of a) and b). $\qquad\qquad\qquad\qquad$ □

The next proposition will be used to prove Theorem 5.1.4 below and will also be referred to in the subsequent section.

5.1.3 Proposition. *The following statements are true for all functions* g : $A \to B$ *and* $f : B \to C$ *(where* A, B, *and* C *are given nonempty sets):*

    a) *If* f *and* g *are injective, then* $f \circ g$ *is injective as well.*

    b) *If* f *and* g *are surjective, then* $f \circ g$ *is surjective as well.*

    c) *If* f *and* g *are bijective, then* $f \circ g$ *is bijective as well.*

Proof. **a)** Assume that f and g are injective and that $(f \circ g)(x_1) = (f \circ g)(x_2)$ for some $x_1, x_2 \in A$. Then $f(g(x_1)) = f(g(x_2))$ and therefore $g(x_1) = g(x_2)$ because f is injective. Since g is injective as well, it follows that $x_1 = x_2$, as desired.

b) Assume that f and g are surjective and let $z \in C$. Then there exists a $y \in B$ such that $f(y) = z$ because f is surjective, and since g is surjective as well, there also exists an $x \in A$ such that $y = g(x)$. Hence $(f \circ g)(x) = f(g(x)) = f(y) = z$, as desired.

c) This is a trivial consequence of a) and b). $\qquad\qquad\qquad\qquad$ □

5.1.4 Theorem. *If* S *is a nonempty finite set and if* $f : M_n \to S$ *and* $g :$ $M_m \to S$ *are bijective maps for some* $n, m \in \mathbb{N}$, *then* $n = m$.

Proof. The assumption that f and g are bijective in conjunction with Proposition 5.1.3c, implies that $g^{-1} \circ f : M_n \to M_m$ is bijective as well (see also Exercise 5.1.10). Consequently, Lemma 5.1.2c allows us to infer that $m = n$, as desired. $\qquad\qquad\qquad\qquad$ □

Given the statement of Theorem 5.1.4, we are now justified in introducing the following definition:

5.1.5 Definition. *If* S *is a nonempty finite set and if* $f : M_n \to S$ *is a bijection for some* $n \in \mathbb{N}$, *then* $|S| := n$ *is said to be the* cardinality *of* S. *Furthermore, the* cardinality *of* S *is said to be zero if* $S = \emptyset$.

5.1.6 Theorem. *For all finite nonempty sets* A *and* B *the following statements are equivalent:*

    a) $|A| \le |B|$,

    b) *there exists an injection* $f : A \to B$,

    c) *there exists a surjection* $g : B \to A$.

Proof. Let $n := |A|$ and $m := |B|$. Then, by Definitions 5.1.1 and 5.1.5, there are bijective maps $a : M_n \to A$ and $b : M_m \to B$. In order to prove the theorem, we will show that a) implies b), that b) implies c), and that c) implies a) (see Proposition 1.4.5k).

a) $\Rightarrow$ **b)**: if $|A| = n \le m = |B|$, then $M_n \subset M_m$ and therefore, the map $h : M_n \to M_m$, $h(x) := x$ is injective. Since a^{-1} and b are bijective, Proposition 5.1.3a implies that $b \circ h \circ a^{-1} : A \to B$ is an injection.

b) $\Rightarrow$ **c)**: if $f : A \to B$ is an injection, then, according to Theorem 3.8.9, there exists a function $g : B \to A$ such that $g \circ f = \mathrm{id}_A$, and this in turn implies that g is a surjection.

c) $\Rightarrow$ **a)**: if $g : B \to A$ is a surjection, then, according to Proposition 5.1.3b, the map $a^{-1} \circ g \circ b : M_m \to M_n$ is a surjection as well, and therefore, $|A| = n \le m = |B|$ by Lemma 5.1.2b. $\qquad\square$

5.1.7 Theorem. *Let B be a finite set and let A be a subset of B. Then A is finite as well, $|A| \le |B|$, and $|A| = |B|$ if and only if $A = B$.*

Proof. For clarity we wish to point out that the assumption $A \subset B$ and the attendant obvious conclusion that the inclusion map $i : A \to B$, $i(x) := x$, is injective does not allow us to apply Theorem 5.1.6 to infer that $|A| \le |B|$ because we do not know *a priori* that A is finite. So what we need to do is to use the finiteness of B in conjunction with the assumption $A \subset B$ to find an $m \in \mathbb{N}$ for which there exists a bijection $f : M_m \to A$. To so, we will show first by means of induction that for all $n \in \mathbb{N}$ the assumption $N \subset M_n$ implies that N is finite, and to this end we set

$$P(n) :\Leftrightarrow \forall_{N \subset M_n} N \text{ is finite.}$$

Given this definition, it is clear that $P(1)$ is true because the only subsets of M_1 are the empty set and M_1 itself, and both of these are finite. In order to show that $P(n)$ implies $P(n+1)$, we assume that $P(n)$ is true for some $n \in \mathbb{N}$ and that N is a subset of M_{n+1}. If $N \subset M_n$, then N is finite because $P(n)$ is true. Thus we may assume that N is not a subset of M_n or, equivalently, that $N = (N \cap M_n) \cup \{n + 1\}$. If $N \cap M_n = \emptyset$, then $N = \{n + 1\}$ is finite because the map $f : M_1 \to \{n + 1\}$, $f(1) := n + 1$, is evidently bijective. Alternatively, if $N \cap M_n \ne \emptyset$, then $P(n)$ implies that there is a $k \in \mathbb{N}$ for which there exists a bijection $g : M_k \to N \cap M_n$ because $N \cap M_n \subset M_n$. Consequently, the equation

$$f(x) := \begin{cases} g(x) & \text{if } x \in N \cap M_n, \\ n + 1 & \text{if } x = k + 1 \end{cases}$$

defines a bijection from M_{k+1} to N (this is trivial), and therefore, N is finite. Thus we may conclude that indeed all subsets of M_n are finite for all $n \in \mathbb{N}$. In the light of this conclusion we now can argue as follows: if $A = \emptyset$, the A is finite and $|A| = 0 \le |B| =: n$, as desired. Alternatively, if $A \ne \emptyset$, then

$A \subset B$ implies that $B \neq \emptyset$, and since B is finite, we may infer that there exists a bijection $g : M_n \to B$. Setting $N := g^{-1}(A)$, it follows that N is a nonempty subset of M_n. Hence N is finite and there is an $m \in \mathbb{N}$ for which there exists a bijection $h : M_m \to N$. Consequently, the map $f := g \circ h$ is a bijection from M_m to A, and, by implication, A is finite. Moreover, according to Theorem 5.1.6, it is the case that $|A| \leq |B|$ because the inclusion map i, defined above, is an injection from A to B. Finally, if $A = B$, then, trivially, $|A| = |B|$. Conversely, if $A \neq B$, then $|A| = 0 < |B|$ in the case where $A = \emptyset$ (and $B \neq \emptyset$ by implication). So we only need to show that $|A| < |B|$ in the case where $A \neq \emptyset$ and where, by implication, there exists a bijection $f : M_m \to A$ for some $m \in \mathbb{N}$ (because A is finite). Since $A \neq B$ and $A \subset B$, it follows that $B \setminus A$ is a nonempty subset of B and therefore finite. Consequently, there exists a $j \in \mathbb{N}$ and a bijection $p : M_j \to B \setminus A$. Hence the equation

$$q(x) := \begin{cases} f(x) & \text{if } x \in M_m, \\ p(x - m) & \text{if } x \in M_{m+j} \setminus M_m \end{cases}$$

defines a map q from M_{m+j} to B, and since q is easily seen to be bijective, it follows that $|A| = m < m + j = |B|$, as desired. $\qquad\square$

In order to round up our discussion of finite sets in this section, we will first fill in a gap that we left open earlier in the proof of Proposition 3.3.8, by providing a fully rigorous proof of the fact that the cardinality of the power set of a finite set S is $2^{|S|}$, and then afterwards we will add a fact that is not directly relevant to comparing cardinalities but that is frequently useful in working with maps from a given finite set into itself. For in trying to determine whether such a map is bijective, it is actually sufficient, according to Theorem 5.1.9 below, to prove that it is either injective or surjective.

5.1.8 Proposition. *If S is a finite set, then so is $\mathcal{P}(S)$, and $|\mathcal{P}(S)| = 2^{|S|}$.*

Proof. Since $|\mathcal{P}(\emptyset)| = |\{\emptyset\}| = 1 = 2^0 = 2^{|\emptyset|}$, we only need to show by induction that for every $n \in \mathbb{N}$ and every nonempty set S it is the case that the existence of a bijective map $f : S \to M_n$ implies the existence of a bijective map $g : \mathcal{P}(S) \to M_{2^n}$. Starting with $n = 1$, we observe that the existence of a bijective map $f : S \to M_1 = \{1\}$ implies that $S = \{f^{-1}(1)\}$. Hence $\mathcal{P}(S) = \{\emptyset, S\}$, and therefore, the map $g : \mathcal{P}(S) \to M_{2^1} = M_2$ that assigns to $\emptyset$ the value $1 \in M_2$ and to S the value $2 \in M_2$ is bijective. Having thus established the inductive base, we proceed to assume that we are given an $n \in \mathbb{N}$ for which it is the case that for every set S for which there exists a bijective map $f : S \to M_n$ there also exists a bijective map $g : \mathcal{P}(S) \to M_{2^n}$. Given this inductive assumption, we need to show that the statement constituting this assumption is true as well for $n+1$. So let S be a set for which there exists a bijective map $f : S \to M_{n+1}$. Setting $S_0 := S \setminus \{f^{-1}(n+1)\}$, it follows that $f_0 := f \upharpoonright_{S_0}$ is a bijective map from S_0 to $M_{n+1} \setminus \{n+1\} = M_n$. Using the inductive assumption, we may thus infer that there exists a bijective map $g_0 : \mathcal{P}(S_0) \to M_{2^n}$, and given this map g_0, we may define $g : \mathcal{P}(S) \to M_{2^{n+1}}$ as follows:

$$g(M) := \begin{cases} g_0(M) & \text{if } M \subset S_0 \\ g_0(M \setminus \{f^{-1}(n+1)\}) + 2^n & \text{othherwise.} \end{cases}$$

Since $M \setminus \{f^{-1}(n+1)\} \subset S_0$ for all $M \subset S$, it follows that g is well defined. Furthermore, since

$$\mathcal{P}(S) = \mathcal{P}(S_0) \cup \{M \cup \{f^{-1}(n+1)\} \mid M \in \mathcal{P}(S_0)\},$$

we may infer that g is surjective because

$$g(\mathcal{P}(S)) = g_0(\mathcal{P}(S_0)) \cup \{g_0(M) + 2^n \mid M \in \mathcal{P}(S_0)\}$$
$$= M_{2^n} + \{k + 2^n \mid k \in M_{2^n}\} = M_{2^{n+1}}.$$

In order to prove that g is injective as well, we need to show that $g(M_1) \neq g(M_2)$ whenever $M_1, M_2 \in \mathcal{P}(S)$ and $M_1 \neq M_2$. So let $M_1, M_2 \in \mathcal{P}(S)$ such that $M_1 \neq M_2$. If $M_1, M_2 \in \mathcal{P}(S_0)$, then $g(M_1) = g_0(M_1) \neq g_0(M_2) = g(M_2)$ because g_0 is injective. If $M_1, M_2 \notin \mathcal{P}(S_0)$, then $f^{-1}(n+1)$ is contained in both M_1 and M_2, and therefore, the fact that M_1 and M_2 are not equal implies that $M_1 \setminus \{f^{-1}(n+1)\}$ and $M_2 \setminus \{f^{-1}(n+1)\}$ are not equal either. Thus the injectivity of g_0 again implies that $g(M_1) = g_0(M_1 \setminus \{f^{-1}(n+1)\}) + 2^n \neq g_0(M_2 \setminus \{f^{-1}(n+1)\}) + 2^n = g(M_2)$. Finally, if $M_1 \in \mathcal{P}(S_0)$ and $M_2 \notin \mathcal{P}(S_0)$ (or vice versa), then $g(M_1) \leq 2^n < g(M_2)$ and, in particular, $g(M_1) \neq g(M_2)$, as desired. Hence g is bijective, and the proof is thus complete. $\qquad \square$

5.1.9 Theorem. *If S is a nonempty finite set and if $f : S \to S$ is a given function, then the following conditions are equivalent:*

a) *f is injective,*

b) *f is surjective,*

c) *f is bijective.*

Proof. Evidently, we only need to show that a) implies b) and that b) implies a). So let us assume that f is injective. Then $f(S)$ is finite by Theorem 5.1.7 because S is finite and because $f(S) \subset S$. Consequently, f is a bijection from the nonempty finite set S onto the nonempty finite set $f(S)$, and Theorem 5.1.6 therefore implies that $|S| = |f(S)|$. Hence f is surjective because $S = f(S)$ by Theorem 5.1.7. Conversely, if f is surjective, then $f(S) = S$, and therefore $|f(S)| = |S|$. In order to show that f is injective as well, we give a proof by contradiction: if f were not injective then there would exist distinct elements $x, y \in S$ such that $f(x) = f(y)$. Thus the surjectivity of f would imply that

$f \upharpoonright_{S \smallsetminus \{y\}}$ is a surjective map from the proper subset $S \smallsetminus \{y\}$ of S onto S. Hence it would follow that $|S| \leq |S \smallsetminus \{y\}|$ by Theorem 5.1.6 and that $|S \smallsetminus \{y\}| < |S|$ by Theorem 5.1.7 which is impossible. $\qquad\square$

Exercises

5.1.10. Use Theorems 3.8.9 and 3.8.11 to argue that the inverse of a bijective function is bijective.

5.1.11. What is the number of bijections from a finite set S into itself?

5.1.12. What is the number of surjections from a finite set S into itself?

5.1.13. Let $m, n \in \mathbb{N}$ such that $m \leq n$. What is the number of injections from M_m to M_n?

5.1.14. ★ What is the number of bijections f from a finite set S into itself that satisfy the equation $f(x) \neq x$ for all $x \in S$ and what is the limit of that number divided by $|S|!$ as $|S|$ tends to infinity?

5.1.15. Assume that A and B are finite sets. Show that $A \cup B$, $A \cap B$, $A \smallsetminus B$, and $A \triangle B$ are finite as well.

5.1.16. Assume that A and B are nonempty finite sets. Show that $A \times B$ is finite as well and that $|A \times B| = |A||B|$.

5.2 Orders of Infinity

In the previous section we saw—in Definition 5.1.5 and Theorem 5.1.6—how the problem of comparing the cardinalities of finite sets is related to the question of whether injections, surjections, and bijections do or do not exist. Inspired by this prior discussion, we now introduce the following definition for arbitrary and thus potentially infinite sets:

5.2.1 Definition. Let A and B be nonempty sets.

a) A and B are said to be *cardinally equivalent* or *equipotent* if there exists a bijective map from A to B. In this case we write $|A| = |B|$. (Note: if A is cardinally equivalent to B, then B is also cardinally equivalent to A, because if $f : A \to B$ is bijective, then $f^{-1} : B \to A$ is bijective as well. In other words, $|A| = |B|$ if and only if $|B| = |A|$.)

b) We say that $|A| \leq |B|$ if there exists an injective map from A to B.

c) We say that $|A| \geq |B|$ if there exists a surjective map from A to B.

d) We say that $|A| < |B|$ if $|A| \leq |B|$ and $|A| \neq |B|$ (where $|A| \neq |B| :\Leftrightarrow \neg(|A| = |B|)$).

e) We say that $|A| > |B|$ if $|A| \geq |B|$ and $|A| \neq |B|$.

Remark. The notation used in Definition 5.2.1 suggests that, for instance, the property $|A| = |B|$, asserted in respect of two given sets A and B, depends for its validity not on the sets A and B themselves but only on the corresponding *cardinalities* $|A|$ and $|B|$. However, when we further ask what might be meant by a 'cardinality' of a potentially infinite set in and of itself, we are naturally led to contemplate the set of all sets whose cardinality is equal to that of a given set A. That is to say, it would be this larger 'set of sets' with which $|A|$ would have to be identified. Unfortunately, though, the embedding universe in which this larger 'set of sets' would be contained would be the set of all sets which in turn, as we saw earlier, is a paradoxical entity. All the same however, the notion that the assertion $|A| = |B|$ depends, not on A and B, but merely on $|A|$ and $|B|$, is still to a certain extent justified because the equality $|A| = |B|$ is well defined in the sense that for all sets A, A', B, and B' the assumption $|A| = |A'|$ and $|B| = |B'|$ implies that $|A| = |B|$ if and only if $|A'| = |B'|$. For if $a : A \to A'$ and $b : B \to B'$ are bijections, then the existence of a bijection $f : A \to B$ implies that $b \circ f \circ a^{-1} : A' \to B'$ is a bijection as well, and conversely, the existence of a bijection $g : A' \to B'$ implies that $b^{-1} \circ g \circ a : A \to B$ is also a bijection. Directly related to this observation is also the statement of Theorem 5.2.2 below, because for any equivalence relation r it is the case that $(A, A'), (B, B') \in r$ implies that $(A, B) \in r$ if and only if $(A', B') \in r$ (see Exercise 3.5.29).

5.2.2 Theorem. *If S is a nonempty set of (nonempty) sets (i.e., the elements of S are sets), then*

$$r := \{(A, B) \in S \times S \mid |A| = |B|\}$$

is an equivalence relation on S. Note: the assumption that the sets in S be nonempty is not necessary but it is convenient because it allows us to disregard certain special cases involving the empty set in the proof that follows.

Proof. r is reflexive because id : $A \to A$ is bijective for all $A \in S$. The symmetry of r was established in the note in Definition 5.2.1, and to establish the transitivity of r, we assume that $|A| = |B|$ and $|B| = |C|$ for some $A, B, C \in S$. Then there exist bijective maps $g : A \to B$ and $f : B \to C$. According to Proposition 5.1.3c, $f \circ g : A \to C$ is bijective as well, and therefore $|A| = |C|$, as desired. $\qquad\square$

For ordinary numbers $x, y \in \mathbb{R}$, the inequalities $x \leq y$ and $y \geq x$ are completely identical in meaning. By implication and in particular, if A and B are finite sets, then $|A| \leq |B|$ is equivalent to $|B| \geq |A|$ because both $|A|$ and $|B|$ are in this case integers in $\mathbb{N} \cup \{0\}$. But if A and B are arbitrary, potentially infinite sets, then $|A| \leq |B|$ is on the surface quite distinct in its meaning from $|B| \geq |A|$, because the former relation signifies the existence of an injection from A to B and the latter that of a surjection from B to A. Hence the following theorem is clearly worth stating:

5.2.3 Theorem. *If A and B are nonempty sets, then $|A| \leq |B|$ if and only if $|B| \geq |A|$.*

Proof. '$\Rightarrow$' If $|A| \leq |B|$, then there exists an injective map $f : A \to B$. In order to prove that $|B| \geq |A|$, we need to prove that there exists a surjective map $g : B \to A$. Since f is injective, we may apply Theorem 3.8.9 to infer that there exists a map $g : B \to A$ such that $g \circ f = \mathrm{id}_A$, and this in turn implies that g is surjective, as desired.
'$\Leftarrow$' If $|B| \geq |A|$, then there exists a surjection $f : B \to A$, and Lemma 3.8.10 (in conjunction with the remark following it) therefore implies that there exists an injective map $g : A \to B$. Thus $|A| \leq |B|$, as desired. $\square$

In the light of Theorem 5.2.2, it is natural to ask whether for a given nonempty set S of nonempty sets the relation

$$r := \{(A, B) \in S \times S \mid |A| \leq |B|\} \tag{5.1}$$

is an order relation on S, but the answer is evidently, "no," because the antisymmetry property is not satisfied. After all, the fact that $|\{1\}| \leq |\{2\}|$ and $|\{2\}| \leq |\{1\}|$ (because $|\{1\}| = |\{2\}| = 1$) does not imply that $\{1\} = \{2\}$. However, the following theorem shows that (5.1) defines a relation that is transitive and in the next section we will see that the less-than-or-equal-to relation for cardinalities induces a linear order on the set of equivalence classes of the equivalence relation r as defined in Theorem 5.2.2.

5.2.4 Theorem. *For all nonempty sets A, B, and C, the following statements are valid:*

  a) *If $|A| = |B|$, then $|A| \leq |B|$.*

  b) *If $A \subset B$, then $|A| \leq |B|$.*

  c) *If $|A| \leq |B|$ and $|B| \leq |C|$, then $|A| \leq |C|$.*

  d) *If $|A| \leq |B|$ and $|B| = |C|$, then $|A| \leq |C|$.*

  e) *If $|A| = |B|$ and $|B| \leq |C|$, then $|A| \leq |C|$.*

Proof. Property a) follows from the fact that every bijective map is also injective, b) is true because the inclusion map $i : A \to B$, $x \mapsto x$ is injective whenever $A \subset B$, c) is an immediate consequence of Proposition 5.1.3a, and d) and e) are implied by a) and c). $\square$

5.2.5 Definition. A set S is said to be *infinite* if it is not finite.

5.2.6 Theorem. *The following statements are equivalent for any set S:*

a) S *is infinite,*

b) $|\mathbb{N}| \leq |S|$,

c) $|S| \geq |\mathbb{N}|$,

d) $|M_n| \leq |S|$ *for all* $n \in \mathbb{N}$,

e) $|S| \geq |M_n|$ *for all* $n \in \mathbb{N}$.

Proof. The pairwise equivalence of b) and c) and of d) and e) directly follows from Theorem 5.2.3. Thus we only need to prove the equivalence of a), b), and d). In order to show that a) implies b), we will inductively define—under the assumptiion that S is infinite—a sequence of injective maps $f_n : M_n \to S$ that satisfy the condition $f_{n+1} \upharpoonright_{M_n} = f_n$ for all $n \in \mathbb{N}$. For $n = 1$ we pick an element $x_1 \in S$ (which is possible because S, by being infinite, is also nonempty) and define $f_1 : \{1\} \to S$ via the equation $f_1(1) := x_1$. Assuming now that an injective map $f_n : M_n \to S$ has been defined, we observe that the assumption that S is infinite implies that f_n is not surjective (for otherwise M_n would be cardinally equivalent to S, and S would thus be finite). Consequently, there is an element $x_{n+1} \in S \setminus f_n(M_n)$. Setting

$$f_{n+1}(k) := \begin{cases} f_n(k) & \text{if } k \in M_n, \\ x_{n+1} & \text{if } k = n+1, \end{cases}$$

it follows that $f_{n+1} : M_{n+1} \to S$ is well defined and that f_n is equal to $f_{n+1} \upharpoonright_{M_n}$. Furthermore, the injectivity of f_n implies that f_{n+1} is injective as well, because the assumption $x_{n+1} \in S \setminus f(M_n)$ implies that $f_{n+1}(n+1) \neq f_n(k) = f_{n+1}(k)$ for all $k \in M_n$. To proceed, we consider each f_n to be a subset of $M_n \times S$, and observe that the restriction condition $f_{n+1} \upharpoonright_{M_n}$ is equivalent to the inclusion relation $f_n \subset f_{n+1}$. Thus Lemma 4.3.13 allows us to infer that

$$f := \bigcup_{n \in \mathbb{N}} f_n$$

is a function from $\mathbb{N} = \bigcup_{n \in \mathbb{N}} M_n$ to S. So in order to prove that $|\mathbb{N}| \leq |S|$, we only need to show that f is injective. To do so we pick $k_1, k_2 \in \mathbb{N}$ such that $k_1 \neq k_2$. Assuming w.l.o.g. that $k_1 < k_2$, it follows that $k_1, k_2 \in M_{k_2}$, and therefore, $f(k_1) = f_{k_2}(k_1) \neq f_{k_2}(k_2) = f(k_2)$ because f_{k_2} is injective. Hence f is indeed injective, and therefore, a) implies b). In order to show that b) implies d), we assume that we are given an injection $f : \mathbb{N} \to S$. Then $f \upharpoonright_{M_n} : M_n \to S$ is an injection as well for all $n \in \mathbb{N}$, and therefore $|M_n| \leq |S|$ for all $n \in \mathbb{N}$, as desired. Finally, to give a proof by contradiction of the implication d)$\Rightarrow$a), we assume that d) is satisfied and that S is finite. Then there exists a bijection $f : M_n \to S$ for some $n \in \mathbb{N}$, and according to d), there also exists an injection $g : M_{n+1} \to S$. Using Proposition 5.1.3a, it follows that $f^{-1} \circ g : M_{n+1} \to M_n \subset M_{n+1}$ is an injection and therefore a surjection from M_{n+1} onto itself by Theorem 5.1.9. But this is not possible because the fact that $R(f^{-1} \circ g) \subset M_n$ implies that $R(f^{-1} \circ g) \neq M_{n+1}$. □

5.2.7 Theorem. *For any set $S \neq \emptyset$ it is the case that $|S| < |\mathcal{P}(S)|$.*

Proof. If $S = \emptyset$, then $|S| = 0 < 1 = |\{\emptyset\}| = |\mathcal{P}(S)|$, as desired. So we may assume that $S \neq \emptyset$. Then then map $f : S \to \mathcal{P}(S)$, $x \mapsto \{x\}$ is injective, because if $\{x_1\} = \{x_2\}$, then $x_1 = x_2$. Thus $|S| \leq |\mathcal{P}(S)|$. In order to show that $|S| \neq |\mathcal{P}(S)|$, it is sufficient to prove that no map $g : S \to \mathcal{P}(S)$ can be surjective. To do so, we assume that g is a map from S to $\mathcal{P}(S)$ and define $M := \{x \in S \mid x \notin g(x)\}$. Then $M \notin g(S)$, because if there did exist an $x \in S$ with $g(x) = M$, then the assumption $x \in M = g(x)$ would imply that $x \notin M$ (by the definition of M), and the alternative assumption $x \notin M = g(x)$ would imply that $x \in M$ (also by the definition of M), that is, we would arrive at a contradiction in either case. Hence $g(x) \neq M$ for all $x \in S$, and therefore, g is not surjective. □

Remark. The preceding theorem shows that there are infinitely many orders of infinity, because

$$|\mathbb{N}| < |\mathcal{P}(\mathbb{N})| < |\mathcal{P}(\mathcal{P}(\mathbb{N}))| < |\mathcal{P}(\mathcal{P}(\mathcal{P}(\mathbb{N})))| < \ldots ad\ infinitum. \qquad (5.2)$$

However, the orders of infinity are by no means exhausted by this sequence. For if we define inductively $S_1 := \mathbb{N}$ and $S_{n+1} := \mathcal{P}(S_n)$ for all $n \in \mathbb{N}$ and then set

$$S := \bigcup_{n=1}^{\infty} S_n,$$

then $|S_n| < |S|$ for all $n \in \mathbb{N}$. To see this, we argue as follows: since $S_n \subset S$ for all $n \in \mathbb{N}$, we may apply Theorem 5.2.4b to infer that $|S_n| \leq |S|$ for all $n \in \mathbb{N}$. Thus, in order to show that $|S_n| < |S|$ we only need to prove that there does not exist a bijection from S_n to S. Assuming, by way of contradiction, that there does exist such a bijection, that is, that there does exist a bijective map $f : S_n \to S$, we may infer that there also exists a surjection from S_n to S_{n+1}. This is so because $S_{n+1} \subset S$ implies that $|S_{n+1}| \leq |S|$, and since this latter relation is equivalent to $|S| \geq |S_{n+1}|$ (by Theorem 5.2.3), it follows that there exists a surjection $g : S \to S_{n+1}$. Hence $g \circ f : S_n \to S_{n+1}$ is a surjection (by Proposition 5.1.3). But this is impossible because, as we saw in the proof of Theorem 5.2.7, the set S_n cannot be mapped surjectively onto its power set $\mathcal{P}(S_n) = S_{n+1}$. Furthermore, having shown that the cardinality of S is strictly greater than all the cardinalities in (5.2), we can create another such sequence of strictly increasing cardinalities by taking the power sets of the power sets of S:

$$|S| < |\mathcal{P}(S)| < |\mathcal{P}(\mathcal{P}(S))| < |\mathcal{P}(\mathcal{P}(\mathcal{P}(S)))| < \ldots ad\ infinitum.$$

Thus it becomes apparent that there really is no limit whatsoever to the range of infinities that the cardinalities of infinite sets can represent.

Exercises

5.2.8. Assume that A, B, A', and B' are sets such that $|A| = |A'|$ and $|B| = |B'|$. Prove the following statements:

a) $|A| = |B|$ if and only if $|A'| = |B'|$.

b) $|A| \leq |B|$ if and only if $|A'| \leq |B'|$.

c) $|A| \geq |B|$ if and only if $|A'| \geq |B'|$.

d) $|A| < |B|$ if and only if $|A'| < |B'|$.

e) $|A| > |B|$ if and only if $|A'| > |B'|$.

5.2.9. Use induction to prove that a nonempty set A is finite if there exists an $n \in \mathbb{N}$ such that $|A| \leq |M_n|$.

5.2.10. Assume that A and B are cardinally equivalent and that A is infinite. Show that B is infinite as well.

5.2.11. Assume that $|A| = |\mathbb{N}|$ and that B is finite. Prove that $|A \cup B| = |\mathbb{N}|$.

5.3 ★ Antisymmetry and Linearity

As we saw in the previous section, the relation r, defined in (5.1), is not antisymmetric because the inequalities $|A| \leq |B|$ and $|B| \leq |A|$ do not in general imply that A equals B. However, there is a somewhat modified antisymmetry property that is in fact valid, namely the fact that $|A| \leq |B| \wedge |B| \leq |A|$ always implies that $|A| = |B|$. In other words, the assumption that A can be mapped injectively into B and vice versa or, equivalently (see Theorem 5.2.3), that A can be mapped injectively into B and also surjectively onto B implies that A can also be mapped bijectively onto B. So the existence of an injection $f : A \to B$ and the *separate* existence of a surjection $g : A \to B$ together imply the existence of a bijection $h : A \to B$. In order to establish this claim, we need the following lemma:

5.3.1 Lemma. *If $g : A \to B$ is a surjection and N a nonempty subset of B, then there exists a set $M \subset g^{-1}(N)$ such that $|M \cap g^{-1}(\{y\})| = 1$ for all $y \in N$.*

Proof. According to Lemma 3.8.10, the assumption of surjectivity on g implies that there exists a function $h : B \to A$ such that $h(y) \in g^{-1}(\{y\})$ for all $y \in B$. Setting $M := h(N)$, it follows that for all $x \in M$ there exists a $y \in N$ such that

$x = h(y) \in g^{-1}(\{y\}) \subset g^{-1}(N)$. Hence $M \subset g^{-1}(N)$. In order to prove that $|M \cap g^{-1}(\{y\})| = 1$ for all $y \in N$, we observe to begin with that the condition $x \in M \cap g^{-1}(\{y\})$ implies that $g(x) = y$ (because $x \in g^{-1}(\{y\})$) and that $h(z) = x$ for some $z \in N$ (by the definition of M). Since $h(z) \in g^{-1}(\{z\})$, it follows that $z = g(h(z)) = g(x) = y$, and therefore, $x = h(z) = h(y)$. In other words, what we have here shown is that $x \in M \cap g^{-1}(\{y\}) \Rightarrow x \in \{h(y)\}$ for all $x \in A$, or, equivalently, that $M \cap g^{-1}(\{y\}) \subset \{h(y)\}$. But since $h(y) \in g^{-1}(\{y\})$ (by the definition of h) and $h(y) \in M$ (because $y \in N$), it also is the case that $\{h(y)\} \subset M \cap g^{-1}(\{y\})$, and therefore, $\{h(y)\} = M \cap g^{-1}(\{y\})$. Hence $|M \cap g^{-1}(\{y\})| = 1$, as desired. □

Remark. The conditions on M in Lemma 5.3.1 trivially imply that $g(M) = N$, because if $M \subset g^{-1}(N)$, then $g(M) \subset N$, and the reverse inclusion is implied by the fact that $|M \cap g^{-1}(\{y\})| = 1$ for all $y \in N$, because given this fact and given an element $y \in N$, it follows that there exists exactly one element $x \in M$ such that $y = g(x) \in g(M)$.

In order to intuitively grasp how Lemma 5.3.1 will enable us to establish the claim that two nonempty sets A and B are cardinally equivalent whenever $|A| \leq |B| \wedge |A| \geq |B|$ (see the introductory paragraph above), we assume that we are given an injection $f : A \to B$ and a surjection $g : A \to B$. How can we use f and g to construct a bijection $h : A \to B$? Well, if f happens to be, not only injective, but also surjective, then f is bijective, and we can simply define h to be equal to f. But what if f is not surjective? In that case we do have a problem because the set $N_1 := B \setminus f(A)$ is nonempty and its elements cannot be reached by f. Fortunately, though, they can be reached by g because g is surjective, and this is where Lemma 5.3.1 comes in. For according to this lemma, there is a set $M_1 \subset g^{-1}(N_1)$ such that $|M_1 \cap g^{-1}(\{y\})| = 1$ for all $y \in N_1$. Consequently, if we define $h(x)$ to be equal to $g(x)$ for all $x \in M_1$, then h covers N_1 because $h(M_1) = g(M_1) = N_1$ (see the remark above). Furthermore, h is injective on M_1 because if $h(x_1) = h(x_2)$ for some $x_1, x_2 \in M_1$, then, in setting $y := h(x_1)$, we find that both x_1 and x_2 are contained in $M_1 \cap g^{-1}(\{y\})$, and this in turn implies that $x_1 = x_2$ because $|M_1 \cap g^{-1}(\{y\})| = 1$. So by means of this construction we can injectively cover the set $B \setminus f(A) = N_1$, as desired, but the problem here is that in defining h to be equal to g on M_1 we now punch a hole into the range of f because the range of $f \restriction_{A \setminus M_1}$ is not $f(A)$ but $f(A) \setminus f(M_1)$. Consequently, we need to keep going and again use g to patch up the 'hole' $N_2 := f(M_1)$. To do so, we apply Lemma 5.3.1 to infer the existence of a set $M_2 \subset g^{-1}(N_2)$ such that $|M_2 \cap g^{-1}(\{y\})| = 1$ for all $y \in N_2$ and define $h(x) := g(x)$ for all $x \in M_2$. Again we readily observe that h is injective on M_2 and that $f(M_2) = N_2$, but unfortunately, there is yet another hole opening up in the range of f because $f(A \setminus (M_1 \cap M_2))$ is not equal to $f(A) \setminus f(M_1)$ but rather to $(f(A) \setminus f(M_1)) \setminus f(M_2)$. Hence we set $N_3 := f(M_2)$ and use Lemma 5.3.1

to find a set $M_3 \subset g^{-1}(N_3)$ such that $|M_3 \cap g^{-1}(\{y\})| = 1$ for all $y \in N_3$. Proceeding in this way *ad infinitum*, we produce an infinite sequence of sets N_n and M_n such that $N_{n+1} = f(M_n)$ for all $n \in \mathbb{N}$, and in setting $M := \bigcup_{n=1}^{\infty} M_1$, we are ultimately left with a bijective map $h : A \to B$ that is defined to be equal to g on M and equal to f on $A \smallsetminus M$.

5.3.2 Theorem. *Let A and B be nonempty sets. If there exists an injection $f : A \to B$ and a surjection $g : A \to B$, then there also exists a bijection $h : A \to B$. In other words, if $|A| \le |B|$ and $|A| \ge |B|$, then $|A| = |B|$.*

Proof. To get started, we apply Lemma 5.3.1 (and the Axiom of Choice) to associate with each non-empty subset N of B a set $M(N) \subset g^{-1}(N)$ such that $|g^{-1}(\{y\}) \cap M(N)| = 1$ for all $y \in N$. If f is surjective, then $h := f$ is a bijection. Assuming therefore that f is not surjective, it follows that the sets $N_1 := B \smallsetminus f(A)$ and $M_1 := M(N_1)$ are both nonempty. To inductively define two sequences of nonempty sets $N_n \subset B$ and $M_n \subset A$, we assume that nonempty sets N_n and M_n have been defined for some $n \in \mathbb{N}$ and set

$$N_{n+1} := f(M_n)$$

and

$$M_{n+1} := M(N_{n+1}).$$

Given these definitions, we further set

$$M := \bigcup_{n=1}^{\infty} M_n,$$

and

$$h(x) := \begin{cases} f(x) & \text{if } x \in A \smallsetminus M, \\ g(x) & \text{if } x \in M. \end{cases}$$

According to the remark on p.145, it is the case that $g(M_n) = N_n$ for all $n \in \mathbb{N}$, and therefore, we may apply Theorem 3.8.19a to infer that

$$g(M) = g\left(\bigcup_{n=1}^{\infty} M_n\right) = \bigcup_{n=1}^{\infty} g(M_n) = \bigcup_{n=1}^{\infty} N_n$$

$$= (B \smallsetminus f(A)) \cup \bigcup_{n=2}^{\infty} N_n = (B \smallsetminus f(A)) \cup \bigcup_{n=1}^{\infty} f(M_n) \qquad (5.3)$$

$$= (B \smallsetminus f(A)) \cup f\left(\bigcup_{n=1}^{\infty} M_n\right) = (B \smallsetminus f(A)) \cup f(M).$$

Consequently, Proposition 3.8.23 in conjunction with the assumed injectivity of f implies that

$$h(A) = h(M \cup (A \smallsetminus M)) = h(M) \cup h(A \smallsetminus M) = g(M) \cup f(A \smallsetminus M)$$
$$= (B \smallsetminus f(A)) \cup f(M) \cup (f(A) \smallsetminus f(M)) = (B \smallsetminus f(A)) \cup f(A) = B.$$

This proves that h is surjective. To demonstrate that h is also injective, we will pick two elements $x_1 \neq x_2$ and show that $h(x_1) \neq h(x_2)$.

Case 1: $x_1, x_2 \in A \setminus M$.

In this case the definition of h, the injectivity of f, and the assumption $x_1 \neq x_2$ imply that $h(x_1) = f(x_1) \neq f(x_2) = h(x_2)$, as desired.

Case 2: $x_1 \in A \setminus M$ and $x_2 \in M$.

Since

$$
\begin{aligned}
f(A \setminus M) \cap g(M) &= f(A \setminus M) \cap ((B \setminus f(A)) \cup f(M)) \quad \text{(by (5.3))} \\
&= (f(A) \setminus f(M)) \cap ((B \setminus f(A)) \cup f(M)) \quad \text{(by Proposition 3.8.23)} \\
&= ((f(A) \setminus f(M)) \cap (B \setminus f(A))) \cup ((f(A) \setminus f(M)) \cap f(M)) \\
&= \emptyset \cup \emptyset = \emptyset,
\end{aligned}
$$

it follows that $h(x_1) = f(x_1) \neq g(x_2) = h(x_2)$, as desired.

Case 3: $x_1, x_2 \in M$.

To begin with we will prove via mathematical induction that

$$
M_{n+1} \cap \bigcup_{k=1}^{n} M_k = \emptyset = N_{n+1} \cap \bigcup_{k=1}^{n} N_k \tag{5.4}
$$

for all $n \in \mathbb{N}$. For $n = 1$ we have $N_2 = f(M_1) \subset f(A)$ and therefore $N_2 \cap N_1 = N_2 \cap (B \setminus f(A)) = \emptyset$. This observation also shows that $M_1 \cap M_2 = \emptyset$, because $M_1 = M(N_1) \subset g^{-1}(N_1)$ and $M_2 = M(N_2) \subset g^{-1}(N_2)$. Assuming now (5.4) to be satisfied for a given $n \in \mathbb{N}$, we use the assumed injectivity of f in conjunction with Theorem 3.8.21 to infer that

$$
\begin{aligned}
N_{n+2} \cap \bigcup_{k=1}^{n+1} N_k &= f(M_{n+1}) \cap \left((B \setminus f(A)) \cup \bigcup_{k=1}^{n} f(M_k) \right) \\
&= f(M_{n+1}) \cap \bigcup_{k=1}^{n} f(M_k) = f(M_{n+1}) \cap f \left(\bigcup_{k=1}^{n} M_k \right) \\
&= f \left(M_{n+1} \cap \bigcup_{k=1}^{n} M_k \right) = f(\emptyset) = \emptyset.
\end{aligned}
$$

Using this result, we also find that

$$
\begin{aligned}
M_{n+2} \cap \bigcup_{k=1}^{n+1} M_k &= M(N_{n+2}) \cap \bigcup_{k=1}^{n+1} M(N_k) \subset g^{-1}(N_{n+2}) \cap \bigcup_{k=1}^{n+1} g^{-1}(N_k) \\
&= g^{-1}(N_{n+2}) \cap g^{-1} \left(\bigcup_{k=1}^{n+1} N_k \right) = g^{-1} \left(N_{n+2} \cap \bigcup_{k=1}^{n+1} N_k \right) \\
&= g^{-1}(\emptyset) = \emptyset,
\end{aligned}
$$

as desired. Moreover, given the assumption $x_1, x_2 \in M$, it follows that there are integers m and n such $x_1 \in M_m$ and $x_2 \in M_n$. Assuming w.l.o.g. that $m \le n$, we first consider the case $m = n$. Then there are elements y_1 and y_2 in N_n such that $x_1 \in g^{-1}(\{y_1\})$ and $x_2 \in g^{-1}(\{y_2\})$. Since $|M_n \cap g^{-1}(\{y\})| = 1$ for all $y \in N_n$ and $x_1 \ne x_2$, we may conclude that $y_1 \ne y_2$, and therefore $h(x_1) = g(x_1) = y_1 \ne y_2 = g(x_2) = h(x_2)$. Finally, if $m < n$, then (5.4) allows us to infer that

$$\{g(x_2)\} \cap \{g(x_1)\} \subset g(M_n) \cap g\left(\bigcup_{k=1}^{n-1} M_k\right) = g(M_n) \cap \bigcup_{k=1}^{n-1} g(M_k)$$

$$= N_n \cap \bigcup_{k=1}^{n-1} N_k = \emptyset.$$

This shows that $h(x_2) = g(x_2) \ne g(x_1) = h(x_1)$ and hence completes the proof. $\square$

Given this result, we will now prove some additional transitivity properties involving the strictly-less-than relation.

5.3.3 Theorem. *For all nonempty sets A, B, and C, the following statements are valid:*

a) *If $|A| < |B|$ and $|B| < |C|$, then $|A| < |C|$.*

b) *If $|A| \le |B|$ and $|B| < |C|$, then $|A| < |C|$.*

c) *If $|A| < |B|$ and $|B| \le |C|$, then $|A| < |C|$.*

d) *If $|A| = |B|$ and $|B| < |C|$, then $|A| < |C|$.*

e) *If $|A| < |B|$ and $|B| = |C|$, then $|A| < |C|$.*

Proof. Properties a) and d) are implied by b), and e) is implied by c). Thus we only need to prove b) and c), but the latter of these we will leave as an exercise to the reader (see Exercise 5.3.18). To establish b), we assume that $|A| \le |B|$ and $|B| < |C|$. Then $|A| \le |C|$ by Theorem 5.2.4c. In order to prove, by way of contradiction, that $|A| \ne |C|$, we assume that $|A| = |C|$. Then $|C| = |A|$ and $|A| \le |B|$, and therefore, Theorem 5.2.4e implies that $|C| \le |B|$. Hence $|B| \ge |C|$ (by Theorem 5.2.3), and therefore, $|C| = |B|$ (by Theorem 5.3.2) because $|B| \le |C|$ by assumption. Since this result contradicts the assumed inequality $|B| < |C|$, the proof is complete. $\square$

Having addressed the issue of antisymmetry, we now turn to the property of linearity: for given nonempty sets A and B it is the case that either $|A| \le |B|$ or $|B| \le |A|$. To prove this claim, we need to use a prominent result of set theory which is known as *Zorn's Lemma* and which among all the major results of set theory is perhaps the one most frequently employed in other branches of mathematics. Hence we introduce the following definitions:

5.3.4 Definition. A partial order relation $r \subset S \times S$ (see Definition 3.5.7) is said to be *reflexive* if $(x, x) \in r$ for all $x \in S$. In order to visually convey the fact that reflexivity is a property that r possesses, we agree to denote r by the common less-than-or-*equal* sign, that is,

$$x \leq y :\Leftrightarrow (x, y) \in r$$

for all $x, y \in S$. Furthermore, we also agree that

$$x < y :\Leftrightarrow x \leq y \wedge x \neq y$$

for all $x, y \in S$.

Remark. For clarity, we wish to point out that our use of the less-than-or-equal sign in Definition 5.3.4 is inherently arbitrary. For we could have just as well agreed to use the greater-than-or-equal sign instead. That is to say, if r was the common greater-than-or-equal relation on $\mathbb{N}$ or $\mathbb{R}$, then Definition 5.3.4 would tell us to denote it by '$\leq$' rather than the familiar '$\geq$'. This may seem counterintuitive at first, but the greater-than-or-equal relation, as a mathematical object, is of course not affected in any way whatever by the symbol that we use to denote it. So it really doesn't matter: greater than is less than and less than is greater than, and up is down and down is up, and math is still the same regardless.

5.3.5 Lemma. *If '$\leq$' is a reflexive partial order on a set S, then a set $A \subset S$ is linearly ordered by '$\leq$' if and only if $\forall_{x,y \in A} \, x \leq y \vee y \leq x$.*

Proof. Since $\forall_{x,y \in A} \, x \leq y \vee y \leq x$ is a stronger statement than $\forall_{x,y \in A} \, x \leq y \vee y \leq x \vee x = y$, we only need to prove that the latter statement implies the former. So let us assume that $x \leq y \vee y \leq x \vee x = y$ is true for any given $x, y \in A$. If $x = y$, then $x \leq y$ by reflexivity, and if $x \neq y$, then $x \leq y \vee y \leq x$ by assumption. So in either case $x \leq y \vee y \leq x$ is true. $\square$

5.3.6 Definition. If '$\leq$' is a reflexive partial order on a set S, then $x \in S$ is said to be a *maximal element* of a subset A of S if $x \in A$ and if there is no element $y \in A$ such that $x < y$. Similarly, x is said to be a *minimal element* of A if $x \in A$ and if there is no $y \in A$ such that $y < x$.

5.3.7 Example. In the partial order case minimal and maximal elements can be unique but do not need to be. To provide an example, we assume that U is a nonempty set that containns two nonempty disjoint subsets V and W (i.e., $V, W \subset U$, $V, W \neq \emptyset$, and $V \cap W = \emptyset$). If '$\leq$' denotes the subset relation on $S := \mathcal{P}(U)$ (i.e., $A \leq B :\Leftrightarrow A \subset B$ for all $A, B \in S = \mathcal{P}(U)$), then V and W are distinct maximal elements of $A := \mathcal{P}(V) \cup \mathcal{P}(W) \subset \mathcal{P}(U) = S$, and the empty set is the unique minimal element of A (see Exercise 5.3.19).

5.3.8 Lemma. *Let '$\leq$' be a reflexive partial order on a set S. If $A \subset S$ is linearly ordered by '$\leq$', then x is a maximal element of A if and only if $x \in A \land \forall_{y \in A}\, y \leq x$, and x is a minimal element of A if and only if $x \in A \land \forall_{y \in A}\, x \leq y$. Furthermore, maximal and minimal elements of linearly ordered subsets of S are unique.*

Proof. To begin with, we observe that, by definition, x is maximal in A if $x \in A \land \neg \exists_{y \in A}\, x < y$, that is, if $x \in A \land \forall_{y \in A}\, \neg(x < y)$. Hence it is sufficient to show that $y \leq x \Leftrightarrow \neg(x < y)$ or, equivalently, that $y \leq x \Leftrightarrow (x \leq y \Rightarrow x = y)$. In order to show that $y \leq x \Rightarrow (x \leq y \Rightarrow x = y)$, we assume that $y \leq x$ and $x \leq y$ and use the antisymmetry of '$\leq$' to infer that $x = y$, as desired. To prove the reverse implication $y \leq x \Leftarrow (x \leq y \Rightarrow x = y)$, we notice that

$$(y \leq x \Leftarrow (x \leq y \Rightarrow x = y)) \Leftrightarrow ((\neg(x \leq y) \lor x = y) \Rightarrow y \leq x)$$

and assume that $\neg(x \leq y) \lor x = y$ is true. If $x = y$, then $y \leq x$ by reflexivity, and if $\neg(x \leq y)$, then $y \leq x$ by linearity and Lemma 5.3.5. Thus we find that indeed x is maximal in A if and only if $x \in A \land \forall_{y \in C}\, y \leq x$. In a completely analogous fashion it can also be shown that x is minimal in A if and only if $x \in A \land \forall_{y \in A}\, x \leq y$. Moreover, under the present assumption of linearity on A, maximal and minimal elements are unique, because if x and z are, say, both maximal in A, then $z \leq x$ (because x is maximal) and $x \leq z$ (because z is maximal), and therefore, $x = z$ by antisymmetry. $\qquad \square$

5.3.9 Definition. If '$\leq$' is a reflexive partial order on a set S, then a set $A \subset S$ is said to be *well ordered* by '$\leq$' if A is linearly ordered by '$\leq$' and if every nonempty subset of C has a minimal element, that is, if

$$(\forall_{x,y \in A}\, x \leq y \lor y \leq x) \land (\forall_{C \subset A}\, C \neq \emptyset \Rightarrow \exists_{x \in C} \forall_{y \in C}\, x \leq y).$$

Remark. The less-than-or-equal relation on $\mathbb{N}$ is a well-ordering by Theorems 4.2.9 and 4.3.9. Moreover and very amazingly, it can be shown that a well ordering can be defined on any given set S regardless of its nature or structure. In other words, if S is an entity whose existence is guaranteed by and consistent with the axioms of set theory, then S can be well ordered (see for instance the exposition in [JW]).

5.3.10 Definition. If '$\leq$' is a reflexive partial order on a set S, then a set $A \subset S$ is said to be *closed* in a set $B \subset S$ if

$$A \subset B \land \forall_{x \in A} \forall_{y \in B}\, y \leq x \Rightarrow y \in A,$$

and in this case we write $A \leq B$. Furthermore, if $A \leq B$ and $A \neq B$, then we write $A < B$.

5.3.11 Lemma. *Let '≤' be a reflexive partial order on a set S. If B is a well ordered subset of S and $A < B$, then A consists of all elements y that are strictly less than the unique minimal element of $B \smallsetminus A$, that is,*

$$A = \{y \in B \mid y < \min(B \smallsetminus A)\}.$$

Proof. If $A < B$, then $B \smallsetminus A \neq \emptyset$, and therefore, $x := \min(B \smallsetminus A)$ is well defined and unique (by Lemma 5.3.8) because B is well ordered. If $y < x$, then $y \notin B \smallsetminus A$ because otherwise x would not be the minimal element of $B \smallsetminus A$. Hence $y \in A$, and this shows that $\{y \in B \mid y < x\} \subset A$. In order to show that the reverse inclusion relation is valid as well, we pick a $y \in A$ and observe that $y \neq x$ because $x \in B \smallsetminus A$. Using Lemma 5.3.5, it further follows that $y \leq x \vee x \leq y$, and therefore, we only need to show that $x \leq y$ is impossible, for in the alternative case—$y \leq x$—we find that $y < x$ (because $y \neq x$), as desired. So why can it not be the true that $x \leq y$? Because if it were true, then the fact that A is closed in B would imply that x is contained in A (see Definition 5.3.10). $\qquad\square$

5.3.12 Lemma. *Let '≤' be a reflexive partial order on a set S. If $W \subset \mathcal{P}(S)$ is a collection of well ordered subsets of S such that $A \leq B \vee B \leq A$ for all $A, B \in W$, then $E := \bigcup_{A \in W} A$ is well ordered.*

Proof. we need to show that

$$(\forall_{x,y \in E}\ x \leq y \vee y \leq x) \wedge (\forall_{C \subset E}\ C \neq \emptyset \Rightarrow \exists_{x \in C} \forall_{y \in C}\ x \leq y).$$

So let $x, y \in E$. Then there are sets $A, B \in W$ such that $x \in A$ and $y \in B$. Since $A \leq B \vee B \leq A$, we may assume w.l.o.g. that $A \leq B$ (as the case $B \leq A$ is completely analogous). Then both x and y are contained in B, and since B is linearly ordered (in fact, B is well ordered), we may infer that $x \leq y \vee y \leq x$, as desired. Having thus established linearity, we proceed to assume that C is a nonempty subset of E. Then, by the definition of E, there exists an $A \in W$ such that $C \cap A \neq \emptyset$. Since A is well ordered, it follows that $x := \min(C \cap A)$ is well defined. Hence it is sufficient to show that x is the minimal element of C or, equivalently, that $x \leq y$ for all $y \in C$. So let y be in C. Then there is a $B \in W$ such that $y \in B$. If $B \leq A$, then $y \in C \cap B \subset C \cap A$, and therefore, $x \leq y$ because x is minimal in $C \cap A$. Alternatively, if $A \leq B$, then $x, y \in B$, and the fact that B is linearly ordered therefore implies that $x \leq y \vee y \leq x$. Since $x \leq y$ is the relation that we wish to infer, we only need to consider the latter case where $y \leq x$. So let this latter inequality be valid. Then $A \leq B$ implies that $y \in A$ (by Definition 5.3.10), and, as before, we find that $y \in C \cap A$ and that $x \leq y$ by implication (in fact, here we even find that $x = y$ by antisymmetry). Thus the proof is complete. $\qquad\square$

5.3.13 Definition. *If '≤' is a reflexive partial order on a nonempty set S, then a subset A of S is said to be bounded above if there exists an $x \in S$ such* that $y \leq x$ for all $y \in A$. In this case x is said to be an *upper bound* of A. Similarly, A is said to be *bounded below*, if A has a *lower bound*, that is, if there exists an $x \in S$ such that $x \leq y$ for all $y \in A$.

5.3.14 Theorem. *(Zorn's Lemma, First Version) If '$\leq$' is a reflexive partial order on a nonempty set S, and if every well ordered subset of S is bounded above, then S has a maximal element.*

Proof. Let us assume that the statement is false. Then there is a nonempty set S equipped with a reflexive partial order '$\leq$' such that every well ordered subset of S is bounded above and such that S has no maximal element. This means that for every well ordered subset $A \subset S$ there exists an upper bound in $S \setminus A$. For suppose that there did exist a well ordered subset A of S for which there did not exist an upper bound in $S \setminus A$. Then A would have an upper bound $x \in A$ because A, by being well ordered, would have an upper bound by assumption. Furthermore, for any $z \in S \setminus A$ we would have $\neg(x < z)$ because if x were strictly less than z, then, by transitivity, z would be an upper bound of A that is contained in $S \setminus A$. Consequently, x would in fact be a maximal element of S, because all elements in A would be less than or equal to x and no element in $S \setminus A$ would be strictly greater than x, and therefore, no element in S would be strictly greater than x. Since this conclusion would contradict our initial assumption that S does not have a maximal element, we may infer that our secondary assumption concerning the existence of a well ordered set A for which there does not exist an upper bound in $S \setminus A$ must have been false. Hence we may apply the Axiom of Choice to identify for every well ordered set $A \subset S$ an upper bound $g(A) \in S \setminus A$. To proceed, we agree to call $A \subset S$ a g-set if A is well ordered and

$$x = g(\{y \in A \mid y < x\}) \tag{5.5}$$

for all $x \in A$. In order to help the reader understand what a g-set really is, we may want to assume that $A \subset S$ is a nonempty g-set. Then A is well ordered and therefore has a unique minimal element x_0. Consequently, $\{y \in A \mid y < x_0\} = \emptyset$, and (5.5) therefore implies that $x_0 = g(\emptyset)$. So the minimal element x_0 of any nonempty g-set is necessarily $g(\emptyset)$. Furthermore, if $A \setminus \{g(\emptyset)\}$ happens to be nonempty as well, then the well-ordering property of A implies that $A \setminus \{g(\emptyset)\}$ has a minimal element x_1 and that x_0 is the only element in A that is strictly smaller than x_1. Hence $\{y \in A \mid y < x_1\} = \{x_0\}$, and therefore, (5.5) allows us to infer that $x_1 = g(\{x_0\}) = g(\{g(\emptyset)\})$. Continuing in this fashion, we readily find that

$$A = \{g(\emptyset), g(\{g(\emptyset)\}), g(\{g(\emptyset), g(\{g(\emptyset)\})\}), \ldots\}.$$

This is not to say by any means that the elements of a g-set can always be listed like the elements of $\mathbb{N}$, but it is to say that any initial string of finitely many elements in A is uniquely determined by (5.5). Having thus made the notion of a g-set a little more transparent, we move on to claim that for any given g-sets $A, B \subset S$ it always is the case that

$$A \leq B \vee B \leq A. \tag{5.6}$$

To prove this claim, we set

$$V := \{C \subset S \mid C \leq A \wedge C \leq B\}.$$

Given this definition, it follows that $\emptyset \in V$ (by Definition 5.3.10), and therefore, $V \neq \emptyset$. Thus we may further define

$$D := \bigcup_{C \in V} C,$$

and observe that $D \leq A$ and $D \leq B$ (see Exercise 5.3.20). If it were the case that $D = A$ or $D = B$, then the claim would be proven. Hence we only need to show that $D < A \wedge D < B$ cannot be true. To give a proof of this latter claim by way of contradiction, we assume that $D < A \wedge D < B$ is true and use Lemma 5.3.11 to conclude that

$$\{y \in A \mid y < \min(A \smallsetminus D)\} = D = \{y \in B \mid y < \min(B \smallsetminus D)\}.$$

Since A and B are g-sets, it follows that

$$z := \min(A \smallsetminus D) = g(\{y \in A \mid y < \min(A \smallsetminus D)\})$$
$$= g(\{y \in B \mid y < \min(B \smallsetminus D)\}) = \min(B \smallsetminus D).$$

Given this observation and given this definition of z, we wish to show that $D \cup \{z\} \leq A$ and $D \cup \{z\} \leq B$. To do so, we assume that $y \leq x$ for some $x \in D \cup \{z\}$ and some $y \in A$. If $x \in D$, then $y \in D$ because $D \leq A$, and if $x = z$, then there are two cases: if $y = x$, then $y = z \in D \cup \{z\}$, and if $y \neq x$, then $y \in \{y \in A \mid y < x\} = D$. So in all cases we find that $y \in D \cup \{z\}$, and therefore, we may indeed infer that $D \cup \{z\} \leq A$. Since a completely analogous argument can be used to show that $D \cup \{z\} \leq B$, it follows that $D \cup \{z\} \in V$. Hence

$$D \cup \{z\} \subset \bigcup_{C \in V} C = D$$

and therefore, $z \in D$. Since this conclusion contradicts the fact that $z = \min(A \smallsetminus D) \in A \smallsetminus D$, it follows that $D < A \wedge D < B$ is false, as desired. In order to complete the proof, we now set

$$W := \{A \subset S \mid A \text{ is a } g\text{-set}\}$$

and

$$E := \bigcup_{A \in W} A$$

and claim that E itself is a g-set. Since E is well-ordered by (5.6) and Lemma 5.3.12, we only need to show that $x = g(\{y \in E \mid y < x\})$ for all $x \in E$. So let x be in E. Then there exists an $A \in W$ such that $x \in A$. Since A is a g-set, it follows that $x = g(\{y \in A \mid y < x\})$, and therefore, it is sufficient to show that $\{y \in A \mid y < x\} = \{y \in E \mid y < x\}$. But since $A \subset E$ trivially implies that $\{y \in A \mid y < x\} \subset \{y \in E \mid y < x\}$, we really need to prove only the reverse inclusion $\{y \in E \mid y < x\} \subset \{y \in A \mid y < x\}$. So let $y \in E$ such that $y < x$. Then there is a $B \in W$ such that $y \in B$. If $B \leq A$, then $y \in A$, and we are done. Alternatively, if $A \leq B$, then $y \in A$ because A is closed in B, and again we are done. Thus E is a g-set and $g(E)$ is well defined. Since $g(E) \notin E$ (because $g(E)$ is an upper bound of E that is contained in $S \smallsetminus E$), it is now sufficient to show that $F := E \cup \{g(E)\}$ is a g-set because in that case we arrive at the contradictory conclusion that F, as an element of W, is a subset of E and that therefore $g(E) \in E$. In other words, if we can show that F is a g-set then our initial assumption that S has no maximal element has been proven false, and we are finally done. Using the fact that $x < g(E)$ for all $x \in E$ and that E is well ordered, it is very easy to see that F is well ordered as well (see Exercise 5.3.21). Consequently, all that is left for us to prove is that $x = g(\{y \in F \mid y < x\})$ for all $x \in F$. So let x be in F. If $x = g(E)$, then $\{y \in F \mid y < x\} = E$ (because $x < g(E)$ for all $x \in E$), and therefore, $g(\{y \in F \mid y < x\}) = g(E) = x$, as desired. Alternatively, if $x \in E$, then $\{y \in F \mid y < x\} = \{y \in E \mid y < x\}$ (again because $x < g(E)$ for all $x \in E$), and therefore, $g(\{y \in F \mid y < x\}) = x$ because E is a g-set. $\square$

A slightly weaker but more commonly encountered version of Zorn's Lemma, which also is more easily applied, is the following:

5.3.15 Theorem. *(Zorn's Lemma, Second Version) If '$\leq$' is a reflexive partial order on a nonempty set S, and if every linearly ordered subset of S is bounded above, then S has a maximal element.*

Proof. Since every well ordered subset of S is linearly ordered as well, Theorem 5.3.15 is a direct consequence of Theorem 5.3.14. $\square$

Having established Zorn's Lemma, we now are ready to prove that the cardinal less-than-or-equal relation is linear.

5.3.16 Theorem. *If A and B are given sets, then $|A| \leq |B|$ or $|B| \leq |A|$.*

Proof. If A is empty, then $|A| = 0 \leq |B|$. Thus we may assume that both A and B are nonempty. Given this assumption, we need to show that there exists either an injection from A to B or a surjection from A to B. To do so, we set

$$S := \{f \subset A \times B \mid \exists_{C \subset A} f \text{ is an injective function from } C \text{ to } B\}.$$

Since A and B are nonempty, it trivially follows that S is nonempty as well and that

$$f \leq g :\Leftrightarrow f \subset g$$

defines a reflexive partial order on S. In order to be able to apply the second version of Zorn's Lemma, we need to prove that every linearly ordered subset of S is bounded above. So let $M \subset S$ be linearly ordered and let us denote by C_f the domain of any given $f \in M$. We wish to show that $h := \bigcup_{f \in M} f$ is an injective function from $C := \bigcup_{f \in M} C_f$ into B. Since $h \subset C \times B$ is evidently a relation with domain C, it follows that h is a function from C to B if $\forall_{x \in C} \forall_{y_1, y_2 \in B} ((x, y_1) \in h \wedge (x, y_2) \in h) \Rightarrow y_1 = y_2$. So let $x \in C$ and $y_1, y_2 \in B$ such that $(x, y_1) \in h \wedge (x, y_2) \in h$. Then, by the definition of h, there are functions f_1 and f_2 in M such that $(x, y_1) \in f_1$ and $(x, y_2) \in f_2$. Since M is reflexively linearly ordered, we may infer that $f_1 \leq f_2$ or $f_2 \leq f_1$. Assuming w.l.o.g. that $f_2 \leq f_1$ or, equivalently, that $f_2 \subset f_1$, it follows that both (x, y_1) and (x, y_2) are contained in f_1, and since f_1 is a function, it further follows that $y_1 = y_2$, as desired. In order to prove that h is injective, we assume that $h(x_1) = h(x_2)$ for some $x_1, x_2 \in C$ and set $y := h(x_1)$. Using again the definition of h, we can find functions $f_1, f_2 \in M$ such that $(x_1, y) \in f_1$, $(x_2, y) \in f_2$, and $f_2 \subset f_1$ (w.l.o.g.). Thus (x_1, y) and (x_2, y) are both contained in f_1, and since f_1 is injective, it follows that $x_1 = x_2$ and that h is injective. Hence h is an element of S, as desired. Furthermore, h is an upper bound of M because trivially $f \subset h$ for all $f \in M$, and therefore, we may apply Zorn's Lemma to conclude that S has a maximal element f_0. If the domain of f_0 is A, then f_0 is an injection from A to B and we are done. Alternatively, if the domain of f_0 is a proper subset C of A (i.e., $C \subset A$ and $C \neq A$), then there are two possibilities: if f_0 happens to be surjective, then we pick an arbitrary element y_0 in B and observe that the equation

$$g(x) := \begin{cases} f_0(x) & \text{if } x \in C, \\ y_0 & \text{if } x \in A \setminus C \end{cases}$$

defines a surjection from A to B. Since this implies that $|B| \leq |A|$, we again are done. So all we have left to do is to consider the alternative possibility that f_0 is not surjective (and that C is a proper subset of A). Given this case, we pick an $x_0 \in A \setminus C$ and a $y_0 \in B \setminus f_0(C)$ and define a function $f_1 : C \cup \{x_0\} \to B$ via the equation

$$f_1(x) := \begin{cases} f_0(x) & \text{if } x \in C, \\ y_0 & \text{if } x = x_0. \end{cases}$$

Since f_0 is injective and since y_0 is not contained in the range of f_0, it trivially follows that f_1 is injective as well and thus contained in S. Furthermore, since f_0 is evidently a proper subset of f_1, it follows that $f_0 < f_1$. But this is impossible because f_0 is maximal in S. Consequently, the case where f_0 is not surjective and C is not equal to A cannot occur, and hence the proof is complete. □

5.3.17 Theorem. *If S is a nonempty set of (nonempty) sets, and if R is the set of equivalence classes $[A]$ with respect to the equivalence relation r defined in Theorem 5.2.2, then*

$$s := \{([A], [B]) \in R \times R \mid |A| \leq |B|\}$$

is a well defined linear order on R. Note: as in the case of Theorem 5.2.2, the assumption that the sets in S be nonempty is not necessary but it is again convenient for the same reason.

Proof. s is well defined by Exercise 5.2.8b, it is antisymmetric by Theorem 5.3.2, it is transitive by Theorem 5.2.4c, and it is linear by Theorem 5.3.16. □

Exercises

5.3.18. Prove Theorem 5.3.3c.

5.3.19. Prove the assertion made in Example 5.3.7.

5.3.20. Let '$\leq$' be a reflexive partial order on a set S and let $\{C_i\}_{i \in I}$ be a collection of subsets of S that are each closed in a given set $A \subset S$ (i.e., $C_i \leq A$ for all $i \in I$). Prove that $\bigcup_{i \in I} C_i \leq A$ and $\bigcap_{i \in I} C_i \leq A$.

5.3.21. Let '$\leq$' be a reflexive partial order on a set S and let A be a well ordered subset of S. Prove the following statement: if x is an upper bound of A, then $A \cup \{x\}$ is well ordered as well.

5.3.22. ★ Assume that A and B are cardinally equivalent and that A is infinite. Show that $|A| = |A \cup B|$. *Hint:* use Zorn's Lemma to prove that every infinite set is a disjoint union of sets that are each cardinally equivalent to $\mathbb{N}$.

Chapter 6

A Story of Creation

6.1 The Integers

The purpose of this concluding chapter is to build up the familiar number systems $\mathbb{Z}$, $\mathbb{Q}$, $\mathbb{R}$, and $\mathbb{C}$ from the set $\mathbb{N}$ of positive counting numbers. Starting with the first of these—the set $\mathbb{Z}$ of positive and negative integers—we introduce the following definition:

$$r_{\mathbb{Z}} := \{((m, n), (k, l)) \in (\mathbb{N} \times \mathbb{N}) \times (\mathbb{N} \times \mathbb{N}) \mid m + l = k + n\}. \tag{6.1}$$

We will show that $r_{\mathbb{Z}}$ is an equivalence relation, and the reason for doing so will be that—intuitively speaking—each equivalence class $[(m, n)]_{\mathbb{Z}}$ represents the integer value $m - n$ which can be positive, zero, or negative. For if $(k, l) \in [(m, n)]_{\mathbb{Z}}$, then $m + l = k + n$ or, equivalently, $m - n = k - l$. In other words, (m, n) and (k, l) are equivalent if and only if the difference values represented by these two pairs are equal.

In order to prove that $r_{\mathbb{Z}}$, as defined in (6.1), is an equivalence relation on $\mathbb{N} \times \mathbb{N}$, we observe that $r_{\mathbb{Z}}$ is reflexive, because the equation $m + n = m + n$ trivially implies that $((m, n), (m, n)) \in r_{\mathbb{Z}}$ for all $(m, n) \in \mathbb{N} \times \mathbb{N}$. Furthermore, $r_{\mathbb{Z}}$ is symmetric because if $((m, n), (k, l)) \in r_{\mathbb{Z}}$, then $m + l = k + n$, and therefore $((k, l), (m, n)) \in r_{\mathbb{Z}}$. Finally, in order to show that $r_{\mathbb{Z}}$ is transitive, we assume that $((m, n), (k, l)) \in r_{\mathbb{Z}}$ and $((k, l), (i, j)) \in r_{\mathbb{Z}}$. Then $m + l = k + n$ and $k + j = i + l$, and adding $i + j$ to both sides of the first equation yields $m + j + i + l = k + j + i + n$. Substituting $k + j$ for $i + l$ and using the notation introduced in Lemma 4.2.4, we find that

$$u_{k+j}(m + j) = m + j + (k + j) = i + n + (k + j) = u_{k+j}(i + n).$$

Since u_{k+j} is injective (by Lemma 4.2.4), it follows that $m + j = i + n$, and therefore, $((m, n), (i, j)) \in r_{\mathbb{Z}}$, as desired. Having thus established that $r_{\mathbb{Z}}$ is an equivalence relation, we proceed to define the set of integers as follows:

$$\boxed{\mathbb{Z} := \{[(m, n)]_{\mathbb{Z}} \mid (m, n) \in \mathbb{N} \times \mathbb{N}\}.}$$

Intuitively speaking, this definition means that

$$\vdots$$
$$-2 = 1 - 3 = [(1, 3)]_{\mathbb{Z}},$$
$$-1 = 1 - 2 = [(1, 2)]_{\mathbb{Z}},$$
$$0 = 1 - 1 = [(1, 1)]_{\mathbb{Z}},$$
$$1 = 2 - 1 = [(2, 1)]_{\mathbb{Z}},$$
$$2 = 3 - 1 = [(3, 1)]_{\mathbb{Z}},$$
$$\vdots$$

ad infinitum. However, in order to rigorously justify this listing, we need to define addition and multiplication on $\mathbb{Z}$ in such a way that the familiar rules of arithmetic are valid. To this end we assume that we are given 'integers'

$$x = [(m, n)]_{\mathbb{Z}}$$

and

$$y = [(k, l)]_{\mathbb{Z}}$$

in $\mathbb{Z}$ that may be thought to represent the difference values $m - n$ and $k - l$, respectively. Since the sum and the product of these difference values are $(m+k) - (n+l)$ and $(mk+nl) - (ml+nk)$, respectively, and since $-(m-n) = n - m$, it is natural to introduce the following definitions:

$$-x := [(n, m)]_{\mathbb{Z}},$$
$$x + y := [(m + k, n + l)]_{\mathbb{Z}}, \qquad (6.2)$$
$$x - y := x + (-y) = [(m + l, n + k)]_{\mathbb{Z}}$$

and

$$xy = x \cdot y := [(mk + nl, ml + nk)]_{\mathbb{Z}}. \qquad (6.3)$$

In order to convince ourselves that these operations are well defined, we need to show that they are dependent only on the equivalence classes x and y and not on the particular pairs of integers that represent them. For instance, in order to prove that addition is well defined, we need to show that

$$[(m, n)]_{\mathbb{Z}} + [(k, l)]_{\mathbb{Z}} = [(m + k, n + l)]_{\mathbb{Z}} = [(m' + k', n' + l')]_{\mathbb{Z}}$$
$$= [(m', n')]_{\mathbb{Z}} + [(k', l')]_{\mathbb{Z}}$$

whenever $[(m, n)]_{\mathbb{Z}} = [(m', n')]_{\mathbb{Z}}$ and $[(k, l)]_{\mathbb{Z}} = [(k', l')]_{\mathbb{Z}}$, that is, whenever $m + n' = m' + n$ and $k + l' = k' + l$. Since the latter two equations imply that

$$m + k + n' + l' = m' + k' + n + l,$$

we may infer that indeed

$$[(m + k, n + l)]_{\mathbb{Z}} = [(m' + k', n' + l')]_{\mathbb{Z}}.$$

The proof of the fact that the remaining operations of sign inversion and multiplication are well defined too will be left as an exercise to the reader (see Exercise 6.1.6).

6.1.1 Lemma. *Recalling the description of subtraction in Theorem 4.2.7, it is the case that*

$$[(m, n)]_{\mathbb{Z}} = \begin{cases} \{(m - n + k, k) \mid k \in \mathbb{N}\} & \text{if } m > n, \\ \{(k, k) \mid k \in \mathbb{N}\} & \text{if } m = n, \\ \{(k, n - m + k) \mid k \in \mathbb{N}\} & \text{if } m < n \end{cases}$$

for all $m, n \in \mathbb{N}$, and in particular,

$$[(m, n)]_{\mathbb{Z}} = [(m + k, n + k)]_{\mathbb{Z}}$$

for all $k, m, n \in \mathbb{N}$.

Proof. If $m = n$, then

$$(i, j) \in [(m, n)]_{\mathbb{Z}} \Leftrightarrow (i, j) \in [(m, m)]_{\mathbb{Z}} \Leftrightarrow m + j = m + i \Leftrightarrow u_m(j) = u_m(i)$$
$$\Leftrightarrow i = j \quad \text{(because } u_m \text{ is injective)}$$
$$\Leftrightarrow (i, j) \in \{(k, k) \mid k \in \mathbb{N}\}.$$

Furthermore, if $m = n$, then $m + k = n + k$, and therefore, $(m + k, n + k) \in \{(k, k) \mid k \in \mathbb{N}\}$. Hence $[(m, n)]_{\mathbb{Z}} = [(m + k, n + k)]_{\mathbb{Z}}$, as desired. Since the remaining two cases—$m > n$ and $m < n$—are analogous, we may assume w.l.o.g. that $m > n$. Using Theorem 4.2.7, we find that

$$(i, j) \in [(m, n)]_{\mathbb{Z}} \Leftrightarrow m + j = n + i \Leftrightarrow (m - n) + n + j = n + i$$
$$\Leftrightarrow u_n((m - n) + j) = u_n(i)$$
$$\Leftrightarrow (m - n) + j = i \quad \text{(because } u_n \text{ is injective)}$$
$$\Leftrightarrow (i, j) \in \{(m - n + k, k) \mid k \in \mathbb{N}\},$$

and therefore, $[(m, n)]_{\mathbb{Z}} = \{(m - n + k, k) \mid k \in \mathbb{N}\}$. Consequently, $[(m, n)]_{\mathbb{Z}} = [(m - n + k, k)]_{\mathbb{Z}}$ for all $k \in \mathbb{N}$ (by Lemma 3.6.2), and in replacing k by $n + k$, we find that

$$[(m, n)]_{\mathbb{Z}} = [(m - n + n + k, n + k)]_{\mathbb{Z}}$$
$$= [(m + k, n + k)]_{\mathbb{Z}} \quad \text{(by Theorem 4.2.7)},$$

as desired. $\qquad \square$

6.1.2 Theorem. *For all $x, y, z \in \mathbb{Z}$ it is the case that*

a) $x + (y + z) = (x + y) + z$,

b) $x + 0 = x$, *where* $0 = [(1,1)]_{\mathbb{Z}} = -0$,

c) $x - x = -x + x = 0$,

d) $x + y = y + x$,

e) $x(yz) = (xy)z$,

f) $xy = yx$,

g) $x = 1 \cdot x$ *and* $-x = (-1) \cdot x$, *where* $1 = [(2,1)]_{\mathbb{Z}}$ *and* $-1 = [(1,2)]_{\mathbb{Z}}$,

h) $xy = 0 \Leftrightarrow (x = 0 \vee y = 0)$,

i) $x(y + z) = xy + xz$, *and*

j) $x(y - z) = xy - xz$.

Proof. We will be content to prove only g), h), and i) and leave the proofs of the remaining properties as exercises to the reader (see Exercise 6.1.7). To prove g), we pick a number $x = [(m, n)] \in \mathbb{Z}$ and observe that

$$1 \cdot x = [(2,1)]_{\mathbb{Z}}[(m,n)]_{\mathbb{Z}} = [(2m + 1n, 2n + 1m)]_{\mathbb{Z}}$$
$$= [((1+1)m + n, (1+1)n + m)]_{\mathbb{Z}}$$
$$\quad \text{(because } 2 = 1 + 1 \text{, by definition, and } 1 \cdot n = n \text{, by Theorem 4.4.2)}$$
$$= [(m + m + n, n + n + m)]_{\mathbb{Z}} \quad \text{(by Theorem 4.4.2)}$$
$$= [(m, n)]_{\mathbb{Z}} = x \quad \text{(by Lemma 6.1.1),}$$

as desired. Similarly, we find that

$$(-1)x = [(1,2)]_{\mathbb{Z}}[(m,n)]_{\mathbb{Z}} = [(1m + 2n, 2m + 1n)]_{\mathbb{Z}}$$
$$= [(m + (1+1)n, (1+1)m + n)]_{\mathbb{Z}}$$
$$= [(n + m + n, m + m + n)]_{\mathbb{Z}} = [(n, m)]_{\mathbb{Z}} = -x \quad \text{(by Lemma 6.1.1).}$$

Moving on to h), we argue as follows: if $x = [(m,n)]_{\mathbb{Z}}$ is a given integer and $y = [(1,1)]_{\mathbb{Z}} = 0$, then

$$xy = [(m,n)]_{\mathbb{Z}}[(1,1)]_{\mathbb{Z}} = [(1m + 1n, 1m + 1n)]_{\mathbb{Z}} = [(1,1)]_{\mathbb{Z}} = 0$$

by Lemma 6.1.1, and this shows that $(x = 0 \vee y = 0) \Rightarrow xy = 0$. To prove the reverse implication by contraposition, we assume that we are given non-zero integers $x = [(m,n)]_{\mathbb{Z}}$, $y = [(k,l)]_{\mathbb{Z}}$. Then there are four cases: $n < m \wedge l < k$, $n < m \wedge k < l$, $m < n \wedge l < k$, and $m < n \wedge l < k$. Since the first of these cases is analogous to the fourth and the second to the third, we only need to consider the first and the second. If $n < m \wedge l < k$, then $x = [(m - n + 1, 1)]_Z$ and $y = [(k - l + 1, 1)]_Z$ (by Lemma 6.1.1), and therefore,

$$
\begin{aligned}
xy &= [((m - n + 1)(k - l + 1) + 1 \cdot 1, 1(m - n + 1) + 1(k - l + 1))]_Z \\
&= [((m - n)(k - l) + m - n + k - l + 1, m - n + k - l + 1 + 1)]_Z \\
&= [((m - n)(k - l) + 1, 1)]_Z \quad \text{(by Lemma 6.1.1)} \\
&\neq 0
\end{aligned}
$$

because $1 < (m - n)(k - l) + 1$. Similarly, if $n < m \wedge k < l$, then $x = [(m - n + 1, 1)]_Z$ and $y = [(1, l - k + 1)]_Z$, and, by implication,

$$
\begin{aligned}
xy &= [(1(m - n + 1) + 1(l - k + 1), (m - n + 1)(l - k + 1) + 1 \cdot 1)]_Z \\
&= [(m - n + l - k + 1 + 1, (m - n)(l - k) + m - n + l - k + 1)]_Z \\
&= [(1, (m - n)(k - l) + 1)]_Z \quad \text{(by Lemma 6.1.1)} \\
&\neq 0.
\end{aligned}
$$

Finally, in order to prove i), we pick integers $x = [(m, n)]_Z$, $y = [(k, l)]_Z$, and $z = [(i, j)]_Z$ and observe that

$$
\begin{aligned}
x(y + z) &= [(m, n)]_Z([(k, l)]_Z + [(i, j)]_Z) = [(m, n)]_Z[(k + i, l + j)]_Z \\
&= [(m(k + i) + n(l + j), m(l + j) + n(k + i))]_Z \\
&= [(mk + mi + nl + nj, ml + mj + nk + ni)]_Z \quad \text{(by Theorem 4.4.2)} \\
&= [(mk + nl, ml + nk)]_Z + [(mi + nj, mj + ni)]_Z \\
&= [(m, n)]_Z[(k, l)]_Z + [(m, n)]_Z[(i, j)]_Z \\
&= xy + xz,
\end{aligned}
$$

as claimed. $\qquad\square$

In the light of this theorem, we are now able to further refine and clarify the central result discussed in this text, namely the claim that $1 + 1 = 2$. Previously, we interpreted this equation to mean that 2, by definition equals $\varphi(1) = p(1, 1)$, but now we can dig a little deeper and observe that $2 = [(3, 1)]_Z = [(\varphi(\varphi(1)), 1)]_Z$ and that $1 = [(2, 1)]_Z = [(\varphi(1), 1)]_Z$. So what we really mean by saying that one plus one equals two is that

$$
\boxed{[(\varphi(1), 1)]_Z + [(\varphi(1), 1)]_Z = [(\varphi(\varphi(1)), 1)]_Z,}
$$

where

$$
[(\varphi(1), 1)]_Z = \{(m, n) \in \mathbb{N} \times \mathbb{N} \mid \varphi(1) + n = 1 + m\}
$$

and

$$
[(\varphi(\varphi(1)), 1)]_Z = \{(m, n) \in \mathbb{N} \times \mathbb{N} \mid \varphi(\varphi(1)) + n = 1 + m\}.
$$

There is no doubt, this insight is momentous.

Our next objective in this section is to define and explore the natural order on $\mathbb{Z}$: for integers $x = [(m, n)]_\mathbb{Z}$ and $y = [(k, l)]_\mathbb{Z}$ we say that x is strictly less than y if $m + l < k + n$ (see Theorem 4.2.5) or, in logic notation,

$$x < y :\Leftrightarrow m + l < k + n. \tag{6.4}$$

Furthermore, we say that x is less than or equal to y if $x < y$ or $x = y$:

$$x \leq y :\Leftrightarrow (x < y \lor x = y).$$

In order to show that '$<$' is well defined, we need to prove that $m' + l' < k' + n'$ whenever $m + l < k + n$, $[(m.n)]_\mathbb{Z} = [(m', n')]_\mathbb{Z}$, and $[(k, l)]_\mathbb{Z} = [(k', l')]_\mathbb{Z}$, that is, whenever $m + l < k + n$, $m + n' = m' + n$, and $k + l' = l + k'$. So let us assume that $m + l < k + n$, $m + n' = m' + n$, and $k + l' = l + k'$. Then $m + l + n' + k' < k + n + n' + k'$ (by Proposition 4.2.10b) and therefore, $m' + n + l' + k < k + n + n' + k'$ (because $m + n' = m' + n$ and $k + l' = l + k'$). Thus $m' + l' < n' + k'$, as desired (by Proposition 4.2.10d).

Remark. The relation "$<$' is strictly reflexive in the sense that $x < y \Rightarrow x \neq y$ for all $x, y \in \mathbb{Z}$. This is so because if $x = [(m.n)]_\mathbb{Z}$, $y = [(k, l)]_\mathbb{Z}$, and $m + l < k + n$, then $m + l \neq k + n$ (by Theorem 4.2.5), and therefore, $[(m, n)]_\mathbb{Z} \neq [(k, l)]_\mathbb{Z}$. Furthermore, for all $x = [(m, n)]_\mathbb{Z} \in \mathbb{Z}$ it is the case that

$$\begin{aligned} x < 0 &\text{ if and only if } m < n, \\ x = 0 &\text{ if and only if } m = n, \text{ and} \\ 0 < x &\text{ if and only if } n < m. \end{aligned} \tag{6.5}$$

For if $m = n$, then $x = 0 = [(1, 1)]_\mathbb{Z}$ (by Lemma 6.1.1), and if $m < n$, then

$$\begin{aligned} 0 = [(1, 1)]_\mathbb{Z} &= [(m + n, n + m)]_\mathbb{Z} \quad \text{(by Lemma 6.1.1)} \\ &= [(m, n)]_\mathbb{Z} + [(n, m)]_\mathbb{Z} = x + [(n, m)]_\mathbb{Z}, \end{aligned}$$

and, by implication, $x < 0$, as claimed (because $m < n$). Conversely, if $x = 0 = [(1, 1)]_\mathbb{Z}$, then $m + 1 = 1 + n$, and therefore, $m = n$ (by Lemma 4.2.4); and if $x < 0 = [(1, 1)]_\mathbb{Z}$, then $m + 1 < 1 + n$ (by (6.4)), and therefore, $m < n$ (by Proposition 4.2.10d). Finally, since the case $n < m$ is completely analogous to the case $m < n$, we may agree that (6.5) is indeed valid.

6.1.3 Lemma. *For all $x, y \in \mathbb{Z}$ it is the case that $x < y$ if and only if there exists a $z \in \mathbb{Z}$ such that $0 < z$ and $x + z = y$.*

Proof. If $x = [(m, n)]_\mathbb{Z}$, $y = [(k, l)]_\mathbb{Z}$, and $x < y$, then $m + l < k + n$, and therefore, $m + l + i = k + n$ for some $i \in \mathbb{N}$ (by Theorem 4.2.5). Thus $m + l + i + 1 = k + n + 1$ (by Proposition 4.2.10b), and in setting $z := [(i + 1, 1)]_\mathbb{Z}$, it follows that $x + z = [(m + i + 1, n + 1)]_\mathbb{N} = [(k, l)]_\mathbb{N} = y$ and $0 < z$ (by (6.5)) because $1 < i + 1$). Conversely, if $x + z = y$ for some $z = [(i, j)]_\mathbb{Z} \in \mathbb{Z}$ with $j < i$, then $m + i + l = k + n + j$ and $j + p = i$ for some $p \in \mathbb{N}$. Consequently, $m + j + p + l = k + n + j$, and therefore, $m + p + l = k + n$ (by Lemma 4.2.4). Hence $m + l < k + n$ (by Theorem 4.2.5), and, by implication, $x = [(m, n)]_\mathbb{Z} < [(k, l)]_\mathbb{Z} = y$, as claimed. $\qquad\square$

To proceed, we turn to the problem of establishing that both the strictly-less-than and the less-than-or-equal-to relations are linear orders on $\mathbb{Z}$. According to Lemma 4.2.3, it is sufficient to prove this claim only with respect to the strictly-less-than relation. To do so, we observe that '$<$' is transitive, because if $x + [(i, j)]_\mathbb{Z} = y$ and $y + [(k, l)]_\mathbb{Z} = z$ for some $i, j, k, l \in \mathbb{N}$ with $j < i$ and $l < k$ (see Lemma 6.1.3), then

$$z = x + [(i, j)]_\mathbb{Z} + [(k, l)]_\mathbb{Z} = x + [(i + k, j + l)]_\mathbb{Z},$$

and therefore, $x < z$ because $j + l < i + k$ (see Exercise 6.1.8). In order to show that '$<$' is antisymmetric as well, we assume that we are given integers $x = [(m, n)]_\mathbb{Z}$ and $y = [(k, l)]_\mathbb{Z}$ such that $x < y$ and $y < x$. Then, in contradiction to the last statement in Theorem 4.2.5, we find that $m + l < k + n$ and $k + n < m + l$. Consequently, $x < y \wedge y < x$ is false, and this proves that '$<$' is indeed antisymmetric. Finally, since '$<$' is a linear order on $\mathbb{N}$, we may infer that for all $x = [(m, n)]_\mathbb{Z}$ and $y = [(k, l)]_\mathbb{N}$ it is the case that $m + l < k + n \vee k + n < m + l \vee m + l = k + n$. Hence it also is the case that $x < y \vee y < x \vee x = y$, and therefore, '$<$' is a linear order on $\mathbb{Z}$.

Furthermore, in the light of (6.5), we may naturally identify $\mathbb{N}$ with the positive integers

$$\mathbb{Z}_+ = \{[(m, n)]_\mathbb{Z} \mid n < m\} = \{[(n + 1, 1)]_\mathbb{Z} \mid n \in \mathbb{N}\}.$$

That is to say, the natural identification map

$$\begin{aligned} i : \mathbb{N} &\to \mathbb{Z}_+ \\ n &\mapsto [(n + 1, 1)]_\mathbb{Z} \end{aligned} \tag{6.6}$$

is a bijection that preserves addition and thereby also subtraction, multiplication, and the strictly-less-than relation (i.e., $i(m + n) = i(m) + i(n)$, $i(mn) = i(m)i(n)$, $m < n \Rightarrow i(n - m) = i(n) - i(m)$, and $m < n \Leftrightarrow i(m) < i(n)$ for all $m, n \in \mathbb{N}$ (see Exercise 6.1.9)). Consequently, it will be permissible from here on onwards to simply consider $\mathbb{N}$ to be equal to the subset $\mathbb{Z}_+$ of $\mathbb{Z}$:

$$\boxed{\mathbb{N} = \{n \in \mathbb{Z} \mid 0 < n\} = \mathbb{Z}_+ \subset \mathbb{Z}.}$$

To complete the present section, we add the following propositions for later reference:

6.1.4 Proposition. *For all $x, y \in \mathbb{Z}$ the following statements are valid:*

a) $0 < xy$ if and only if $0 < x \wedge 0 < y$ or $x < 0 \wedge y < 0$,

b) $xy < 0$ if and only is $0 < x \wedge y < 0$ or $x < 0 \wedge 0 < y$.

Proof. Since '$<$' is a linear order, it follows that $0 < x \vee x < 0 \vee 0 = x$ and $0 < y \vee y > 0 \vee 0 = y$ for all $x, y \in \mathbb{Z}$. So there are a total of nine pairings as each of the three options for x—positive, negative, or zero—can be paired up with one of the three corresponding options for y. In five of these pairings, either x or y equals zero, and therefore, so does xy (by Theorem 6.1.2h). Consequently, it is actually sufficient to show that $0 < xy$ whenever $0 < x \wedge 0 < y$ or $x < 0 \wedge y < 0$ and that $xy < 0$ whenever $0 < x \wedge y < 0$ or $x < 0 \wedge 0 < y$. That is to say, the fact that the nine pairings just referred to are exhaustive implies that the whenever-statements in the previous sentence are in fact assertions of equivalence. For if, for example, $0 < xy$, then the five pairings in which either $x = 0$ or $y = 0$ can be excluded, because xy would equal zero in each of them, and the two pairings listed in b) can be excluded as well because xy would be negative in each of them. So the assumption $0 < xy$ is here indeed seen to imply that $0 < x \wedge 0 < y$ or $x < 0 \wedge y < 0$, and similarly, the assumption $xy < 0$ implies that $0 < x \wedge y < 0$ or $x < 0 \wedge 0 < y$. To proceed, we will prove only that $0 < x \wedge 0 < y$ implies that $0 < xy$ and leave it to the reader to show that $(x < 0 \wedge y < 0) \Rightarrow 0 < xy$ and $((0 < x \wedge y < 0) \vee (x < 0 \wedge 0 < y)) \Rightarrow xy < 0$ (see Exercise 6.1.10). So let us assume that we are given integers $x, y \in \mathbb{Z}$ such that $0 < x \wedge 0 < y$. Then there are elements $m, n, k, l \in \mathbb{N}$ such that $x = [(m, n)]_\mathbb{Z}$, $y = [(k, l)]_\mathbb{Z}$, $n < m$, and $l < k$. Consequently, Lemma 6.1.1 allows us to infer that

$$
\begin{aligned}
xy &= [(m - n + 1, 1)]_\mathbb{Z}[(k - l + 1, 1)]_\mathbb{Z} \\
&= [((m - n + 1)(k - l + 1) + 1, m - n + 1 + k - l + 1)]_\mathbb{Z} \\
&= [((m - n)(k - l) + m - n + k - l + 1 + 1, m - n + k - l + 1 + 1)]_\mathbb{Z} \\
&= [((m - n)(k - l) + 1, 1)]_\mathbb{Z}.
\end{aligned}
$$

Hence $0 < xy$ because $1 < (m - n)(k - l) + 1$ (and this in turn is so because $(m - n)(k - l) \in \mathbb{N}$). $\qquad\square$

6.1.5 Proposition. The following statements are true for all $x, y \in \mathbb{Z}$:

a) $x < y \Leftrightarrow 0 < y - x$ and $x \le y \Leftrightarrow 0 \le y - x$,

b) $x < y \Rightarrow \forall_{z \in \mathbb{Z}}\, x + z < y + z$,

c) $x \le y \Rightarrow \forall_{z \in \mathbb{Z}}\, x + z \le y + z$,

d) $(\exists_{z \in \mathbb{Z}}\, x + z < y + z) \Rightarrow x < y$,

e) $(\exists_{z \in \mathbb{Z}}\, x + z \le y + z) \Rightarrow x \le y$,

f) $(y \le x \le y + 1) \Rightarrow (x = y \vee x = y + 1)$,

g) $\forall_{z \in \mathbb{Z}}\, 0 < z \Rightarrow x < x + z$,

h) $\forall_{z \in \mathbb{Z}} \, 0 \le z \Rightarrow x \le x + z$,

i) $x < y \Leftrightarrow -y < -x$,

j) $x \le y \Leftrightarrow -y \le -x$,

k) $x < y \Rightarrow \forall_{z \in \mathbb{Z}} \, 0 < z \Rightarrow xz < yz$,

l) $x < y \Rightarrow \forall_{z \in \mathbb{Z}} \, z < 0 \Rightarrow yz < xz$,

m) $x \le y \Rightarrow \forall_{z \in \mathbb{Z}} \, 0 \le z \Rightarrow xz \le yz$,

n) $x \le y \Rightarrow \forall_{z \in \mathbb{Z}} \, z \le 0 \Rightarrow yz \le xz$.

o) $(\exists_{z \in \mathbb{Z}} \, 0 < z \wedge xz < yz) \Rightarrow x < y$,

p) $(\exists_{z \in \mathbb{Z}} \, z < 0 \wedge yz < xz) \Rightarrow x < y$,

q) $(\exists_{z \in \mathbb{Z}} \, 0 < z \wedge xz \le yz) \Rightarrow x \le y$,

r) $(\exists_{z \in \mathbb{Z}} \, z < 0 \wedge yz \le xz) \Rightarrow x \le y$.

Proof. We will prove a), b), k), and o) and leave the proofs of the remaining statements as exercises to the reader (see Exercise 6.1.11). To establish a), we apply Theorem 6.1.2a,b,c,d, (6.5), and Lemma 6.1.3 to infer that for all $x, y \in \mathbb{Z}$ it is the case that

$$x < y \Leftrightarrow \exists_{u \in \mathbb{Z}} \, 0 < u \wedge x + u = y \Leftrightarrow \exists_{u \in \mathbb{Z}} \, 0 < u \wedge 0 + u = y - x \Leftrightarrow 0 < y - x,$$

and

$$x \le y \Leftrightarrow (x < y \vee x = y) \Leftrightarrow (0 < y - x \vee 0 = y - x) \Leftrightarrow 0 \le y - x,$$

as desired. Moving on to b), we assume that we are given integers $x, y, z \in \mathbb{Z}$ such that $x < y$. Then a) and Theorem 6.1.2 imply that $0 < y - x = y + z - z - x = (y + z) - (x + z)$ and another application of a) therefore allows us to conclude that $x + z < y + z$. To prove k), we assume that we are given integers $x, y, z \in \mathbb{Z}$ such that $0 < z$ and $x < y$. Then $0 < y - x$ (by a)), and therefore, Proposition 6.1.4a in conjunction with Theorem 6.1.2 implies that $0 < (y - x)z = yz - xz$. Hence $xz < yz$ by b). Finally, in order to establish o), we assume that we are given integers $x, y, z \in \mathbb{Z}$ such that $0 < z$ and $xz < yz$. Then $0 < yz - xz = (y - x)z$ (by a) and Theorem 6.1.2), and therefore, $0 < y - x$ (by Proposition 6.1.4a because $0 < z$). Thus $x < y$ (by a)), as desired. $\qquad\square$

Exercises

6.1.6. Show that the operations $x \mapsto -x$ and $(x, y) \mapsto xy$ in (6.2) and (6.3) are well defined.

6.1.7. Prove Theorem 6.1.2a,b,c,d,e,f,j.

6.1.8. Let $i, j, k, l \in \mathbb{N}$ such that $j < i$ and $l < k$. Show that $j + l < i + k$.

6.1.9. Show that the natural identification map i, defined in (6.6), is bijective and satisfies the following properties for all $m, n \in \mathbb{N}$:

   a) $i(m + n) = i(m) + i(n)$,

   b) $i(mn) = i(m)i(n)$,

   c) $m < n \Rightarrow i(n - m) = i(n) - i(m)$, and

   d) $m < n \Leftrightarrow i(m) < i(n)$.

6.1.10. Complete the proof of Theorem 6.1.4.

6.1.11. Prove Proposition 6.1.5c,d,e,f,g,h,i,j,l,m,n,p,q,r.

6.2   The Rationals

Concerning the construction of the rational numbers from the integers, we already did some preparatory work in Chapter 3 where we showed, in Example 3.5.15, that

$$r_{\mathbb{Q}} := \{((m, n), (p, q)) \in (\mathbb{Z} \times \mathbb{N}) \times (\mathbb{Z} \times \mathbb{N}) \mid mq = np\}$$

is an equivalence relation on $\mathbb{Z} \times \mathbb{N}$. Furthermore, in Example 3.6.3 we went on to suggest that the set of rational numbers ought to be defined to be the set of equivalence classes of this equivalence relation, that is,

$$\boxed{\mathbb{Q} := \{[(m, n)]_{\mathbb{Q}} \mid (m, n) \in \mathbb{Z} \times \mathbb{N}\}.}$$

In order to properly describe the equivalence classes $[(m, n)]_{\mathbb{Q}}$, we need to discuss some elementary divisibility properties of the positive integers. To this end, we recall our discussion in Example 1.5.4 and agree to say that $m \in \mathbb{N}$ is a *divisor* of $n \in \mathbb{N}$—written as $m|n$—if there is a $k \in \mathbb{N}$ such that $n = km$. Using quantification notation, we can write this definition as follows:

$$m|n :\Leftrightarrow \exists_{k \in \mathbb{N}} \, n = mk.$$

Furthermore, the *greatest common divisor* of m and n we denote by $\gcd(m, n)$, that is,

$$\gcd(m, n) := \max\{k \in \mathbb{N} \mid k|m \wedge k|n\}.$$

Note: the definition of divisibility above implies that $1|n$ for all $n \in \mathbb{N}$ because $1n = n$, and this observation in turn implies that the set of common divisors of two given integers $m, n \in \mathbb{N}$ can never be empty. Moreover, since $k \leq kj$ for all $k, j \in \mathbb{N}$ (by Exercise 4.4.4 and Proposition 6.1.5m), we may conclude that a divisor is always less than or equal to the number it divides. Hence

$$1 \leq \gcd(m, n) \leq \min\{m, n\}.$$

6.2.1 Lemma. *If $m|n$ for some $m, n \in \mathbb{N}$, then the integer $k \in \mathbb{N}$ that satisfies the equation $n = km$ is uniquely determined by m and n.*

Proof. If $n = k_1 m = k_2 m$ for some $k_1, k_2 \in \mathbb{N}$, then $0 = (k_1 - k_2)n$ (by Theorem 6.1.2c,h,j). Hence $k_1 - k_2 = 0$ by Theorem 6.1.2g because $n \neq 0$, and therefore, $k_1 = k_2$ (by Theorem 6.1.2a,b,c), as desired. $\square$

In the light of this lemma, it is meaningful and feasible to denote by $k(m, n)$ the unique integer in $\mathbb{N}$ that satisfies the equation

$$mk(m, n) = n$$

whenever $m, n \in \mathbb{N}$ and $m|n$.

6.2.2 Theorem. *(The Euclidean Algorithm) For all integers $m \in \mathbb{N}$ and $n \in \mathbb{Z}$ there exist unique integers $p, q \in \mathbb{Z}$ such that $0 \leq q \leq m - 1$, and $n = pm + q$.*

Proof. In order to prove uniqueness, we assume that $p_1 m + q_1 = p_2 m + q_2$ for some integers $p_1, p_2, q_1, q_2 \in \mathbb{Z}$ that satisfy the inequalities $0 \leq q_1, q_2 \leq m - 1$, and, w.l.o.g., $q_1 \leq q_2$. Then $0 \leq q_2 - q_1 \leq q_2 \leq m - 1$ (by Proposition 6.1.5a,c) and $(p_1 - p_2)m = p_1 m - p_2 m = q_2 - q_1$. Consequently, if it were the case that q_1 is strictly less than q_2, then $q_2 - q_1$ would be strictly greater than zero and so would be $p_1 - p_2$ by implication (by Propositions 6.1.5a and 6.1.4). Thus we would find that $m \leq (p_1 - p_2)m = q_2 - q_1 \leq m - 1$ (by Proposition 6.1.5m). Since this is impossible, we may infer that $q_1 = q_2$, and this in turn implies that $0 = (p_1 - p_2)m$. Since $m \neq 0$ (because $m \in \mathbb{N}$), we may apply Theorem 6.1.2g to deduce that $p_1 - p_2 = 0$ or, equivalently, that $p_1 = p_2$. Hence p and q are indeed uniquely determined by the conditions $0 \leq q \leq m-1$ and $n = pm+q$. In order to establish the existence of p and q, we assume to begin with that $1 \leq n$ and observe that in this case $n \leq mn$ (by Proposition 6.1.5m). Hence the set $M := \{k \in \mathbb{N} \mid n \leq mk\}$ is nonempty, and using the Well Ordering Principle, it follows that $k_0 := \min M$ is well defined. Since this definition of k_0 implies in particular that $k_0 \in M$, we may infer that $n \leq mk_0$. If $n = mk_0$, then the statement of the theorem is satisfied for $p := k_0$ and $q := 0$. Alternatively, if $n < mk_0$, then there are again two possibilities: if $k_0 = 1$, then $n < m$ and the statement of the theorem is satisfied for $p = 0$ and $q = n$. Finally, if $1 < k_0$, then $k_0 - 1 \in \mathbb{N}$, and therefore, $m(k_0 - 1) < n$ (because $k_0 - 1 < k_0$ and $k_0 = \min M$). Consequently, in setting $p := k_0 - 1$ and $q := n - pm$ it follows that $n = pm+q$ and that $0 < 1 \leq q$ (because $pm = m(k_0-1) < n$ implies that $n - pm \in \mathbb{N}$). Hence we only need to show that $q < m$. Using the inequality $n < mk_0 = m(p + 1) = pm + m$ in conjunction with Proposition 6.1.5b, we find that

$$q = n - pm < pm + m - pm = m,$$

as desired. Furthermore, if $n \leq -1$, then $1 \leq -n$, and therefore, there are integers $p_0, q_0 \in \mathbb{Z}$ such that $0 \leq q_0 \leq m-1$ and $-n = p_0 m + q_0$ or, equivalently,

$$n = -(p_0 m + q_0) = (-1)(p_0 m + q_0) = (-p_0)m + (-q_0).$$

If $q_0 = 0$, then $-q_0 = 0$, and therefore, $n = pm+q$ for $p := -p_0$ and $q := 0$, and we are done. Alternatively, if $0 < q_0$, then we set $p := -p_0 - 1$ and $q := m - q_0$ and observe that

$$pm + q = (-p_0 - 1)m + m - q_0 = (-p_0)m + (-q_0) = n.$$

Thus in this latter case we only need to show that $0 \le q < m$. Using Proposition 6.1.5i,j in conjunction with the fact that $0 < q_0 \le m - 1$, we may infer that $-m + 1 = -(m - 1) \le -q_0 < -0 = 0$, and therefore, $0 < 1 = (-m+1)+m \le -q_0 + m = q < 0+m = m$ (by Proposition 6.1.5b,c), as desired. Finally, if $n = 0$, then $n = pm + q$ for $p := q := 0$, and again we are done.   □

Remark. The reason why the preceding theorem is referred to as the Euclidean *Algorithm* is that it gives rise to an algorithmic method for the computation of greatest common divisors. For instance, if we wish to determine the greatest common divisor of 294 and 1400—which is easily seen to be 14—we repeatedly apply Theorem 6.2.2 to generate the following sequence of divisions with a remainder:

$$1400 = 4 \cdot 294 + 224$$
$$294 = 1 \cdot 224 + 70$$
$$224 = 3 \cdot 70 + 14$$
$$70 = 5 \cdot 14 + 0.$$

So what we see here is that the greatest common divisor—in this case 14—is the first divisor that produces the remainder zero. In order to understand why this algorithmic procedure always works, we may argue as follows: starting with two integers $m, n \in \mathbb{N}$, we apply the Euclidean Algorithm to infer that

$$n = p_1 m + q_1$$
$$m = p_2 q_1 + q_2$$
$$q_1 = p_3 q_2 + q_3$$
$$\vdots$$

where $q_1 < m$, $q_2 < q_1$, $q_3 < q_2$, etc. So the sequence of the remainders q_i is strictly decreasing, and therefore, there must exist a k such that $q_k = 0$. If $k = 1$, then $m | n$ and $\gcd(m, n) = m$, as desired. In order to see what happens if $k > 1$, we observe that the equation $n = p_1 m + q_1$ trivially implies that every common divisor of m and n is also a common divisor of m and q_1 and vice versa. So the pairs (m, n) and (m, q_1) have exactly the same common divisors, and this in turn implies that $\gcd(m, n) = \gcd(m, q_1)$. Similarly, we find that $\gcd(m, q_1) = \gcd(q_1, q_2) = \gcd(q_2, q_3) = \cdots = \gcd(q_{k-2}, q_{k-1})$. But since $q_k = 0$, it follows that $q_{k-2} = p_k q_{k-1}$. Hence $q_{k-1} | q_{k-2}$, and therefore, $\gcd(m, n) = \gcd(q_{k-2}, q_{k-1}) = q_{k-1}$, as claimed.

6.2.3 Proposition. *Setting* $I(m, n) := \{xm + yn \mid x, y \in \mathbb{Z}\}$ *for any given integers* $m, n \in \mathbb{N}$, *it is the case that* $I(m, n) \cap \mathbb{N} \neq \emptyset$ *and*

$$\gcd(m, n) = \min I(m, n) \cap \mathbb{N}.$$

Proof. If $m, n \in \mathbb{N}$ are given positive integers, then $m = 1m + 0n \in I(m, n) \cap \mathbb{N}$, and therefore, $I(m, n) \cap \mathbb{N} \neq \emptyset$. Consequently, $k_0 := \min I(m, n) \cap \mathbb{N}$ is well defined by the Well Ordering Principle. Since the definition of k_0 implies that $k_0 \in I(m, n)$, there are integers $x_0, y_0 \in \mathbb{Z}$ such that $k_0 = x_0 m + y_0 n$, and therefore, $jk_0 = jx_0 m + jy_0 n \in I(m, n)$ for all $j \in \mathbb{Z}$. This shows that $\{jk_0 \mid j \in \mathbb{Z}\} \subset I(m, n)$. In order to prove that the reverse inclusion relation is valid as well, we pick a value $xm + yn \in I(m, n)$ and use the Euclidean Algorithm to infer that there are integers $p, q \in \mathbb{Z}$ such that $xm + yn = pk_0 + q$ and $0 \leq q \leq k_0 - 1$. Hence $0 \leq q = (x - px_0)m + (y - py_0)n \in I(m, n)$, and therefore, $q = 0$ because $q < k_0 = \min(I(m, n) \cap \mathbb{N})$. Thus $xm + yn = pk_0 \in \{jk_0 \mid j \in \mathbb{Z}\}$, and therefore, $I(m, n) = \{jk_0 \mid j \in \mathbb{Z}\}$. Since $m, n \in I(m, n) \cap \mathbb{N}$, it follows that there are $i, j \in \mathbb{N}$ such that $m = ik_0$ and $n = jk_0$, and this shows that k_0 is a common divisor of m and n. Consequently, in order to show that k_0 is the greatest common divisor of m and n, we only need to show that any given common divisor l of m and n is necessarily less than or equal to k_0. So let $l \in \mathbb{N}$ be a common divisor of m and n. Then there are integers $a, b \in \mathbb{N}$ such that $m = al$ and $n = bl$. Hence $k_0 = x_0 m + y_0 n = (ax_0 + by_0)l$. Since this equation implies that $0 < ax_0 + by_0$ or, equivalently, that $1 \leq ax_0 + by_0$ (by Proposition 6.1.4 because $k_0, l \in \mathbb{N}$), we may apply Proposition 6.1.5m to infer that indeed $l \leq k_0$. $\qquad\square$

Before we can use the preceding proposition to describe the equivalence classes of $r_{\mathbb{Q}}$, we need to introduce one more definition: for a given $x \in \mathbb{Z}$, the *absolute value* of x is

$$|x| := \begin{cases} x & \text{if } 0 \leq x, \\ -x & \text{if } x < 0. \end{cases} \tag{6.7}$$

According to Proposition 6.1.5i, this definition readily implies that $0 \leq |x|$ for all $x \in \mathbb{Z}$ and that $0 < |x|$ if and only if $x \neq 0$.

6.2.4 Lemma. *Let* $m \in \mathbb{Z}$ *and* $n \in \mathbb{N}$ *be given integers. If* $m = 0$, *then*

$$[(m, n)]_{\mathbb{Q}} = \{(0, j) \mid j \in \mathbb{N}\},$$

and if $m \neq 0$, *then, in setting* $a := k(\gcd(|m|, n), |m|)$ *and* $b := k(\gcd(|m|, n), n)$, *we find that*

$$[(m, n)]_{\mathbb{Q}} = \begin{cases} \{(ja, jb) \mid j \in \mathbb{N}\} & \text{if } 0 < m, \\ \{(-ja, jb) \mid j \in \mathbb{N}\} & \text{if } m < 0. \end{cases}$$

Proof. Since $((m, n), (p, q)) \in r_{\mathbb{Q}}$ if and only if $mq = np$, it follows that $((m, n), (p, q)) \in r_{\mathbb{Q}}$ if and only if $0 = np$ whenever $m = 0$. However, since $n \neq 0$ (because $n \in \mathbb{N}$), the equation $0 = np$ is satisfied if and only if $p = 0$ (by Theorem 6.1.2g), and therefore, $[(0, n)]_{\mathbb{Q}} = \{(0, q) \mid q \in \mathbb{N}\}$, as desired. To proceed we consider the case $0 < m$. In order to show that $[(m, n)]_{\mathbb{Q}} \subset \{(ja, jb) \mid j \in \mathbb{N}\}$, we pick a pair $(p, q) \in [(m, n)]_{\mathbb{Q}}$. Then $\gcd(m, n)aq = mq = np = \gcd(m, n)bq$, and therefore, $\gcd(m, n)(aq - bp) = 0$. Since $\gcd(m, n) \in \mathbb{N}$, we may again apply Theorem 6.1.2g to infer that $aq - bp = 0$, or equivalently, $aq = bp$. If $c \in \mathbb{N}$ is a common divisor of a and b, then $\gcd(m, n)c$ is a common divisor of m and n, and therefore, $\gcd(m, n)c \leq \gcd(m, n)$. Thus $c = 1$, because if 1 were strictly less than c, then Proposition 6.1.5l would imply that $\gcd(m, n) < \gcd(m, n)c$. Hence $\gcd(a, b) = 1$, and according to Proposition 6.2.3, it therefore follows that there are integers $x, y \in \mathbb{Z}$ such that $1 = xa + yb$. Consequently, in setting $j := xp + yq$, we find that $q = xaq + ybq = xbp + ybq = b(xp + yq) = jb$, and $p = xap + ybp = xap + yaq = a(xp + yq) = ja$, and this proves that $(p, q) \in \{(ja, jb) \mid j \in \mathbb{N}\}$, as desired. Conversely, if $(p, q) = (ja, jb)$ for some $j \in \mathbb{N}$, then $mq = \gcd(m, n)ajb = \gcd(m, n)bja = np$, and therefore, $(p, q) \in [(m, n)]_{\mathbb{Q}}$. Thus $[(m, n)]_{\mathbb{Q}} = \{(ja, jb) \mid j \in \mathbb{N}\}$, as claimed. Given this result, it further follows that $[(|m|, n)]_{\mathbb{Q}} = \{(ja, jb) \mid j \in \mathbb{N}\}$ whenever $m < 0$. Thus it is sufficient to show that

$$(p, q) \in [(m, n)]_{\mathbb{Q}} \Leftrightarrow (-p, q) \in [(|m|, n)]_{\mathbb{Q}}$$

whenever $m < 0$. So let m be strictly less than zero. Then $-m = |m|$, and therefore,

$$(-p, q) \in [(|m|, n)]_{\mathbb{Q}} \Leftrightarrow |m|q = -pn \Leftrightarrow -|m|q = np \Leftrightarrow mq = np$$
$$\Leftrightarrow (p, q) \in [(m, n)]_{\mathbb{Q}},$$

as desired. $\square$

As we now introduce the standard quotient notation

$$\frac{m}{n} := [(m, n)]_{\mathbb{Q}},$$

we may apply Lemma 6.2.4 to infer that

$$\frac{m}{n} = \frac{k(\gcd(m, n), m)}{k(\gcd(m, n), n)}$$

for all $m, n \in \mathbb{N}$. In particular, if $m|n$, then $\gcd(m, n) = m$, and therefore,

$$\frac{m}{n} = \frac{k(m, n)}{1}. \tag{6.8}$$

To proceed, we define the standard operations of sign inversion, addition, subtraction, multiplication, and division as follows:

$$-\frac{m}{n} := \frac{-m}{n},$$

$$\frac{m}{n} + \frac{p}{q} := \frac{mq + np}{nq},$$

$$\frac{m}{n} - \frac{p}{q} := \frac{m}{n} + \left(-\frac{p}{q}\right) = \frac{mq - np}{nq},$$

$$\frac{m}{n} \cdot \frac{p}{q} := \frac{mp}{nq},$$

for all $(m, n), (p, q) \in \mathbb{Z} \times \mathbb{N}$, and if $p \neq 0$, then

$$\frac{m/n}{p/q} := \begin{cases} \frac{m}{n} \cdot \frac{q}{p} = \frac{mq}{np} & \text{if } p > 0, \\ -\frac{m}{n} \cdot \frac{q}{|p|} = \frac{-mq}{n|p|} & \text{if } p > 0. \end{cases}$$

In order to show that these operations are well defined, we assume that we are given integers $m', p' \in \mathbb{Z}$ and $n', q' \in \mathbb{N}$ such that $m/n = m'/n'$ and $p/q = p'/q'$. Using Lemma 6.2.4, we may infer that there are integers $a, b \in \mathbb{Z}$ and $c, d, i, i', j, j' \in \mathbb{N}$ such that $m = ia$, $n = ib$, $m' = i'a$, $n' = i'b$, $p = jc$, $q = jd$, $p' = j'c$, and $q' = j'd$. Hence

$$\frac{m}{n} + \frac{p}{q} = \frac{mq + np}{nq} = \frac{ij(ad + bc)}{ijbc}$$

and

$$\frac{m'}{n'} + \frac{p'}{q'} = \frac{m'q' + n'p'}{n'q'} = \frac{i'j'(ad + bc)}{i'j'bc}.$$

Since evidently

$$ij(ad + bc)i'j'bc = i'j'(ad + bc)ijbc,$$

it follows that

$$\frac{m}{n} + \frac{p}{q} = \frac{m'}{n'} + \frac{p'}{q'},$$

and this shows that the operation of addition is well defined. As in the previous section, we will leave it to the reader to prove that the remaining operations of sign inversion, subtraction, multiplication, and division are well defined as well (see Exercise 6.2.8).

Furthermore, the statement

$$\frac{m}{n} < \frac{p}{q} :\Leftrightarrow mq < np$$

defines a linear order on $\mathbb{Q}$, and according to Lemma 4.2.3, so does the statement

$$\frac{m}{n} \leq \frac{p}{q} :\Leftrightarrow \left(\frac{m}{n} < \frac{p}{q} \vee \frac{m}{n} = \frac{p}{q} \right).$$

In order to prove that the strictly-less-than relation is well defined, we need to show that $m'q' < n'p'$ whenever $m/n = m'/n'$, $p'/q' = p/q$, and $mq < np$. If $m/n = m'/n'$ and $p'/q' = p/q$ then, according to Lemma 6.2.4, there are integers $a, c \in \mathbb{Z}$ and $b, d, j, j', k, k' \in \mathbb{N}$ such that $m/n = (ja)/(jb)$, $m'/n' = (j'a)/(j'b)$, $p/q = (kc)/(kd)$, and $p'/q' = (k'c)/(k'd)$. (Note: a is positive, zero, or negative if both m and m' are positive, zero, or negative, and the same is true for c, p, and p'. So Lemma 6.2.4 implies, in particular, that m and m' and also p and p' have the same sign.) Thus the inequality $mq < np$ is equivalent to $jakd < jbkc$, and according to Theorem 6.1.5o, this latter inequality allows us to infer that $ad < bc$ (because $jk \in \mathbb{N}$ implies that $jk > 0$). Using Theorem 6.1.5k, it follows that $j'ak'd < j'bk'c$ because $j'k' > 0$, and therefore, $m'q' < n'p'$, as desired.

Turning to the problem of establishing that '$<$' is a linear order, we observe that '$<$' is antisymmetric because the conjunction $mq < np \wedge np < mq$ is false and because, by implication, the conjunction $m/n < p/q \wedge p/q < m/n$ is false as well. Regarding transitivity, we notice that the assumption $m/n < p/q \wedge p/q < x/y$ is equivalent to $mq < np \wedge py < qx$. Consequently, using Proposition 6.1.5k, we find that $mqy < npy < nqx$, and therefore, $0 < q(nx - my)$ (by Proposition 6.1.5a). Since $0 < q$ (because $q \in \mathbb{N}$), it follows that $0 < nx - my$ (by Proposition 6.1.4a). Hence $my < nx$ (by Proposition 6.1.5a), and this implies that $m/n < x/y$, as desired. Finally, since '$<$' is linear on $\mathbb{Z}$, we may infer that for given fractions $m/n, p/q \in \mathbb{Q}$ it is the case that $mq < np \vee np < mq \vee mq = np$, and therefore, it also is the case that

$$\frac{m}{n} < \frac{p}{q} \vee \frac{p}{q} < \frac{m}{n} \vee \frac{m}{n} = \frac{p}{q}.$$

Remark. The order relation '$<$' on $\mathbb{Q}$ is strictly reflexive in the sense that $x < y \Rightarrow x \neq y$ for all $x = m/n$ and $y = p/q$ in $\mathbb{Q}$. For if $x < y$, then $mq < pn$, and therefore $mq \neq pn$ (by the remark on p.162).

Moreover, in analogy to the identification of $\mathbb{N}$ with $\mathbb{Z}_+$, we can presently identify $\mathbb{Z}$ with the fractions of the form $m/1$. That is to say, the natural identification map

$$i : \mathbb{Z} \to \mathbb{Q}$$
$$m \mapsto \frac{m}{1}$$

is injective and satisfies the following statements for all $m, n \in \mathbb{Z}$ (see Exercise 6.2.9):

$$i(-m) = -i(m) \wedge i(m+n) = i(m) + i(n) \wedge i(mn) = i(m)i(n) \qquad (6.9)$$

and
$$m < n \Leftrightarrow i(m) < i(n). \tag{6.10}$$

In the light of this identification, it is natural to write m instead of $m/1$ for all $m \in \mathbb{Z}$ and we will do so from here on onwards.

6.2.5 Theorem. *For all $x, y, z \in \mathbb{Q}$ it is the case that*

    a) $x + (y + z) = (x + y) + z$,

    b) $x + 0 = x$ *(where $0 = 0/1 = 0/n = -0$ for all $n \in \mathbb{N}$)*,

    c) $x - x = -x + x = 0$,

    d) $x + y = y + x$,

    e) $x(yz) = (xy)z$,

    f) $xy = yx$,

    g) $x = 1 \cdot x$ *and* $-x = (-1) \cdot x$ *where* $1 = 1/1$,

    h) $(1/x)x = x/x = 1$ *whenever* $x \neq 0$,

    i) $xy = 0 \Leftrightarrow (x = 0 \lor y = 0)$,

    j) $x(y + z) = xy + xz$, *and*

    k) $x(y - z) = xy - xz$.

Proof. We will prove only h) and i) and leave the proofs of the remaining statements as exercises to the reader (see Exercise 6.2.10). To prove h), we assume that $x \neq 0$. Then $x = m/n$ for some $n \in \mathbb{N}$ and some $m \in \mathbb{Z} \smallsetminus \{0\}$. If $m > 0$, then $(1/x)x = (nm)/(mn) = 1$ (by (6.8)), and if $0 < m$, then $(1/x)x = (-nm)/(|m|n) = (n|m|)/(|m|n) = 1$ (again by (6.8)), as desired. In order to establish i), it is evidently sufficient to show that $x = 0 \Rightarrow xy = 0$ and that $(xy = 0 \land x \neq 0) \Rightarrow y = 0$. To do so we argue as follows: if $x = 0 = 0/1$ and $y = p/q$, then $xy = 0/(mq) = 0$, and therefore, if $xy = 0 \land x \neq 0$, then $y = 1y = ((1/x)x)y = (1/x)(xy) = (1/x)0 = 0$. $\qquad \square$

The next two propositions are almost completely analogous to Propositions 6.1.4 and 6.1.5, and their proofs will again be mostly left as exercises to the reader.

6.2.6 Proposition. *For all $x, y \in \mathbb{Q}$ the following statements are valid:*

    a) $0 < xy$ *if and only if $0 < x \land 0 < y$ or $x < 0 \land y < 0$*,

    b) $xy < 0$ *if and only is $0 < x \land y < 0$ or $x < 0 \land 0 < y$.*

Proof. We will prove a) and leave the proof of b) as an exercise to the reader (see Exercise 6.2.11). If $x = m/n > 0 = 0/1$ and $y = p/q > 0 = 0/1$ for some $m, p \in \mathbb{Z}$ and $n, q \in \mathbb{N}$, then $m = 1 \cdot m > 0 \cdot n = 0$, and similarly, $p > 0$. Hence $xy = (mp)/(nq) > 0$ because $mp > 0$ (by Proposition 6.1.4a). Furthermore, if

$x < 0$ and $y < 0$, then $m < 0$ and $p < 0$, and therefore, $xy = (mp)/(nq) > 0$ because $mp > 0$ (by Proposition 6.1.4a). To prove the reverse implication, we assume that $0/1 = 0 < xy = (mp)/(nq)$. Since this assumption is equivalent to the inequality $0 < mp$, we may apply Proposition 6.1.4a to infer that $0 < m \wedge 0 < p$ or $0 > m \wedge 0 > p$. In the former case it follows that $0 < x \wedge 0 < y$ and in the latter that $0 > x \wedge 0 > y$, as desired.   □

6.2.7 Proposition. The following statements are true for all $x, y \in \mathbb{Q}$:

a) $x < y \Leftrightarrow 0 < y - x$ and $x \leq y \Leftrightarrow 0 \leq y - x$,

b) $x < y \Rightarrow \forall_{z \in \mathbb{Q}} \, x + z < y + z$,

c) $x \leq y \Rightarrow \forall_{z \in \mathbb{Q}} \, x + z \leq y + z$,

d) $(\exists_{z \in \mathbb{Q}} \, x + z < y + z) \Rightarrow x < y$,

e) $(\exists_{z \in \mathbb{Q}} \, x + z \leq y + z) \Rightarrow x \leq y$,

f) $\forall_{z \in \mathbb{Q}} \, 0 < z \Rightarrow x < x + z$,

g) $\forall_{z \in \mathbb{Q}} \, 0 \leq z \Rightarrow x \leq x + z$,

h) $x < y \Leftrightarrow -y < -x$,

i) $x \leq y \Leftrightarrow -y \leq -x$,

j) $x < y \Rightarrow \forall_{z \in \mathbb{Q}} \, 0 < z \Rightarrow xz < yz$,

k) $x < y \Rightarrow \forall_{z \in \mathbb{Q}} \, z < 0 \Rightarrow yz < xz$,

l) $x \leq y \Rightarrow \forall_{z \in \mathbb{Q}} \, 0 < z \Rightarrow xz \leq yz$,

m) $x \leq y \Rightarrow \forall_{z \in \mathbb{Q}} \, z < 0 \Rightarrow yz \leq xz$.

n) $(\exists_{z \in \mathbb{Q}} \, 0 < z \wedge xz < yz) \Rightarrow x < y$,

o) $(\exists_{z \in \mathbb{Q}} \, z < 0 \wedge yz < xz) \Rightarrow x < y$,

p) $(\exists_{z \in \mathbb{Q}} \, 0 < z \wedge xz \leq yz) \Rightarrow x \leq y$,

q) $(\exists_{z \in \mathbb{Q}} \, z < 0 \wedge yz \leq xz) \Rightarrow x \leq y$.

r) $0 < x < y \Leftrightarrow 0 < 1/y < 1/x$,

s) $0 < x \leq y \Leftrightarrow 0 < 1/y \leq 1/x$,

t) $1 < x \Leftrightarrow 0 < 1/x < 1$,

u) $1 \leq x \Leftrightarrow 0 < 1/x \leq 1$,

v) $0 < x < y \Rightarrow 1 < y/x$.

w) $0 < x \leq y \Rightarrow 1 \leq y/x$.

Proof. We will prove a), b), j), n), and v) and leave the proofs of the remaining properties as exercises to the reader (see Exercise 6.2.12). Starting with a), we assume that $x = m/n$ and $y = p/q$ are given rational numbers. Then

$$x < y \Leftrightarrow mq < np \Leftrightarrow 0 < np - mq \Leftrightarrow 0 < \frac{np - mq}{nq} \Leftrightarrow 0 < y - x,$$

and the proof of the equivalence $x \leq y \Leftrightarrow 0 \leq y - x$ is an exact copy of the corresponding proof in the case of Proposition 6.1.5a. Furthermore, the proofs of b), j), and n) are completely analogous to the proofs of Proposition 6.1.5 b), k), and o), respectively, and to prove v), we assume that $0 < x < y$ for some $x, y \in \mathbb{Q}$ with $x = m/n$. Then $0 < m$, and, by implication, $0 < n/m = 1/x$. Thus, $1 = (1/x)x < (1/x)y = y/x$ (by j) and Theorem 6.2.5h). $\qquad\square$

To conclude this section, we wish to mention that the domain of the absolute value function, defined in (6.7), can be easily extended from $\mathbb{Z}$ to $\mathbb{Q}$:

$$|x| := \begin{cases} x & \text{if } 0 \leq x, \\ -x & \text{if } x < 0. \end{cases} \qquad (6.11)$$

for all $x \in \mathbb{Q}$. Furthermore, this function not only satisfies the triangle-type inequalities in Proposition 2.3.1 and Corollary 2.3.2, but also the equation

$$|xy| = |x||y| \qquad (6.12)$$

for all $x, y \in \mathbb{Q}$ as well as the equation

$$\left|\frac{x}{y}\right| = \frac{|x|}{|y|} \qquad (6.13)$$

whenever $y \neq 0$ (see Exercise 6.2.13).

Exercises

6.2.8. Show that the operations of sign inversion, subtraction, multiplication, and division on $\mathbb{Q}$ are well defined.

6.2.9. Prove (6.9) and (6.10) and explain why i is injective.

6.2.10. Prove Theorem 6.2.5a,b,c,d,e,f,g,j,k.

6.2.11. Prove Proposition 6.2.6b.

6.2.12. Prove Proposition 6.2.7b,c,d,f,g,h,i,j,k,l,n,p,q,r,s,t,u,v,w,x,z.

6.2.13. Prove the equations (6.12) and (6.13).

6.3 The Reals

In Section 2.3, p.49, we exhibited a sequence of rational numbers a_n that converges to the irrational limit $\sqrt{2}$. In doing so we highlighted the fact that the set of rational numbers $\mathbb{Q}$ is not, as we say, *complete*. For there are Cauchy sequences of rational numbers—such as the sequence $(a_n)_{n \in \mathbb{N}}$ here in question—that are not convergent in $\mathbb{Q}$ because the limit that they converge to happens to be an irrational number in $\mathbb{R} \setminus \mathbb{Q}$.

Speaking informally, this observation is the key that allows us to actually define the real numbers in such a way that the completeness condition just suggested—the convergence of all Cauchy sequences—is satisfied. To further clarify this idea, it is helpful to note (as on p.49) that real numbers are commonly represented by infinite decimal expansions:

$$\sqrt{2} = 1.4142\ldots,$$
$$\pi = 3.1415\ldots,$$
$$1/3 = 0.3333\ldots,$$
$$1/2 = 0.5000\ldots.$$

For in the light of this expandability, it is evidently possible to assign to every real number x a sequence $(a_n)_{n \in \mathbb{N}}$ in $\mathbb{Q}$ such that $\lim_{n \to \infty} a_n = x$ (see again the discussion on p.49). After all, if

$$x = m + 0.x_1 x_2 x_3 \ldots$$

for some $m \in \mathbb{Z}$ and $x_1, x_2, x_3, \cdots \in \{0, \ldots, 9\}$, then the sequence

$$a_1 := m + 0.x_1,$$
$$a_2 := m + 0.x_1 x_2,$$
$$a_3 := m + 0.x_1 x_2 x_3$$
$$\vdots$$

is rational and converges to x. Consequently, since every convergent sequence is in particular a Cauchy sequence (by Theorem 2.3.7), it may occur to us to simply define $\mathbb{R}$ to be the set of all Cauchy sequences of rational numbers. But here we need to be careful because a Cauchy sequence that converges to a real number x is not uniquely determined by x. That is to say, for any given x there can—and do—exist *distinct* rational Cauchy sequences $(a_n)_{n \in \mathbb{N}}$ and $(b_n)_{n \in \mathbb{N}}$ whose common limit is x. However, if a_n and b_n get closer and closer to the same number x as n increases to infinity, then the difference $a_n - b_n$ must converge to zero. By implication, it does seem feasible to identify the limit x of $(a_n)_{n \in \mathbb{N}}$ with the set of all rational Cauchy sequences $(b_n)_{n \in \mathbb{N}}$ for which it is the case that $\lim_{n \to \infty}(a_n - b_n) = 0$. In order to convert this idea into a rigorous definition of $\mathbb{R}$, we need to establish the following theorems:

6.3.1 Theorem. *If $(a_n)_{n \in \mathbb{N}}$ and $(b_n)_{n \in \mathbb{N}}$ are convergent rational sequences and if $\lambda \in \mathbb{Q}$, then*

a) $\lim_{n\to\infty}(\lambda a_n) = \lambda \lim_{n\to\infty} a_n$, and

b) $\lim_{n\to\infty}(a_n + b_n) = \lim_{n\to\infty} a_n + \lim_{n\to\infty} b_n$.

Proof. **a)** If $\lambda = 0$, then, trivially, $\lim_{n\to\infty}(\lambda a_n) = 0 = \lambda \lim_{n\to\infty} a_n$, and if $\lambda \neq 0$, then for a given rational number $\varepsilon > 0$ we can find an $N \in \mathbb{N}$ such that $|a_n - L| < \varepsilon/|\lambda|$ for all $n > N$, where $L := \lim_{n\to\infty} a_n$. Using (6.12), it follows that $|\lambda a_n - \lambda L| = |\lambda||a_n - L| < \varepsilon$ for all $n > N$, as desired.
b) Setting $L_a := \lim_{n\to\infty} a_n$ and $L_b := \lim_{n\to\infty} b_n$ and picking a rational number $\varepsilon > 0$, we can find values $N_a, N_b \in \mathbb{N}$ such that $|a_n - L_a| < \varepsilon/2$ for all $n > N_a$ and $|b_n - L_b| < \varepsilon/2$ for all $n > N_b$. Thus for $N := \max\{N_a, N_b\}$ it is the case that $|a_n + b_n - (L_a + L_b)| \leq |a_n - L_a| + |b_n - L_b| < \varepsilon/2 + \varepsilon/2 = \varepsilon$ for all $n > N$ (by Proposition 2.3.1), and therefore, $\lim_{n\to\infty}(a_n + b_n) = L_a + L_b$, as claimed. $\square$

Remark. As indicated in the preceding proof, we need to restrict the set of values from which ε can be chosen in the definitions of convergent sequences and Cauchy sequences (Definitions 1.6.4 and 2.3.6) to the set of positive *rational* numbers as the irrational numbers $\mathbb{R} \setminus \mathbb{Q}$ have yet to be defined. However, this restriction obviously leaves Definitions 1.6.4 and 2.3.6 invariant in terms of their content because for every positive real number ε there exists—as we shall see in the next section (see the remark on p.187)—a positive rational number that is less than ε.

6.3.2 Theorem. *Setting*

$$C := \{(a_n)_{n\in\mathbb{N}} \mid (a_n)_{n\in\mathbb{N}} \text{ is a rational Cauchy sequence}\},$$

it follows that

$$r_{\mathbb{R}} := \left\{((a_n)_{n\in\mathbb{N}}, (b_n)_{n\in\mathbb{N}}) \in C \times C \mid \lim_{n\to\infty}(a_n - b_n) = 0\right\}$$

is an equivalence relation on C.

Proof. $r_{\mathbb{R}}$ is reflexive because $\lim_{n\to\infty}(a_n - a_n) = 0$ for all $(a_n)_{n\in\mathbb{N}} \in C$. Furthermore, $r_{\mathbb{R}}$ is symmetric, because if $\lim_{n\to\infty}(a_n - b_n) = 0$, then $\lim_{n\to\infty}(b_n - a_n) = \lim_{n\to\infty}(-1)(a_n - b_n) = (-1)\lim_{n\to\infty}(a_n - b_n) = 0$ (by Theorem 6.3.1a). Fiinally, in order to show that $r_{\mathbb{R}}$ is transitive, we assume that we are given two pairs $((a_n)_{n\in\mathbb{N}}, (b_n)_{n\in\mathbb{N}}), ((b_n)_{n\in\mathbb{N}}, (c_n)_{n\in\mathbb{N}}) \in r_{\mathbb{R}}$. Given this assumption, it follows that $\lim_{n\to\infty}(a_n - b_n) = \lim_{n\to\infty}(b_n - c_n) = 0$, and therefore, Theorem 6.3.1b implies that $\lim_{n\to\infty}(a_n - c_n) = \lim_{n\to\infty}(a_n - b_n + (b_n - c_n)) = \lim_{n\to\infty}(a_n - b_n) + \lim_{n\to\infty}(b_n - c_n) = 0$, as desired. $\square$

Having proven that $r_{\mathbb{R}}$ is an equivalence relation, we proceed to define the set of real numbers as the set of equivalence classes of $r_{\mathbb{R}}$:

$$\boxed{\mathbb{R} := \{[(a_n)_{n\in\mathbb{N}}]_{\mathbb{R}} \mid (a_n)_{n\in\mathbb{N}} \in C\}.}$$

6.3.3 Example. Since

$$\lim_{n\to\infty} \frac{n+1}{n} = \lim_{n\to\infty} \frac{n^2+3n+1}{n^2+6} = 1$$

and

$$\lim_{n\to\infty} \frac{2n-3}{n+5} = 2$$

(see Exercise 6.3.17), it follows that the equation

$$1+1=2$$

can now be written in the form

$$\left[\left(\frac{n+1}{n}\right)_{n\in\mathbb{N}}\right]_{\mathbb{R}} + \left[\left(\frac{n^2+3n+1}{n^2+6}\right)_{n\in\mathbb{N}}\right]_{\mathbb{R}} = \left[\left(\frac{2n-3}{n+5}\right)_{n\in\mathbb{N}}\right]_{\mathbb{R}}$$

or even more lucidly in the form

$$\left[\left([(n+1,n)]_{\mathbb{Q}}\right)_{n\in\mathbb{N}}\right]_{\mathbb{R}} + \left[\left([(n^2+3n+1,n^2+6)]_{\mathbb{Q}}\right)_{n\in\mathbb{N}}\right]_{\mathbb{R}}$$
$$= \left[\left([(2n-3,n+5)]_{\mathbb{Q}}\right)_{n\in\mathbb{N}}\right]_{\mathbb{R}} .$$

In order to properly define the standard algebraic operations on $\mathbb{R}$, we need to first establish several facts concerning Cauchy sequences.

6.3.4 Proposition. *Every sequence in C is bounded in $\mathbb{Q}$.*

Proof. If $(a_n)_{n\in\mathbb{N}} \in C$, then $(a_n)_{n\in\mathbb{N}}$ is a Cauchy sequence, and therefore, there exists an $N \in \mathbb{N}$ such that $|a_m - a_n| < 1$ for all $m, n > N$. In particular, using Corollary 2.3.2a, we find that

$$|a_m| = |a_m| - |a_{N+1}| + |a_{N+1}| \le |a_m - a_{N+1}| + |a_{N+1}| < 1 + |a_{N+1}|$$

for all $m > N$. Consequently, for $M := \max\{|a_1|, \ldots, |a_N|, 1+|a_{N+1}|\}$ it is the case that $M \in \mathbb{Q}$ and $|a_m| \le M$ for all $n \in \mathbb{N}$, and this proves that $(a_n)_{n\in\mathbb{N}}$ is bounded in $\mathbb{Q}$. $\qquad\square$

6.3.5 Lemma. *If $(a_n)_{n\in\mathbb{N}}$ converges to zero and $(b_n)_{n\in\mathbb{N}}$ is bounded, then $(a_n b_n)_{n\in\mathbb{N}}$ converges to zero.*

Proof. Assume that $\lim_{n\to\infty} a_n = 0$ and that $|b_n| \le M$ for all $n \in \mathbb{N}$ and some $M > 0$. Then, for a given $\varepsilon > 0$, we can find an $N \in \mathbb{N}$ such that $|a_n| < \varepsilon/M$ for all $n > N$, and therefore, $|a_n b_n| = |a_n||b_n| \le M|a_n| < M\varepsilon/M = \varepsilon$ for all $n > N$, as desired. $\qquad\square$

6.3.6 Theorem. *For all sequences $(a_n)_{n\in\mathbb{N}}, (b_n)_{n\in\mathbb{N}} \in C$ it is the case that $(a_n + b_n)_{n\in\mathbb{N}}$ and $(a_n b_n)_{n\in\mathbb{N}}$ are in C as well.*

Proof. Let $\varepsilon > 0$ be a given rational number. Then there are values $N_a, N_b \in \mathbb{N}$ such that $|a_m - a_n| < \varepsilon/2$ for all $m, n > N_a$ and $|b_m - b_n| < \varepsilon/2$ for all $m, n > N_b$. Hence $|(a_m + b_m) - (a_n + b_n)| \leq |a_m - a_n| + |b_m - b_n| < \varepsilon/2 + \varepsilon/2 = \varepsilon$ for all $m, n > N := \max\{N_a, N_b\}$, and therefore, $(a_n + b_n)_{n \in \mathbb{N}} \in C$. In order to show that $(a_n b_n)_{n \in \mathbb{N}}$ is a Cauchy sequence as well, we apply Proposition 6.3.4 to infer that there are positive rational numbers M_a and M_b such that $|a_n| \leq M_a$ and $|b_n| \leq M_b$ for all $n \in \mathbb{N}$. Setting $M := \max\{M_a, M_b\}$, it follows that $M > 0$ and $|a_n|, |b_n| \leq M$ for all $n \in \mathbb{N}$. Since $(a_n)_{n \in \mathbb{N}}$ and $(b_n)_{n \in \mathbb{N}}$ are Cauchy sequences, we can find values N_a and N_b in $\mathbb{N}$ such that $|a_m - a_n| < \varepsilon/(2M)$ for all $m, n > N_a$ and $|b_m - b_n| < \varepsilon/(2M)$ for all $m, n > N_b$. Hence

$$
\begin{aligned}
|a_m b_m - a_n b_n| &= |a_m b_m - a_n b_m + a_n b_m - a_n b_n| \\
&\leq |b_m(a_m - a_n)| + |a_n(b_m - b_n)| \\
&= |b_m||a_m - a_n| + |a_n||b_m - b_n| < \frac{\varepsilon M}{2M} + \frac{\varepsilon M}{2M} = \varepsilon
\end{aligned}
$$

for all $m, n > N := \max\{N_a, N_b\}$, and this shows that $(a_n b_n)_{n \in \mathbb{N}}$ is indeed a Cauchy sequence. $\qquad\square$

6.3.7 Proposition. *If $(a_n)_{n \in \mathbb{N}} \in C$ and if $(a_n)_{n \in \mathbb{N}}$ does not converge to zero, then there exists an $N \in \mathbb{N}$ and a $\delta > 0$ such that $a_n > \delta$ for all $n > N$ or $a_n < -\delta$ for all $n > N$, and such that the sequence $(a_{N,n}^{-1})_{n \in \mathbb{N}}$ that is defined by the equation*

$$
a_{N,n}^{-1} := \begin{cases} 1 & \text{if } n \leq N, \\ 1/a_n & \text{if } n > N \end{cases}
$$

is a Cauchy sequence.

Proof. If the sequence $(a_n)_{n \in \mathbb{N}} \in C$ does not converge to zero, then there exists a rational number $\varepsilon_0 > 0$ such that for all $M \in \mathbb{N}$ there exists an $m \in \mathbb{N}$ such that $m > M$ and $|a_m| \geq \varepsilon_0$. Furthermore, since $(a_n)_{n \in \mathbb{N}}$ is a Cauchy sequence, there also exists an $N \in \mathbb{N}$ such that $N > M$ and $|a_m - a_n| < \varepsilon_0/2$ for all $m, n > N$. To proceed, we pick an integer $m_0 > N$ (so that $|a_{m_0}| \geq \varepsilon_0$) and set $\delta := \varepsilon_0/2$. If $a_{m_0} > 0$, then

$$
a_n = -(a_{m_0} - a_n) + a_{m_0} \geq -|a_{m_0} - a_n| + a_{m_0} > -\frac{\varepsilon_0}{2} + \varepsilon_0 = \delta
$$

for all $n > N$, and similarly, if $a_{m_0} < 0$, then

$$
a_n = -(a_{m_0} - a_n) + a_{m_0} \leq |a_{m_0} - a_n| + a_{m_0} < \frac{\varepsilon_0}{2} - \varepsilon_0 = -\delta
$$

for all $n > N$, as desired. In order to show that $(a_{N,n}^{-1})_{n \in \mathbb{N}}$ is a Cauchy sequence, we choose an arbitrary rational number $\varepsilon > 0$ and a $K > N$ such that $|a_m - a_n| < \varepsilon_0^2 \varepsilon / 4$ for all $m, n > K$. Then for all $m, n > K$ it is the case that

$$|a_{N,m}^{-1} - a_{N,n}^{-1}| = \left| \frac{1}{a_m} - \frac{1}{a_n} \right| = \frac{|a_m - a_n|}{|a_m||a_n|} \quad \text{(by (6.12) and (6.13))}$$

$$< \frac{\varepsilon_0^2 \varepsilon / 4}{\delta^2} = \frac{\varepsilon_0^2 \varepsilon / 4}{(\varepsilon_0/2)^2} = \varepsilon,$$

as desired.   □

If $x = [(a_n)_{n \in \mathbb{N}}]_{\mathbb{R}}$ and $y = [(b_n)_{n \in \mathbb{N}}]_{\mathbb{R}}$ are real numbers, that is, elements of $\mathbb{R}$, then $(a_n + b_n)_{n \in \mathbb{N}}$, and $(a_n b_n)_{n \in \mathbb{N}}$ are contained in C (by Theorem 6.3.6) and so is, trivially, the sequence $(-a_n)_{n \in \mathbb{N}}$. Thus we may attempt the following definitions:

$$-x := [(-a_n)_{n \in \mathbb{N}}]_{\mathbb{R}},$$
$$x + y := [(a_n + b_n)_{n \in \mathbb{N}}]_{\mathbb{R}},$$
$$x - y := x + (-y) = [(a_n - b_n)_{n \in \mathbb{N}}]_{\mathbb{R}},$$
$$xy := [(a_n b_n)_{n \in \mathbb{N}}]_{\mathbb{R}}.$$

Furthermore, if $(b_n)_{n \in \mathbb{N}}$ does not converge to zero, and if $N \in \mathbb{N}$ and $(b_{N,n}^{-1})_{n \in \mathbb{N}}$ are chosen as in Proposition 6.3.7, then

$$\frac{x}{y} := [(a_n b_{N,n}^{-1})_{n \in \mathbb{N}}]_{\mathbb{R}}.$$

In order to show that these operations are well defined, we assume that we are given sequences $(a_n')_{n \in \mathbb{N}}, (b_n')_{n \in \mathbb{N}} \in C$ such that $[(a_n')_{n \in \mathbb{N}}]_{\mathbb{R}} = [(a_n)_{n \in \mathbb{N}}]_{\mathbb{R}}$ and $[(b_n')_{n \in \mathbb{N}}]_{\mathbb{R}} = [(b_n)_{n \in \mathbb{N}}]_{\mathbb{R}}$. Given this assumption, it follows that $\lim_{n \to \infty}(a_n' - a_n) = \lim_{n \to \infty}(b_n' - b_n) = 0$, and therefore,

$$\lim_{n \to \infty} ((a_n' + b_n') - (a_n + b_n)) = \lim_{n \to \infty} (a_n' - a_n) + \lim_{n \to \infty} (b_n' - b_n) = 0$$

by Theorem 6.3.1. Hence $[(a_n' + b_n')_{n \in \mathbb{N}}]_{\mathbb{R}} = [(a_n + b_n)_{n \in \mathbb{N}}]_{\mathbb{R}}$, and this shows that the operation of addition is well defined. Turning from addition to multiplication, we may apply Theorem 6.3.1 in conjunction with Proposition 6.3.4 and Lemma 6.3.5 to infer that

$$\lim_{n \to \infty} (a_n' b_n' - a_n b_n) = \lim_{n \to \infty} (a_n' b_n' - a_n' b_n + a_n' b_n - a_n b_n)$$

$$= \lim_{n \to \infty} a_n' (b_n' - b_n) + \lim_{n \to \infty} (a_n' - a_n) b_n = 0 + 0 = 0.$$

Consequently, $[(a_n' b_n')_{n \in \mathbb{N}}]_{\mathbb{R}} = [(a_n b_n)_{n \in \mathbb{N}}]_{\mathbb{R}}$, and therefore, multiplication is well defined as well. In order to show that the same is true for the operation of division, we assume that $(b_n)_{n \in \mathbb{N}}$ and $(b_n')_{n \in \mathbb{N}}$ do not converge to zero and choose corresponding values $N, N' \in \mathbb{N}$ and $\delta, \delta' > 0$ as described in Proposition 6.3.7. Setting $N_0 := \max\{N, N'\}$ and $\delta_0 := \min\{\delta, \delta'\}$, we may infer that $|b_n|, |b'_n| > \delta_0$ for all $n > N_0$, and therefore, the sequences $(b_{N,n}^{-1})_{n \in \mathbb{N}}$ and $(a'_n b_{N,n}^{-1} b'_{N',n}^{-1})_{n \in \mathbb{N}}$ are bounded (because $(a'_n)_{n \in \mathbb{N}}$ is bounded by Proposition 6.3.4). Since

$$a_n b_{N,n}^{-1} - a'_n b'^{-1}_{N',n} = \frac{a_n}{b_n} - \frac{a'_n}{b'_n} = \frac{a_n - a'_n}{b_n} + \frac{a'_n(b'_n - b_n)}{b_n b'_n}$$

$$= (a_n - a'_n)b_{N,n}^{-1} + (b'_n - b_n)a'_n b_{N,n}^{-1} b'^{-1}_{N',n}$$

for all $n > N_0$, we may thus apply Theorem 6.3.1 and Lemma 6.3.5 to conclude that

$$\lim_{n \to \infty}(a_n b_{N,n}^{-1} - a'_n b'^{-1}_{N',n}) = \lim_{n \to \infty}(a_n - a'_n)b_{N,n}^{-1} + \lim_{n \to \infty}(b'_n - b_n)a'_n b_{N,n}^{-1} b'^{-1}_{N',n}$$

$$= 0 + 0 = 0,$$

as desired. Finally, the fact that the operations of sign inversion and subtraction are also well defined is left for the reader to prove as an exercise (see Exercise 6.3.11).

Having established the standard algebraic operations, we move on to define the order on $\mathbb{R}$: for real numbers $x = [(a_n)_{n \in \mathbb{N}}]_{\mathbb{R}}$ and y we set

$$0 < x :\Leftrightarrow \exists_{\varepsilon \in \mathbb{Q}}(\varepsilon > 0 \wedge \exists_{N \in \mathbb{N}}\forall_{n \in \mathbb{N}} \, n > N \Rightarrow \varepsilon < a_n),$$
$$0 = x :\Leftrightarrow \lim_{n \to \infty} a_n = 0,$$
$$x < 0 :\Leftrightarrow 0 < -x,$$

and

$$x < y :\Leftrightarrow 0 < y - x,$$

and, as usual,

$$x \leq y :\Leftrightarrow (x < y \vee x = y).$$

In order to prove that the statements '$0 < x$' and '$0 = x$' are well defined, we assume that $x = [(a_n)_{n \in \mathbb{N}}]_{\mathbb{R}} = [(a'_n)_{n \in \mathbb{N}}]_{\mathbb{R}}$. Then $\lim_{n \to \infty}(a'_n - a_n) = 0$, and therefore, the assumption that $\lim_{n \to \infty} a_n = 0$ implies that $\lim_{n \to \infty} a'_n = \lim_{n \to \infty}((a'_n - a_n) + a_n) = 0 + 0 = 0$ (by Theorem 6.3.1b). Consequently, the statement '$0 = x$' is well defined. Turning to the statement '$0 < x$', we assume that there is a rational number $\varepsilon > 0$ and an $N \in \mathbb{N}$ such that $\varepsilon < a_n$ for all $n > N$. Since $(a'_n - a_n)_{n \in \mathbb{N}}$ converges to zero, it follows that there exists an $M \in \mathbb{N}$ such that $|a'_n - a_n| < \varepsilon/2$ for all $n > M$. Setting $\varepsilon' := \varepsilon/2$ and $N' := \max\{N, M\}$, it follows that $\varepsilon' > 0$ and $a'_n = (a'_n - a_n) + a_n > -\varepsilon/2 + \varepsilon = \varepsilon/2 = \varepsilon'$ for all $n > N'$, as desired.

In order to establish that '$<$' is antisymmetric, it is sufficient to show that the conjunction $x < y \wedge y < x$ is always false. To give a proof by contradiction, we assume that we are given real numbers $x = [(a_n)_{n\in\mathbb{N}}]_\mathbb{R}$ and $y = [(b_n)_{n\in\mathbb{N}}]_\mathbb{R}$ such that $0 < y - x$ and $0 < x - y$. Given this assumption, it follows that there are numbers $\varepsilon_+, \varepsilon_- \in \mathbb{Q}$ and $N_+, N_- \in \mathbb{N}$ such that $0 < \varepsilon_+ < b_n - a_n$ for all $n > N_+$ and $0 < \varepsilon_- < a_n - b_n$ for all $n > N_-$. Consequently, Theorem 6.2.7b implies that $a_n < b_n$ and $b_n < a_n$ for all $n > \max\{N_+, N_-\}$. Since this is impossible (see p.172), we may infer that '$<$' is indeed antisymmetric. In order to show that '$<$' is transitive, we assume that we are given real numbers $x = [(a_n)_{n\in\mathbb{N}}]_\mathbb{R}$, $y = [(b_n)_{n\in\mathbb{N}}]_\mathbb{R}$, and $z = [(c_n)_{n\in\mathbb{N}}]_\mathbb{R}$ such that $0 < y - x$ and $0 < z - y$. Then there are numbers $\delta, \varepsilon \in \mathbb{Q}$ and $M, N \in \mathbb{N}$ such that $0 < \delta < b_n - a_n$ for all $n > M$ and $0 < \varepsilon < c_n - b_n$ for all $n > N$. Hence $0 < \delta + \varepsilon < (c_n - b_n) + (b_n - a_n) = c_n - a_n$ for all $n > \max\{M, N\}$ (by Theorem 6.2.7b), and therefore, $x < z$, as desired. Finally, in order to prove that '$<$' is linear, we need to show that the assumption $\neg(x < y \vee y < x)$ implies that $x = y$. So let $x = [(a_n)_{n\in\mathbb{N}}]_\mathbb{R}$ and $y = [(b_n)_{n\in\mathbb{N}}]_\mathbb{R}$ be real numbers such that neither $y - x$ nor $x - y$ are greater than zero. (Note: $\neg(x < y \vee y < x) \Leftrightarrow (\neg(0 < y - x) \wedge \neg(0 < x - y))$.) Picking an arbitrary rational number $\varepsilon > 0$ and setting $c_n := b_n - a_n$, it follows that $(c_n)_{n\in\mathbb{N}} \in C$ and therefore, there exists an $N \in \mathbb{N}$ such that $|c_n - c_m| < \varepsilon/2$ for all $m, n > N$. Furthermore, since $y - x = [(c_n)_{n\in\mathbb{N}}]_\mathbb{R}$ is not greater than zero, there exists an $m > N$ such that $c_m \leq \varepsilon/2$. Thus $c_n = c_n - c_m + c_m \leq |c_n - c_m| + c_m < \varepsilon/2 + \varepsilon/2 = \varepsilon$ for all $n > N$. Similarly, since $x - y = [(-c_n)_{n\in\mathbb{N}}]_\mathbb{R}$ is not greater than zero either, we can find a $k > N$ such that $-c_k \leq \varepsilon/2$. Hence $-c_n = -c_n + c_k - c_k \leq |c_n - c_k| - c_k < \varepsilon/2 + \varepsilon/2 = \varepsilon$ for all $n > N$. So what we have here shown is that $-\varepsilon < c_n < \varepsilon$ for all $n > N$ or, equivalently, that $|c_n| < \varepsilon$ for all $n > N$. Hence $\lim_{n\to\infty} c_n = 0$, and therefore, $x = y$, as desired.

Remark. Just as the linear orders on $\mathbb{N}$, $\mathbb{Z}$, and $\mathbb{Q}$, the linear order on $\mathbb{R}$ is strictly reflexive in the sense that $x < y \Rightarrow x \neq y$ for all real numbers $x = [(a_n)_{n\in\mathbb{N}}]_\mathbb{R}$ and $y = [(b_n)_{n\in\mathbb{N}}]_\mathbb{R}$. For if $x < y$, then $0 < y - x$ and therefore, $\lim_{n\to\infty}(b_n - a_n) \neq 0$ (by the definition of '$<$').

Having thus established that '$<$' is a well defined linear order on $\mathbb{R}$ (and that the same is true, by implication, for '$\leq$'), our next objective is to describe the natural identification map that allows us to consider $\mathbb{Q}$ to be a subset of $\mathbb{R}$. To do so, we denote by $(x)_{n\in\mathbb{N}}$ the constant sequence $x, x, x, \ldots$ for all $x \in \mathbb{Q}$ and define

$$i : \mathbb{Q} \to \mathbb{R}$$
$$x \mapsto [(x)_{n\in\mathbb{N}}]_\mathbb{R}.$$

Evidently, i is well defined because the constant sequence $(x)_{n\in\mathbb{N}}$ is a Cauchy sequence for all $x \in \mathbb{Q}$. Moreover, i is injective and satisfies the following natural consistency properties (see Exercise 6.3.12):

$$i(-x) = -i(x) \wedge i(x + y) = i(x) + i(y) \wedge i(xy) = i(x)i(y) \qquad (6.14)$$

and

$$x < y \Leftrightarrow i(x) < i(y) \tag{6.15}$$

for all $x, y \in \mathbb{Q}$.

6.3.8 Theorem. *For all $x, y, z \in \mathbb{R}$ it is the case that*

a) $x + (y + z) = (x + y) + z$,

b) $x + 0 = x$ *where* $0 = [(0)_{n \in \mathbb{N}}]_{\mathbb{R}}$,

c) $x - x = -x + x = 0$,

d) $x + y = y + x$,

e) $x(yz) = (xy)z$,

f) $xy = yx$,

g) $x = 1 \cdot x$ *and* $-x = (-1) \cdot x$ *where* $1 = [(1)_{n \in \mathbb{N}}]_{\mathbb{R}}$,

h) $(1/x)x = x/x = 1$ *whenever* $x \neq 0$,

i) $xy = 0 \Leftrightarrow (x = 0 \vee y = 0)$,

j) $x(y + z) = xy + xz$, *and*

k) $x(y - z) = xy - xz$.

Proof. As in the case of Theorem 6.2.5 we will prove only h) and i) and leave the proofs of the remaining properties as exercises to the reader (see Exercise 6.3.13). In order to prove h), we assume that $x = [(a_n)_{n \in \mathbb{N}}]_{\mathbb{R}}$ is a nonzero real number. Then $1/x = [(a_{N,n}^{-1})_{n \in \mathbb{N}}]_{\mathbb{R}}$ and $x/x = [(a_n a_{N,n}^{-1})_{n \in \mathbb{N}}]_{\mathbb{R}}$ for some properly chosen $N \in \mathbb{N}$, as explained in Proposition 6.3.7, and therefore, $(1/x)x = x/x$. Furthermore, since $a_{N,n}^{-1} = 1/a_n$ for all $n > N$, it follows that $a_n a_{N,n}^{-1} = 1$ for all $n > N$, and therefore, $\lim_{n \to \infty}(1 - a_n a_{N,n}^{-1}) = 0$. Hence $x/x = [(a_n a^{-1})_{n \in \mathbb{N}}]_{\mathbb{R}} = [(1)_{n \in \mathbb{N}}]_{\mathbb{R}} = 1$, as desired. Turning to i), we will show, as in the proof of Theorem 6.2.5, that $x = 0 \Rightarrow xy = 0$ and that $(xy = 0 \wedge x \neq 0) \Rightarrow y = 0$. If $x = 0 = [(0)_{n \in \mathbb{N}}]_{\mathbb{R}}$ and $y = [(b_n)_{n \in \mathbb{N}}]_{\mathbb{R}}$, then $xy = [(0b_n)_{n \in \mathbb{N}}]_{\mathbb{R}} = [(0)_{n \in \mathbb{N}}]_{\mathbb{R}} = 0$, and if $xy = 0$ and $x \neq 0$, then $y = 1y = ((1/x)x)y = (1/x)(xy) = (1/x)0 = 0$. □

6.3.9 Proposition. *For all $x, y \in \mathbb{R}$ the following statements are valid:*

a) $0 < xy$ *if and only if* $0 < x \wedge 0 < y$ *or* $x < 0 \wedge y < 0$,

b) $xy < 0$ *if and only is* $0 < x \wedge y < 0$ *or* $x < 0 \wedge 0 < y$.

Proof. Using the same logic as in the proof of Proposition 6.1.4, we only need to show that $((0 < x \wedge 0 < y) \vee (x < 0 \wedge y < 0)) \Rightarrow 0 < xy$ and $((0 < x \wedge y < 0) \vee (x < 0 \wedge 0 < y)) \Rightarrow xy < 0$, and, here as there, we will be content to prove only that $(0 < x \wedge 0 < y) \Rightarrow 0 < xy$ and leave the proofs of the remaining implications as exercises to the reader (see Exercise 6.2.11).

So let us assume that $x = [(a_n)_{n \in \mathbb{N}}]_{\mathbb{R}}$ and $y = [(b_n)_{n \in \mathbb{N}}]_{\mathbb{R}}$ are positive real numbers. Then there are positive numbers $\delta, \varepsilon \in \mathbb{Q}$ and integers $M, N \in \mathbb{N}$ such that $0 < \delta < a_n$ for all $n > M$ and $0 < \varepsilon < b_n$ for all $n > N$. Hence $0 < \delta\varepsilon < a_n b_n$ for all $n > \max\{M, N\}$ (by Proposition 6.2.7o), and therefore, $0 < [(a_n b_n)_{n \in \mathbb{N}}]_{\mathbb{R}} = xy$, as desired. $\qquad \square$

6.3.10 Proposition. The following statements are true for all $x, y \in \mathbb{R}$:

a) $x < y \Rightarrow \forall_{z \in \mathbb{R}} \, x + z < y + z$,

b) $x \leq y \Rightarrow \forall_{z \in \mathbb{R}} \, x + z \leq y + z$,

c) $(\exists_{z \in \mathbb{R}} \, x + z < y + z) \Rightarrow x < y$,

d) $(\exists_{z \in \mathbb{R}} \, x + z \leq y + z) \Rightarrow x \leq y$,

e) $\forall_{z \in \mathbb{R}} \, 0 < z \Rightarrow x < x + z$,

f) $\forall_{z \in \mathbb{R}} \, 0 \leq z \Rightarrow x \leq x + z$,

g) $x < y \Leftrightarrow -y < -x$,

h) $x \leq y \Leftrightarrow -y \leq -x$,

i) $x < y \Rightarrow \forall_{z \in \mathbb{R}} \, 0 < z \Rightarrow xz < yz$,

j) $x < y \Rightarrow \forall_{z \in \mathbb{R}} \, z < 0 \Rightarrow yz < xz$,

k) $x \leq y \Rightarrow \forall_{z \in \mathbb{R}} \, 0 < z \Rightarrow xz \leq yz$,

l) $x \leq y \Rightarrow \forall_{z \in \mathbb{R}} \, z < 0 \Rightarrow yz \leq xz$.

m) $(\exists_{z \in \mathbb{R}} \, 0 < z \wedge xz < yz) \Rightarrow x < y$,

n) $(\exists_{z \in \mathbb{R}} \, z < 0 \wedge yz < xz) \Rightarrow x < y$,

o) $(\exists_{z \in \mathbb{R}} \, 0 < z \wedge xz \leq yz) \Rightarrow x \leq y$,

p) $(\exists_{z \in \mathbb{R}} \, z < 0 \wedge yz \leq xz) \Rightarrow x \leq y$,

q) $0 < x < y \Leftrightarrow 0 < 1/y < 1/x$,

r) $0 < x \leq y \Leftrightarrow 0 < 1/y \leq 1/x$,

s) $1 < x \Leftrightarrow 0 < 1/x < 1$,

t) $1 \leq x \Leftrightarrow 0 < 1/x \leq 1$,

u) $0 < x < y \Rightarrow 1 < y/x$,

v) $0 < x \leq y \Rightarrow 1 \leq y/x$.

Proof. We will prove a), i), m), and u) and leave the proofs of the remaining properties as exercises to the reader (see Exercise 6.3.15). However, since the proofs of a), i), and m) are completely analogous to the proofs of b), j), and n) in Proposition 6.2.7, we may limit ourselves to proving only u). But here again we readily observe that the proof of u) is perfectly analogous to the proof of Proposition 6.2.7v except for the fact that we need to independently establish the implication $0 < x \Rightarrow 0 < 1/x$ for all $x \in \mathbb{R}$. So let $x = [(a_n)_{n \in \mathbb{N}}]_{\mathbb{R}} \in \mathbb{R}$ such that $0 < x$. Choosing N as in Proposition 6.3.7, it follows that $0 < a_n$ for all $n > N$. Moreover, since $(a_n)_{n \in \mathbb{N}}$ is a Cauchy sequence, we may apply Proposition 6.3.4 to infer that there is an $M \in \mathbb{Q}$ such that $a_n < M$ for all $n \in \mathbb{N}$. Thus $0 < a_n < M$ for all $n > N$, and therefore, Proposition 6.2.7r implies that $0 < 1/M < 1/a_n = a_{N,n}^{-1}$ for all $n > N$. Hence $0 < [(a_{N,n}^{-1})_{n \in \mathbb{N}}]_{\mathbb{R}} = 1/x$, as desired. $\qquad\square$

To conclude this section, we wish to point out that the domain of the absolute value function, defined in (6.11), can be extended from $\mathbb{Q}$ to $\mathbb{R}$ as follows:

$$|x| := \begin{cases} x & \text{if } 0 \le x, \\ -x & \text{if } x < 0. \end{cases} \tag{6.16}$$

for all $x \in \mathbb{R}$. Here as there we find that this extended absolute value function not only satisfies the triangle-type inequalities in Proposition 2.3.1 and Corollary 2.3.2, but also the equations

$$|xy| = |x||y| \tag{6.17}$$

for all $x, y \in \mathbb{R}$ as well as the equation

$$\left| \frac{x}{y} \right| = \frac{|x|}{|y|} \tag{6.18}$$

whenever $y \ne 0$ (see Exercise 6.3.16).

Exercises

6.3.11. Prove that the operations of sign inversion and subtraction in $\mathbb{R}$ are well defined.

6.3.12. Prove (6.14) and (6.15) and explain why i is injective.

6.3.13. Prove Theorem 6.3.8a,b,c,d,e,f,g,j,k.

6.3.14. Complete the proof of Proposition 6.3.9.

6.3.15. Prove Proposition 6.2.7b,c,d,f,g,h,i,j,k,l,n,p,q,r,s,t,u,v,w,x,z.

6.3.16. Prove the equations (6.17) and (6.18).

6.3.17. Prove that

$$\left[\left(\frac{n+1}{n} \right)_{n \in \mathbb{N}} \right]_{\mathbb{R}} + \left[\left(\frac{n^2 + 3n + 1}{n^2 + 6} \right)_{n \in \mathbb{N}} \right]_{\mathbb{R}} = \left[\left(\frac{2n - 3}{n + 5} \right)_{n \in \mathbb{N}} \right]_{\mathbb{R}}$$

by demonstrating that

$$\lim_{n\to\infty} \frac{n+1}{n} = \lim_{n\to\infty} \frac{n^2+3n+1}{n^2+6} = 1$$

and

$$\lim_{n\to\infty} \frac{2n-3}{n+5} = 2.$$

6.4   ★ The Completeness and Cardinality of $\mathbb{R}$

As the heading indicates, we will address in this section the issues of the completeness and cardinality of $\mathbb{R}$, but first we will show that $\mathbb{Q}$ is *dense* in $\mathbb{R}$ in the sense that for all $x, y \in \mathbb{R}$ with $x < y$ there exists an $r \in \mathbb{Q}$ such that $x < r < y$ (see Theorem 6.4.2). This fact is interesting in itself, but it will also be helpful in proving Lemma 6.4.5 (on p.188) which in turn will be needed to establish the completeness of $\mathbb{R}$.

6.4.1 Lemma. *For every real number* $x = [(a_n)_{n\in\mathbb{N}}]_\mathbb{R}$ *there exists an integer* $n \in \mathbb{N}$ *such that* $x < n$.

Proof. If $x = [(a_n)_{n\in\mathbb{N}}]_\mathbb{R}$, then $(a_n)_{n\in\mathbb{N}}$ is a Cauchy sequence and hence bounded in $\mathbb{Q}$. So there exists an $M \in \mathbb{Q}$ such that $a_n < M$ for all $n \in \mathbb{N}$ or, equivalently, such that $0 < M - a_n$ for all $n \in \mathbb{N}$. Thus $1 < 1 + M - a_n$ for all $n \in \mathbb{N}$ (by Proposition 6.2.7b), and therefore, $0 < [(1 + M - a_n)_{n\in\mathbb{N}}]_\mathbb{R} = 1 + M - x$. Since $1 + M \in \mathbb{Q}$, there are integers $p \in \mathbb{Z}$ and $q \in \mathbb{N}$ such that $1 + M = p/q$. Consequently, in setting $n := |p|$, it follws that $x < 1 + M \le n$, as desired. □

6.4.2 Theorem. *If* $x, y \in \mathbb{R}$ *and* $x < y$, *then there exists an* $r \in \mathbb{Q}$ *such that* $x < r < y$.

Proof. If $x < 0 < y$, then the inequality $x < r < y$ is satisfied for $r := 0 \in \mathbb{Q}$. Thus we may assume that either $0 \le x < y$ or $x < y \le 0$. Considering first the former case, we find that $0 < y - x$ and apply Lemma 6.4.1 to conclude that there are integers $m, n \in \mathbb{N}$ such that $0 < y < n$ and $0 < 1/(y-x) < m$. Using Proposition 6.3.10q, it follows that $0 < 1/m < y - x$, and since $(nm)/m = n > y$, we may infer that the set $S := \{n \in \mathbb{N} | y \le n/m\}$ is nonempty. Using the Well-Ordering Principle, it follows that S has a minimal element n_0. (Note: $n_0 > 1$ because $1/m < y - x \le y$.) Consequently, $n_0 \in S$ and $n_0 - 1 \notin S$, and this in turn implies that $y \le n_0/m$ and $r := (n_0 - 1)/m < y$. Hence $y \le r + 1/m < r + (y - x)$, and therefore, $x < r < y$, as desired. Finally, if $x < y \le 0$, then $0 \le -y < -x$ (by Proposition 6.3.10g,h), and, according to the statement just proven, there exists an $\tilde{r} \in \mathbb{Q}$ such that $-y < \tilde{r} < -x$. Setting $r := -\tilde{r}$, it follows that $x < r < y$ (by Proposition 6.3.10g). □

Remark. Following up on the remark on p.177, we wish to explain exactly how Theorem 6.4.2 implies that the universe $\mathbb{R}$ can be replaced by $\mathbb{Q}$ in the definition of the Cauchy property and the limit of a sequence. Since the reasoning is completely analogous in both cases, we will consider only the definition of a limit. That is to say, we will consider the assertion that the statement

$$\forall_{\varepsilon \in \mathbb{R}}(\varepsilon > 0 \Rightarrow \exists_{N \in \mathbb{N}} \forall_{n \in \mathbb{N}}(n > N \Rightarrow |a_n - L| < \varepsilon)) \tag{6.19}$$

is equivalent to

$$\forall_{\varepsilon \in \mathbb{Q}}(\varepsilon > 0 \Rightarrow \exists_{N \in \mathbb{N}} \forall_{n \in \mathbb{N}}(n > N \Rightarrow |a_n - L| < \varepsilon)). \tag{6.20}$$

Since $\mathbb{Q} \subset \mathbb{R}$, it is clear that (6.19) implies (6.20), and in order to see why the reverse implication is valid as well, we assume that (6.20) is true and that ε is a given positive real number. Then, according to Theorem 6.4.2, there exists a rational number $\tilde{\varepsilon}$ such that $0 < \tilde{\varepsilon} < \varepsilon$, and according to (6.20), there exists an $N \in \mathbb{N}$ such that $|a_n - L| < \tilde{\varepsilon}$ for all $n > N$. But since $\tilde{\varepsilon} < \varepsilon$, it also is the case that $|a_n - L| < \varepsilon$ for all $n > N$, and this shows that (6.19) is indeed valid.

As we pointed out at the beginning of the preceding section, the rational numbers are *not* complete in the sense that there are Cauchy sequences in $\mathbb{Q}$ that are not convergent in $\mathbb{Q}$. By contrast and very significantly, Theorem 6.4.6 shows that $\mathbb{R}$ is complete, because for every Cauchy sequence $(x_n)_{n \in \mathbb{N}}$ in $\mathbb{R}$ there exists an $x \in \mathbb{R}$ such that $\lim_{n \to \infty} x_n = x$. To prove this fact, we need the following lemmas:

6.4.3 Lemma. *For all* $x = [(a_n)_{n \in \mathbb{N}}]_{\mathbb{R}} \in \mathbb{R}$ *it is the case that* $|x| = [(|a_n|)_{n \in \mathbb{N}}]_{\mathbb{R}}$.

Proof. If $x = 0$, then $\lim_{n \to \infty} a_n = 0$, and therefore, trivially, $\lim_{n \to \infty} |a_n| = 0$, as well. Thus $|x| = 0 = [(|a_n|)_{n \in \mathbb{N}}]_{\mathbb{R}}$. Furthermore, if $x \neq 0$, then, according to Proposition 6.3.7, there exists a positive rational number δ and an $N \in \mathbb{N}$ such that $\delta < a_n$ for all $n > N$ or $a_n < -\delta$ for all $n > N$. In the former case we find that $0 < x$ and $a_n = |a_n|$ for all $n > N$, and therefore, $|x| = x = [(|a_n|)_{n \in \mathbb{N}}]_{\mathbb{R}}$ because $\lim_{n \to \infty} (a_n - |a_n|) = 0$. Moreover, in the latter case we have $x < 0$ and $|a_n| = -a_n$ for all $n > N$. Thus $|x| = -x = [(|a_n|)_{n \in \mathbb{N}}]_{\mathbb{R}}$ because $\lim_{n \to \infty} (-a_n - |a_n|) = 0$. □

6.4.4 Lemma. *If* $x = [(a_n)_{n \in \mathbb{N}}]_{\mathbb{R}} \in \mathbb{R}$, *then for every positive rational number* ε *there exists a* $(b_n)_{n \in \mathbb{N}} \in C$ *such that* $|b_m - b_n| < \varepsilon$ *for all* $n \in \mathbb{N}$ *and* $x = [(b_n)_{n \in \mathbb{N}}]_{\mathbb{R}}$.

Proof. Since $(a_n)_{n \in \mathbb{N}}$ is a Cauchy sequence, there exists an $N \in \mathbb{N}$ such that $|a_m - a_n| < \varepsilon$ for all $m, n > N$. Setting

$$b_n := \begin{cases} a_n & \text{if } n > N, \\ a_{N+1} & \text{if } n \leq N, \end{cases}$$

it follows that $|b_m - b_n| < \varepsilon$ for all $n \in \mathbb{N}$ and $x = [(b_n)_{n\in\mathbb{N}}]_\mathbb{R}$, as desired (because, evidently, $\lim_{n\to\infty}(a_n - b_n) = 0$). □

6.4.5 Lemma. *A real number $x \geq 0$ is equal to zero if $x < \varepsilon$ for all rational numbers $\varepsilon > 0$.*

Proof. To give a proof by contradiction, we assume that x is a real number such that $0 < x < \varepsilon$ for all rational numbers $\varepsilon > 0$. Then, according to Theorem 6.4.2, there exists an $r \in \mathbb{Q}$ such that $0 < r < x$. Setting $\varepsilon := r$, we arrive at the contradictory conclusion that $r < x < r$. □

6.4.6 Theorem. *If $(x_n)_{n\in\mathbb{N}}$ is a Cauchy sequence of real numbers $x_n = [(a_{n,k})_{k\in\mathbb{N}}]_\mathbb{R}$, then there exists a real number $x \in \mathbb{R}$ such that $\lim_{n\to\infty} x_n = x$.*

Proof. According to Lemma 6.4.4, we may assume w.l.o.g. that

$$|a_{n,k} - a_{n,j}| < \frac{1}{n} \tag{6.21}$$

for all $j, k, n \in \mathbb{N}$. That said, our first objective is to show that $(a_{n,k})_{n\in\mathbb{N}}$ is a Cauchy sequence for all $k \in \mathbb{N}$. (Note: $(a_{n,k})_{k\in\mathbb{N}}$ is a Cauchy sequence for all $n \in \mathbb{N}$ by assumption, but the fact that the same is true for the sequence $(a_{n,k})_{n\in\mathbb{N}}$ for all $k \in \mathbb{N}$ was not assumed and therefore needs to be proven.) So let ε be a positive rational number. Since $(x_n)_{n\in\mathbb{N}}$ is a Cauchy sequence in $\mathbb{R}$ by assumption, there exists an $N \in \mathbb{N}$ such that

$$N > \frac{3}{\varepsilon} \tag{6.22}$$

and $|x_m - x_n| < \varepsilon/3$ for all $m, n > N$. Using Lemma 6.4.3, it follows that $0 < \varepsilon/3 - |x_m - x_n| = [(\varepsilon/3 - |a_{m,k} - a_{n,k}|)_{k\in\mathbb{N}}]_\mathbb{R}$ for all $m, n > N$, and therefore, for given integers $m, n > N$ there exists a positive rational number $\delta_{m,n}$ and an integer $K_{m,n} \in \mathbb{N}$ such that

$$0 < \delta_{m,n} < \frac{\varepsilon}{3} - |a_{m,k} - a_{n,k}|$$

for all $k > K_{m,n}$. Hence

$$|a_{m,k} - a_{n,k}| < \frac{\varepsilon}{3} - \delta_{m,n} < \frac{\varepsilon}{3} \tag{6.23}$$

for all $k > K_{m,n}$. To proceed, we pick integers $k \in \mathbb{N}$, $m, n > N$, and $k_0 > K_{m,n}$ and combine (6.21) with (6.22) and (6.23) to infer that

$$|a_{m,k} - a_{n,k}| \leq |a_{m,k} - a_{m,k_0}| + |a_{m,k_0} - a_{n,k_0}| + |a_{n,k_0} - a_{n,k}|$$
$$< \frac{1}{m} + \frac{\varepsilon}{3} + \frac{1}{n} < \frac{2}{N} + \frac{\varepsilon}{3} < \varepsilon.$$

In other words (or symbols), what we have here shown is that

$$\forall_{\varepsilon \in \mathbb{Q}}(\varepsilon > 0 \Rightarrow \exists_{N \in \mathbb{N}} \forall_{k \in \mathbb{N}} \forall_{n,m \in \mathbb{N}}(m, n > N \Rightarrow |a_{m,k} - a_{n,k}| < \varepsilon)), \qquad (6.24)$$

and since this statement evidently implies that

$$\forall_{k \in \mathbb{N}} \forall_{\varepsilon \in \mathbb{Q}}(\varepsilon > 0 \Rightarrow \exists_{N \in \mathbb{N}} \forall_{n,m \in \mathbb{N}}(m, n > N \Rightarrow |a_{m,k} - a_{n,k}| < \varepsilon)),$$

it follows that $(a_{n,k})_{n \in \mathbb{N}}$ is a Cauchy sequence for all $k \in \mathbb{N}$, as claimed. Setting $y_k := [(a_{n,k})_{n \in \mathbb{N}}]_{\mathbb{R}}$ for all $k \in \mathbb{N}$, we wish to show next that $y_k = y_j$ for all $j, k \in \mathbb{N}$ or, equivalently, that $|y_k - y_j| = 0$ for all $j, k \in \mathbb{N}$. Using Lemma 6.4.5, it is sufficient to prove that $|y_k - y_j| < \varepsilon$ for all rational numbers $\varepsilon > 0$ and all $j, k \in \mathbb{N}$. So let $j, k \in \mathbb{N}$ and let ε be a positive rational number. Choosing $N \in \mathbb{N}$ such that $N > 2/\varepsilon$, we may apply (6.21) to infer that

$$|a_{n,k} - a_{n,j}| < \frac{1}{n} < \frac{1}{N} < \frac{\varepsilon}{2}$$

for all $n > N$, and therefore,

$$0 < \frac{\varepsilon}{2} < \varepsilon - |a_{n,k} - a_{k,n}|$$

for all $n > N$. Thus $0 < [(\varepsilon - |a_{n,k} - a_{n,j}|)_{n \in \mathbb{N}}]_{\mathbb{R}} = \varepsilon - |y_k - y_j|$ (by Lemma 6.4.3), and, by implication, $|y_k - y_j| < \varepsilon$, as desired. Consequently, in setting $x := y_1$, it follows that

$$x = y_k$$

for all $k \in \mathbb{N}$, and given this observation, we will now complete the proof by demonstrating that $x = \lim_{n \to \infty} x_n$. To do so, we pick an arbitrary rational number $\varepsilon > 0$ and apply (6.24) to infer that there exists an $N \in \mathbb{N}$ such that $N > 6/\varepsilon$ and

$$|a_{n,k} - a_{m,k}| < \frac{\varepsilon}{6}$$

for all $n, m > N$ and all $k \in \mathbb{N}$. Using this fact in conjunction with (6.21), it follows that for all $n, k > N$ and all $j \in \mathbb{N}$ it is the case that

$$|a_{n,k} - a_{k,n}| \le |a_{n,k} - a_{n,j}| + |a_{n,j} - a_{k,j}| + |a_{k,j} - a_{k,n}|$$
$$< \frac{1}{n} + \frac{\varepsilon}{6} + \frac{1}{k} < \frac{2}{N} + \frac{\varepsilon}{6} < \frac{\varepsilon}{2}$$

or. equivalently,

$$\frac{\varepsilon}{2} < \varepsilon - |a_{n,k} - a_{k,n}|.$$

Consequently, for all $n > N$ we find that

$$0 < [(\varepsilon - |a_{n,k} - a_{k,n}|)_{k \in \mathbb{N}}]_{\mathbb{R}} = \varepsilon - |x_n - y_n| = \varepsilon - |x_n - x|,$$

and therefore,

$$|x_n - x| < \varepsilon,$$

as desired. $\qquad\square$

To conclude our discussion of the real number system, we will now show that $\mathbb{R}$ is cardinally equivalent to the power set of $\mathbb{N}$. As it turns out, the most convenient way to accomplish this is to define and exploit the concept of an infinite series:

6.4.7 Definition. Given a sequence $(a_n)_{n \in \mathbb{N}}$ in $\mathbb{R}$, we say that the infinite sum or *series* $\sum_{k=1}^{\infty} a_k$ is *convergent* if the sequence of the corresponding *partial sums* $S_n := \sum_{k=1}^{n} a_k$ is convergent in $\mathbb{R}$. In that case we write

$$\sum_{k=1}^{\infty} a_k = \lim_{n \to \infty} S_n.$$

6.4.8 Lemma. *Assume that $q \in \mathbb{R}$ and $n, m \in \mathbb{N}$ such that $q \neq 1$ and $m \leq n$. Then*

$$\sum_{k=m}^{n} q^k = \frac{q^m(1 - q^{n-m+1})}{1 - q}.$$

Proof. Since

$$\sum_{k=m}^{n} q^k = \sum_{k=m}^{n} q^m q^{k-m} = q^m \sum_{k=m}^{n} q^{k-m} = q^m \sum_{k=0}^{n-m} q^k,$$

we may use (4.31) to infer that

$$(1 - q) \sum_{k=m}^{n} q^k = q^m(1 - q^{n-m+1}).$$

Since $q \neq 1$, we can divide both sides of this equation by $1 - q$ to obtain the desired result. □

6.4.9 Lemma. *If $-1 < q < 1$ and if $(a_n)_{n \in \mathbb{N}}$ is a sequence of zeroes and ones (i.e., $a_n \in \{0, 1\}$ for all $n \in \mathbb{N}$), then the series $\sum_{k=1}^{\infty} a_k q^k$ is convergent.*

Proof. If $q = 0$, then $S_n = 0$ for all $n \in \mathbb{N}$, and therefore, $\lim_{n \to \infty} S_n = 0$. Thus we may assume that $q \neq 0$. According to Theorem 6.4.6, it is sufficient to show that the sequence of the partial sums $S_n = \sum_{k=1}^{n} a_k q^k$ is a Cauchy sequence. To do so, we pick a (rational) number $\varepsilon > 0$ and assume w.l.o.g. that m and n are positive integers such that $m < n$. Using Lemma 6.4.8, we find that

$$|S_n - S_m| = \left| \sum_{k=m+1}^{n} a_k q^k \right| \leq \sum_{k=m+1}^{n} |a_k||q|^k \leq \sum_{k=m+1}^{n} |q|^k$$

$$= \frac{|q|^{m+1}(1 - |q|^{n-m})}{1 - |q|} \leq \frac{|q|^m}{1 - |q|}.$$

Furthermore, since $q \neq 0$ and $-1 < q < 1$, it follows that $1 < 1/|q|$, and setting $x := 1/|q| - 1 > 0$, we may employ the result of Example 4.5.2 to conclude that

$$|q|^m = \frac{1}{(1+x)^m} \leq \frac{1}{1+mx}.$$

Consequently, if $N \in \mathbb{N}$ such that $N > 1/(x\varepsilon(1-|q|))$, then for all $n, m \in \mathbb{N}$ with $n > m > N$ it is the case that

$$|S_n - S_m| \leq \frac{1}{(1+mx)(1-|q|)} < \frac{1}{mx(1-|q|)} < \frac{1}{Nx(1-|q|)} < \varepsilon,$$

as desired. $\qquad\square$

To proceed, we set

$$A := \{(a_n)_{n\in\mathbb{N}} \mid \forall_{n\in\mathbb{N}}\, a_n \in \{0,1\}\},$$

and observe that A is cardinally equivalent to $\mathcal{P}(\mathbb{N})$, because the map $h : \mathcal{P}(\mathbb{N}) \to A$ that is defined by the equations

$$h(M) := (a_n(M))_{n\in\mathbb{N}} \tag{6.25}$$

and

$$a_n(M) := \begin{cases} 1 & \text{if } n \in M, \\ 0 & \text{if } n \in \mathbb{N} \setminus M \end{cases} \tag{6.26}$$

for all $M \subset \mathbb{N}$ and all $n \in \mathbb{N}$ is easily seen to be a bijection (see Exercise 6.4.12). Consequently, in order to prove that $|\mathbb{R}| = |\mathcal{P}(\mathbb{N})|$, it is sufficient to show that $|\mathbb{R}| = |A|$ (by Theorem 5.2.2). However, since $\mathbb{R}$ in turn is easily seen to be cardinally equivalent to the interval $[0,1]$ (see Exercise 6.4.13), we only need to show that $|A| = |[0,1]|$, and in order to establish this latter statement, it is sufficient to exhibit two maps $f : A \to [0,1]$ and $g : A \to [0,1]$ that are injective and surjective, respectively (by Theorem 5.3.2).

6.4.10 Theorem. *The function $f : A \to [0,1]$ that is defined by the equation*

$$f((a_n)_{n\in\mathbb{N}}) := \sum_{k=1}^{\infty} \frac{a_k}{3^k}$$

is well defined and injective.

Proof. According to Lemma 6.4.9, the series $\sum_{k=1}^{\infty} a_k/3^k$ is convergent for all $(a_n)_{n\in\mathbb{N}} \in A$. Furthermore, using Lemma 6.4.8, we may infer that

$$0 \leq \sum_{k=1}^{n} \frac{a_k}{3^k} \leq \sum_{k=1}^{n} \frac{1}{3^k} = \frac{(1/3)(1-(1/3)^n)}{1-1/3} \leq \frac{1}{2}$$

for all $n \in \mathbb{N}$, and therefore $0 \le f((a_n)_{n \in \mathbb{N}}) \le 1/2 < 1$ (by Exercise 6.4.15). Hence $f((a_n)_{n \in \mathbb{N}})$ is well defined because $f((a_n)_{n \in \mathbb{N}}) \in [0, 1]$. In order to show that f is injective, we assume that we are given sequences $(a_n)_{n \in M}, (b_n)_{n \in \mathbb{N}} \in A$ such that $(a_n)_{n \in M} \ne (b_n)_{n \in \mathbb{N}}$. Then the set $M := \{n \in \mathbb{N} \mid a_n \ne b_n\}$ is nonempty, and according to the Well-Ordering Principle $n_0 := \min M$ is well defined. Assuming w.l.o.g. that $a_{n_0} = 1$ and $b_{n_0} = 0$, it follows that

$$\sum_{k=1}^{n} \frac{a_k}{3^k} - \sum_{k=1}^{n} \frac{b_k}{3^k} = \sum_{k=1}^{n} \frac{a_k - b_k}{3^k} = \frac{1}{3^{n_0}} + \sum_{k=n_0+1}^{n} \frac{a_k - b_k}{3^k}$$

$$\ge \frac{1}{3^{n_0}} - \sum_{k=n_0+1}^{n} \frac{1}{3^k} = \frac{1}{3^{n_0}} - \frac{(1/3)^{n_0+1}(1 - (1/3)^{n-n_0})}{1 - 1/3}$$

$$> \frac{1}{3^{n_0}} - \frac{(1/3)^{n_0+1}}{2/3} = \frac{1}{2 \cdot 3^{n_0}}.$$

Consequently, we may apply Theorem 6.3.1 and Exercise 6.4.15 to infer that

$$0 < \frac{1}{2 \cdot 3^{n_0}} \le \lim_{n \to \infty} \left(\sum_{k=1}^{n} \frac{a_k}{3^k} - \sum_{k=1}^{n} \frac{b_k}{3^k} \right) = \lim_{n \to \infty} \sum_{k=1}^{n} \frac{a_k}{3^k} - \lim_{n \to \infty} \sum_{k=1}^{n} \frac{b_k}{3^k}$$

$$= \sum_{k=1}^{\infty} \frac{a_k}{3^k} - \sum_{k=1}^{\infty} \frac{b_k}{3^k} = f((a_n)_{n \in M}) - f((b_n)_{n \in \mathbb{N}}).$$

Thus $f((a_n)_{n \in M}) \ne f((b_n)_{n \in \mathbb{N}})$, and therefore, f is injective. $\square$

6.4.11 Theorem. *The function* $g : A \to [0, 1]$ *that is defined by the equation*

$$g((a_n)_{n \in \mathbb{N}}) := \sum_{k=1}^{\infty} \frac{a_k}{2^k}$$

is well defined and surjective.

Proof. As in the proof of the preceding theorem, we may apply Lemma 6.4.9 to infer that the series $\sum_{k=1}^{\infty} a_k/2^k$ is convergent for all $(a_n)_{n \in \mathbb{N}} \in A$ and then use Lemma 6.4.8 to conclude that

$$0 \le \sum_{k=1}^{n} \frac{a_k}{2^k} \le \sum_{k=1}^{n} \frac{1}{2^k} = \frac{(1/2)(1 - (1/2)^n)}{1 - 1/2} \le 1$$

for all $n \in \mathbb{N}$. Hence $0 \le g((a_n)_{n \in \mathbb{N}}) \le 1$ (by Exercise 6.4.15), and therefore, $g((a_n)_{n \in \mathbb{N}})$ is well defined because $g((a_n)_{n \in \mathbb{N}}) \in [0, 1]$. In order to prove that g is surjective, we pick a value $x \in [0, 1]$ and inductively define a sequence $(a_n(x))_{n \in \mathbb{N}}$ as follows:

$$a_1(x) := \begin{cases} 1 & \text{if } x > 1/2, \\ 0 & \text{if } x \le 1/2, \end{cases}$$

and

$$a_{n+1}(x) := \begin{cases} 1 & \text{if } x > 1/2^{n+1} + \sum_{k=1}^{n} a_k(x)/2^k, \\ 0 & \text{if } x \le 1/2^{n+1} + \sum_{k=1}^{n} a_k(x)/2^k \end{cases}$$

for all $n \in \mathbb{N}$. To proceed, we will show, by means of induction, that the partial sums

$$S_n(x) := \sum_{k=1}^{n} \frac{a_k(x)}{2^k}$$

satisfy the inequality

$$S_n(x) \le x \le S_n(x) + \frac{1}{2^n} \qquad (6.27)$$

for all $n \in \mathbb{N}$. To establish the inductive base, we observe that for $n = 1$ the inequality above assumes the form

$$\frac{a_1(x)}{2} \le x \le \frac{a_1(x)}{2} + \frac{1}{2},$$

and that the definition of $a_1(x)$ readily implies that this inequality is indeed valid (because $a_1(x) = 0$ if $0 \le x \le 1/2$ and $a_1(x) = 1$ if $1/2 < x \le 1$). As we now assume that (6.27) is satisfied for some $n \in \mathbb{N}$, it remains for us to show that (6.27) is satisfied as well for $n + 1$. Since (6.27) implies that

$$x \in \left[S_n(x), S_n(x) + \frac{1}{2^n} \right],$$

the definition of $a_{n+1}(x)$ allows us to infer that

$$a_{n+1}(x) = 0 \Leftrightarrow S_{n+1}(x) = S_n(x)$$
$$\Leftrightarrow x \in \left[S_n(x), S_n(x) + \frac{1}{2^{n+1}} \right]$$
$$\Leftrightarrow x \in \left[S_{n+1}(x), S_{n+1}(x) + \frac{1}{2^{n+1}} \right]$$

and that

$$a_{n+1}(x) = 1 \Leftrightarrow S_{n+1}(x) = S_n(x) + \frac{1}{2^{n+1}}$$
$$\Leftrightarrow x \in \left[S_n(x) + \frac{1}{2^{n+1}}, S_n(x) + \frac{1}{2^n} \right]$$
$$\Leftrightarrow x \in \left[S_{n+1}(x), S_{n+1}(x) + \frac{1}{2^{n+1}} \right]$$

Consequently, (6.27) is indeed valid for $n + 1$. Finally, in order to show that $(S_n(x))_{n \in \mathbb{N}}$ converges to x, we assume that $\varepsilon > 0$ is a given real or rational number and use the result of Example 4.5.2 to conclude that

$$n < 1 + n \le (1 + 1)^n = 2^n$$

for all $n \in \mathbb{N}$. Given this observation, we pick an $N \in \mathbb{N}$ such that $N > 1/\varepsilon$ and apply (6.27) to infer that

$$|S_n(x) - x| \leq \frac{1}{2^n} < \frac{1}{n} < \frac{1}{N} < \varepsilon$$

for all $n > N$, as desired. Hence

$$x = \lim_{n \to \infty} S_n(x) = \sum_{k=1}^{\infty} \frac{a_k(x)}{2^k} = g((a_n(x))_{n \in \mathbb{N}}),$$

and this shows that g is surjective.   □

Having thus established that $|\mathbb{R}| = |\mathcal{P}(\mathbb{N})| > |\mathbb{N}|$ (by Theorem 5.2.7), there arises the natural question as to whether there exists a set $M \subset \mathbb{R}$ such that $|\mathbb{N}| < |M| < |\mathbb{R}|$. Naively, we might think that $\mathbb{Q}$, for example, could be such a set, as $\mathbb{Q}$ appears to contain 'more' elements than $\mathbb{N}$ and fewer than $\mathbb{R}$. But, as it turns out, $\mathbb{Q}$ is actually cardinally equivalent to $\mathbb{N}$ (see Exercise 6.4.17), and as it further turns out, the very surprising answer to question concerning the existence of such a set M is generally accepted to be that there really is no answer. For the assertion of the non-existence of such a set M—which is known as the *Continuum Hypothesis*—can be shown to be independent of the (commonly accepted) axioms of set theory. That is to say, on Mondays at lunchtime we may assume that no such set M exists, and on Tuesdays during breakfast we may imagine that such a set M does exist. In either case we will not come into conflict with any piece of known mathematics because all of mathematics—as human beings currently construe it—is derivative of the axioms of set theory. Put differently, we can choose freely, according to whim, whether we wish to inhabit a universe in which the Continuum Hypothesis is true or rather one in which it is false—so much for truth in mathematics being well-defined and certain.

Exercises

6.4.12. Prove that h, as defined in (6.25) and (6.26), is a bijection.

6.4.13. Show that $|\mathbb{R}| = |[0,1]|$. *Hint:* consider the function $g : (0,1) \to \mathbb{R}$ that is defined by the equation $g(x) := (2x-1)/(1-(2x-1)^2)$ for all $x \in (0,1)$.

6.4.14. Use the result of Exercise 6.4.13 to show that $|\mathbb{R}| = |[a,b]|$ whenever $a, b \in \mathbb{R}$ and $a < b$.

6.4.15. Prove the following statement: if $(a_n)_{n \in \mathbb{N}}$ is a convergent sequence in $\mathbb{R}$ and if $b \leq a_n \leq c$ for some $b, c \in \mathbb{R}$, then $b \leq \lim_{n \to \infty} a_n \leq c$.

6.4.16. Prove that $|\mathbb{N}| = |\mathbb{Q}|$.

6.4.17. Prove that $|\mathbb{N}| = |\mathbb{Z}|$.

6.5   The Complex Numbers

The common way to introduce the *complex numbers*—denoted by $\mathbb{C}$—is to postulate the existence of an *imaginary number* i whose square is equal to negative one, that is,

$$i^2 = -1.$$

At first sight, this may seem absurd as there really is no such number—at least not if by 'really' we mean 'real' because for all *real* numbers $x \in \mathbb{R}$ it is the case that $x^2 \geq 0$ (by Proposition 6.3.9 and Theorem 6.3.8i). But then again mathematics in its entirety is a product of the human imagination and thus imaginary by nature. So there is nothing wrong *per se* with the addition of an imaginary entity to an imaginary subject.

Moreover, as far as the alignment of number systems with our intuitive grasp of reality is concerned, we already get into trouble when contemplating numbers that are negative. We can eat one apple or two apples or maybe no apple, but eating negative three apples is clearly a challenge. That is to say, the notion of there being something that actually is less than nothing is inherently highly perplexing. But on the other hand negative numbers are so completely familiar to us that the problem of exactly how a negative quantity can be thought to exist doesn't feel very urgent to us. Negative numbers are useful and that is really all there is to it. The equation $x + 1 = 0$ does not have a solution in $\mathbb{N}$ but it does have a solution in $\mathbb{Z}$, and therefore, it makes sense to introduce $\mathbb{Z}$ as an extension of $\mathbb{N}$. Similarly, the equation $x^2 + 1 = 0$ cannot be solved in $\mathbb{R}$ but it can be solved in $\mathbb{C}$ (where its solution is $x = \pm i$), and hence, $\mathbb{C}$ is useful.

That said, however, we also mustn't forget that in mathematics our imagination is not entirely free because it always must be tied to sets and elements of sets. It is in this regard that the proposed introduction of i is somewhat dubious. For i is neither a set nor linked transparently, or so it seems, to any of the axioms that constitute $\mathbb{N}$—and in the context of the present exposition this is simply not acceptable. So what we will do in consequence is to rely on the standard flawed approach, using i, only for inspiration. We will use it as a guide and then turn it into something more rigorous and adequate.

To get started we need to understand that a complex number is commonly construed to be a sum of the form

$$a + ib$$

where both a and b are real numbers. The first observation that follows from this is that a complex number, as a mere object, is exhaustively described by a pair of real numbers $(a, b) \in \mathbb{R}^2$. But this is not to say, of course, that we can somehow construe $\mathbb{C}$ to be identical with $\mathbb{R}^2$ and then just leave it at that. For complex numbers—as numbers—are entities that have number-like properties with respect to the algebraic operations that we naturally perform with them. So $\mathbb{C}$ is equal to $\mathbb{R}^2$ only on condition that the latter set is equipped with proper algebraic structures. And to identify these structures, we need to realize that two complex numbers $a+ib$ and $c+id$, in standard form, are added and multiplied as follows:

$$(a + ib) + (c + id) = (a + c) + i(b + d)$$

and

$$(a + ib)(c + id) = ac + i^2bd + iad + ibc = (ac - bd) + i(ad + bc).$$

Given these rules, it makes sense for us to define addition and multiplication on $\mathbb{R}^2$ by means of the following maps:

$$+ : \mathbb{R}^2 \times \mathbb{R}^2 \to \mathbb{R}^2$$
$$((a, b), (c, d)) \mapsto (a, b) + (c, d) = (a + c, b + d)$$

and

$$\cdot : \mathbb{R}^2 \times \mathbb{R}^2 \to \mathbb{R}^2$$
$$((a, b), (c, d)) \mapsto (a, b) \cdot (c, d) = (ac - bd, ad + bc).$$

So as we now define

$$\boxed{\mathbb{C} := \mathbb{R}^2,}$$

we do so with the understanding that addition and multiplication in $\mathbb{R}^2$ are performed by means of these maps:

$$(a, b) + (c, d) = (a + c, b + d) \tag{6.28}$$

and

$$(a, b) \cdot (c, d) = (ac - bd, ad + bc). \tag{6.29}$$

Remark. Strictly speaking, $\mathbb{C}$ ought to be defined to be the triple $(\mathbb{R}^2, +, \cdot)$. But if we really adopted this approach, many common references to $\mathbb{C}$, concerning subsets or elements (such as, for example, $M \subset \mathbb{C}$ or $(a, b) \in \mathbb{C}$), would have to be re-interpreted to refer only to the set-component $\mathbb{R}^2$ in $\mathbb{C}$ rather than the whole triple $(\mathbb{R}^2, +, \cdot)$.

The natural identification map that allows us to consider $\mathbb{R}$ to be a subset of $\mathbb{C}$ is the map that sends $a \in \mathbb{R}$ into $(a, 0) \in \mathbb{C}$. In other words, $\mathbb{R}$ may here be considered to the set of all complex numbers whose second component is zero:

$$\mathbb{R} = \{(a, b) \in \mathbb{C} \mid b = 0\}.$$

Furthermore, due to this natural identification, the first component of a complex number $(a, b) \in \mathbb{C}$ is referred to as the *real part* of (a, b), and the second is said to be the *imaginary part*:

$$\mathcal{R}e(a, b) := a$$

and

$$\mathcal{I}m(a, b) := b.$$

Given these definitions, the imaginary number $i = 0 + 1i$ is revealed to be the pair $(0, 1)$, and the number $1 \in \mathbb{R}$ is represented by the pair $(1, 0)$. Moreover, using the rules for summation and multiplication in (6.28) and (6.29), we find that the sum $a + ib$ assumes the form

$$(a, 0) + (0, 1)(b, 0) = (a, 0) + (0, b) = (a, b).$$

Having thus established the basic structure of $\mathbb{C}$, our next and already final objective is to state and prove the complex-number analogue of Theorem 6.3.8. And to do so, we first need to define subtraction and division. Concerning the first of these, we agree to define

$$-(a, b) := (-a, -b), \tag{6.30}$$

for all $(a, b) \in \mathbb{C}$ and then set

$$(a, b) - (c, d) := (a, b) + (-(c, d)) = (a - c, b - d)$$

for all $(a, b), (c, d) \in \mathbb{C}$. A trivial consequence of (6.30) is the fact that i^2 equals negative one:

$$i^2 = (0, 1) \cdot (0, 1) = (0 \cdot 0 - 1 \cdot 1, 1 \cdot 0 + 0 \cdot 1) = (-1, 0) = -(1, 0) = -1.$$

Moving on from subtraction to division, we note that the quotient of a complex number $a + ib$ and a real number $c \neq 0$ is naturally construed to be

$$\frac{a + ib}{c} = \frac{a}{c} + i\frac{b}{c}.$$

Furthermore, in defining the *complex conjugate* of a complex number $z = a + ib$ to be

$$\bar{z} := a - ib, \tag{6.31}$$

we observe that the product of z with $\bar{z}$ is always a non-negative real number:

$$z\bar{z} = (a + ib)(a - ib) = a^2 - i^2 b^2 = a^2 + b^2.$$

Consequently, as we further impose—by definition—the natural cancellation requirement that z_1/z_2 be equal to $(z_1 z)/(z_2 z)$ for all $z_1 \in \mathbb{C}$ and $z, z_2 \in$

$\mathbb{C} \smallsetminus \{0\} = \mathbb{C} \smallsetminus \{0 + i0\} = \mathbb{C} \smallsetminus \{(0,0)\}$, we are led to conclude that a complex quotient of the form

$$\frac{a + ib}{c + id}$$

ought to be equal to

$$\frac{(a + ib)(c - id)}{(c + id)(c - id)} = \frac{ac + bd + i(bc - ad)}{c^2 + d^2} = \frac{ac + bd}{c^2 + d^2} + i\frac{bc - ad}{c^2 + d^2}.$$

Thus we define

$$\frac{(a, b)}{(c, d)} = (a, b)(c, d)^{-1} := \left(\frac{ac + bd}{c^2 + d^2}, \frac{bc - ad}{c^2 + d^2} \right)$$

for all $(a, b) \in \mathbb{C}$ and all $(c, d) \in \mathbb{C} \smallsetminus \{(0,0)\}$.

6.5.1 Theorem. *For all $u, v, w \in \mathbb{C}$ it is the case that*

   a) $u + (v + w) = (u + v) + w$,

   b) $u + 0 = u$ where $0 = (0, 0)$,

   c) $u - u = -u + u = 0$,

   d) $u + v = v + u$,

   e) $u(vw) = (uv)w$,

   f) $uv = uv$,

   g) $u = 1 \cdot u$ and $-u = (-1) \cdot u$ where $1 = (1, 0)$,

   h) $(1/u)u = u^{-1}u = u/u = 1$ whenever $u \neq 0$,

   i) $uv = 0 \Leftrightarrow (u = 0 \vee v = 0)$,

   j) $u(v + w) = uv + uw$, and

   k) $u(v - w) = uv - uw$.

Proof. Restricting ourselves to h) and i) and leaving the proofs of the remaining properties as exercises to the reader (see Exercise 6.5.3), we notice that for $u = (a, b) \in \mathbb{C} \smallsetminus \{(0,0)\}$ it is the case that

$$\frac{1}{u} = \frac{(1, 0)}{(a, b)} = \left(\frac{a}{a^2 + b^2}, \frac{-b}{a^2 + b^2} \right)$$

and therefore,

$$\frac{1}{u} \cdot u = \left(\frac{a}{a^2 + b^2}, \frac{-b}{a^2 + b^2} \right) \cdot (a, b) = \left(\frac{a^2 + b^2}{a^2 + b^2}, 0 \right) = (1, 0) = 1$$

and

$$\frac{u}{u} = \frac{(a, b)}{(a, b)} = \left(\frac{a^2 + b^2}{a^2 + b^2}, 0 \right),$$

as desired. Turning to i), we notice that the implication $(u = 0 \lor v = 0) \Rightarrow uv = 0$ is a trivial consequence of (6.29). So we only need to show that for all complex numbers $u = (a, b)$ and $v = (c, d)$ the assumption $uv = 0$ implies that $u = 0 \lor v = 0$ or equivalently, that $(uv = 0 \land u \neq 0) \Rightarrow v = 0$. If $uv = 0$ and $u \neq 0$, then $v = 1v = (1/u)uv = (1/u) \cdot 0 = 0$, as desired. $\square$

Since there is no natural linear order on $\mathbb{C}$, there also is no complex-number analogue of Proposition 6.3.10. Consequently, we now have actually reached the end of our discussion of *Logic, Sets, and Numbers*. There are a few more standard properties of complex numbers listed in the exercises below, but other than that, we are done—except, that is, for one important final add-on: we wish to pay tribute, one last time, to the spirit of the cover image and offer the reader one more version of the equation $1 + 1 = 2$ (see Exercise 6.5.4):

$$\left(\left[([([(n^2, 1)]_{\mathbb{Z}}, [(n^2 + 2, 1)]_{\mathbb{Z}})]_{\mathbb{Q}})_{n \in \mathbb{N}} \right]_{\mathbb{R}}, \left[([([(n, 1)]_{\mathbb{Z}}, [(n^2 + 2.1)]_{\mathbb{Z}})]_{\mathbb{Q}})_{n \in \mathbb{N}} \right]_{\mathbb{R}} \right)$$

$$+ \left(\left[([([(n + 4, 1)]_{\mathbb{Z}}, [(n, 1)]_{\mathbb{Z}})]_{\mathbb{Q}})_{n \in \mathbb{N}} \right]_{\mathbb{R}}, \left[([([(n, 1)]_{\mathbb{Z}}, [(n^2, 2)]_{\mathbb{Z}})]_{\mathbb{Q}})_{n \in \mathbb{N}} \right]_{\mathbb{R}} \right)$$

$$= \left(\left[([([(2n^2, 1)]_{\mathbb{Z}}, [(n^2, 8)]_{\mathbb{Z}})]_{\mathbb{Q}})_{n \in \mathbb{N}} \right]_{\mathbb{R}}, \left[([([(n^2, 1)]_{\mathbb{Z}}, [(n^3, 1)]_{\mathbb{Z}})]_{\mathbb{Q}})_{n \in \mathbb{N}} \right]_{\mathbb{R}} \right).$$

$$(6.32)$$

Exercises

6.5.2. Evaluate the following sums, products, and quotients: $(1 + 2i) + (3 - i)$, $(1 + 2i)(3 - i)$, $(1 + 2i)/(3 - i)$, and $1/i$.

6.5.3. Prove Theorem 6.5.1a,b,c,d,e,f,g,j,k.

6.5.4. Explain why equation (6.32) is equivalent to the all-important assertion that one plus one equals two.

6.5.5. Prove that the operation of complex conjugation, as defined in (6.31) (i.e., $\overline{(a, b)} := (a, -b)$ for all $(a, b) \in \mathbb{C}$), satisfies the following properties for all $u, v \in \mathbb{C}$:

a) $\overline{u + v} = \overline{u} + \overline{v}$,

b) $\overline{u \cdot v} = \overline{u} \cdot \overline{v}$,

c) $\overline{u/v} = \overline{u}/\overline{v}$ whenever $v \neq 0$.

6.5.6. The absolute value of a complex number $(a, b) \in \mathbb{C}$ is defined by the equation

$$|(a, b)| := \sqrt{a^2 + b^2}.$$

Prove that the absolute value satisfies the following properties for all $u, v \in \mathbb{C}$:

a) $|u + v| \le |u| + |v|$,

b) $|uv| = |u||v|$,

c) $|u/v| = |u|/|v|$ whenever $v \ne 0$.

Selected Solutions

Chapter 1

1.1.5 a) and d) are false statements, c) is a true statement, and b) is not a statement because x is not specified.

1.1.11 Colors are not objective qualities but rather are subjective impressions in the human mind. So it is not the sky that is blue but rather the human perception thereof. Consequently, a variant of the given statement that might be considered to be true is the assertion, 'The sky is colorless'. On the other hand, though, it is not at all clear what precisely is meant by the word "sky." Exactly how can this verbal symbol be construed to be representative of a well-defined object to which a quality such as colorlessness can be meaningfully assigned? What part of the physical world is to be considered to be 'the sky' or what exactly is meant by the phrase 'physical world'? If these questions cannot be answered conclusively, then the assertion that the sky is blue is not only false but is not even a statement. In other words, the state of its being a statement is in this case not objective but is instead a subjective opinion conditioned by personal philosophical taste.

1.2.2 The schemata $(P \wedge Q) \wedge \bar{R}$ and $P \wedge (Q \wedge \bar{R})$ are both true if and only if P, Q, and R are each true. So there is no difference between these two schemata, and therefore it is permissible to remove the parantheses and write $P \wedge Q \wedge \bar{R}$.

1.2.4 There are two truth values that a single variable can assume, and for each of these two input values there are two output values. Consequently, the number of truth functions that depend on one variable is $2^2 = 4$. Similarly, the number of truth functions that depend on two variables is $2^{(2^2)} = 16$, and, in general, the number of truth functions that depend on n variables is $2^{(2^n)}$.

1.2.7 Given the truth table below, we notice that the schema $\neg(\bar{P} \wedge \bar{Q})$ is equivalent to the schema $P \vee Q$.

P	Q	$\bar{P}$	$\bar{Q}$	$\bar{P} \wedge \bar{Q}$	$\neg(\bar{P} \wedge \bar{Q})$
T	T	F	F	F	T
T	F	F	T	F	T
F	T	T	F	F	T
F	F	T	T	T	F

201

1.2.10 Given the truth table below, we notice that the schema $\bar{P} \Leftrightarrow \bar{Q}$ is equivalent to the schema $P \Leftrightarrow Q$.

P	Q	$\bar{P}$	$\bar{Q}$	$\bar{P} \Leftrightarrow \bar{Q}$
T	T	F	F	T
T	F	F	T	F
F	T	T	F	F
F	F	T	T	T

1.3.3 a), b), and c) are simple conjunctions of the form $P \wedge Q$. Furthermore, in setting $P :\Leftrightarrow$ 'Grumpy is happy' and $Q :\Leftrightarrow$ 'Stupy is happy', it follows that d) and e) are both equivalent to $P \Rightarrow Q$. Finally, i) and j) are both alternations of the form $P \vee Q$, and l) is ambiguous $((P \Rightarrow Q) \wedge R$ or $P \Rightarrow (Q \wedge R))$.

1.3.4 In comparing i) and j), we notice that, according to i), the reason for Grumpy to get lost is his failure to get up and that j) suggests the reverse causal relation: Grumpy is unable to get up because he gets lost.

1.4.10 If P is true, then the schema $(P \Rightarrow Q) \Leftrightarrow (\bar{Q} \Rightarrow \bar{P})$ simplifies as follows:

$$(T \Rightarrow Q) \Leftrightarrow (\bar{Q} \Rightarrow F)$$
$$Q \Leftrightarrow \neg \bar{Q}$$
$$Q \Leftrightarrow Q$$
$$T.$$

And if P is false, then we find that

$$(F \Rightarrow Q) \Leftrightarrow (\bar{Q} \Rightarrow T)$$
$$T \Leftrightarrow T$$
$$T,$$

as desired.

1.4.11 We will prove b) by means of a truth-value analysis and then establish d) by using b) and (1.8). If P is true, then the schema $\neg(P \vee Q) \Leftrightarrow (\bar{P} \wedge \bar{Q})$ reduces to

$$\neg(T \vee Q) \Leftrightarrow (F \wedge \bar{Q})$$
$$\neg T \Leftrightarrow F$$
$$F \Leftrightarrow F$$
$$T,$$

and if P is false, we find that

$$\neg(F \vee Q) \Leftrightarrow (T \wedge \bar{Q})$$
$$\neg Q \Leftrightarrow \bar{Q}$$
$$T,$$

as desired. Furthermore, to prove d), we argue as follows:

$$\neg(P \Rightarrow Q) \Leftrightarrow \neg(\bar{P} \vee Q) \quad \text{(by (1.8))}$$
$$\Leftrightarrow (\neg\bar{P} \wedge \bar{Q}) \quad \text{(by b))}$$
$$\Leftrightarrow (P \wedge \bar{Q}).$$

1.4.13 a) According to Proposition 1.4.5a, it is the case that

$$\neg(\bar{P} \wedge \bar{Q}) \Leftrightarrow (\neg\bar{P} \vee \neg\bar{Q})$$
$$\Leftrightarrow (P \vee Q).$$

b)

$$\neg(P \wedge \bar{Q}) \Leftrightarrow (\bar{P} \vee \neg\bar{Q}) \quad \text{(by Proposition 1.4.5a)}$$
$$\Leftrightarrow (\bar{P} \vee Q)$$
$$\Leftrightarrow (P \Rightarrow Q) \quad \text{(by (1.8))}.$$

1.5.13 Setting

$$F(x) :\Leftrightarrow \text{'}x \text{ has fur'},$$
$$N(x) :\Leftrightarrow \text{'}x \text{ has a long nose'},$$
$$L_2(x) :\Leftrightarrow \text{'}x \text{ is two-legged'},$$
$$L_4(x) :\Leftrightarrow \text{'}x \text{ is four-legged'}$$

for all x, it follows that statement a) is equivalent to

$$(\neg\exists_x L_4(x)) \Rightarrow \forall_x(\bar{F}(x) \Rightarrow (L_2(x) \wedge N(x)))$$

and that c) is equivalent to

$$\forall_x(L_4(x) \Rightarrow (F(x) \vee N(x)) \wedge \exists_x(L_2(x) \wedge N(x) \wedge \bar{F}(x)).$$

1.5.16 a) $\exists_{x,y\in\mathbb{R}}((x-2)^2 + y^2 = 1 \wedge x^2 + y^2 = 1) \wedge \forall_{x,y\in\mathbb{R}}((x-2)^2 + y^2 = 1 \Rightarrow x^2 + y^2 \geq 1)$
c) $\exists_{n\in\mathbb{N}} 60 = 12n \wedge \exists_{n\in\mathbb{N}} 60 = 10n \wedge \forall_{m\in\mathbb{N}}((\exists_{n\in\mathbb{N}} m = 12n \wedge \exists_{n\in\mathbb{N}} m = 10n) \Rightarrow m \geq 60)$
e) $\exists_{x,y\in\mathbb{R}}((y = -(x-1/\sqrt{2}) + 1/\sqrt{2} \wedge x^2 + y^2 = 1) \wedge \forall_{s,t\in\mathbb{R}}((t = -(s-1/\sqrt{2}) + 1/\sqrt{2} \wedge s^2 + t^2 = 1) \Rightarrow (s = x \wedge t = y))).$
f) $p \in \mathbb{N} \wedge p > 1 \wedge \forall_{n\in\mathbb{N}}((\exists_{m\in\mathbb{N}} mn = p) \Rightarrow (n = p \vee n = 1))$
1.5.17 a) Every odd integer greater than one is a prime number.
b) Every even integer greater than two can be written as a sum of two prime numbers.

1.5.18 a) $\forall_{S\subset\mathbb{R}}((\exists_{x\in S}\forall_{y\in S} x \leq y \wedge \exists_{x\in S}\forall_{y\in S} x \geq y) \Rightarrow ((\forall_{x,y\in S} x = y) \Leftrightarrow \forall_{x,y\in S}((\forall_{z\in S}(x \leq z \wedge y \geq z)) \Rightarrow x = y)))$

b) $\forall_{S\subset\mathbb{R}}(S \neq \emptyset \Rightarrow ((\exists_{M\in\mathbb{R}}\forall_{x\in S}\, x \leq M) \Rightarrow \exists_{N\in\mathbb{R}}(\forall_{x\in S}\, x \leq N \wedge \forall_{M\in S}((\forall_{x\in S}\, x \leq M) \Rightarrow N \leq M))))$

1.5.20 a) $f(x) := x/2$, **c)** $f(x) := 1 - x$.

1.5.21 Here we may define $f(x) := x + 1$ for $x \in [-1,0]$, $f(1) := 0$, and $f(x) := x - 1 + 3/2^n$ for all $n \in \mathbb{N}$ and all $x \in (1 - 1/2^{n-1}, 1 - 1/2^n]$. Note: in order to understand the meaning of this definition, the reader may want to draw the graph of f on $[-1,0]$, on $(1 - 1/2^{n-1}, 1 - 1/2^n]$ for $n = 1$ and $n = 2$, and at $x = 1$.

1.6.11 Let $\varepsilon > 0$ and $N > 1/\varepsilon$. Then for all $n > N$ it is the case that

$$\left|\frac{2n+3}{n+1} - 2\right| = \frac{1}{n+1} < \frac{1}{n} < \frac{1}{N} < \varepsilon,$$

as desired.

1.6.13 a) The statement is false because the schema $n > N \Rightarrow 1/n < \varepsilon$ is false for $\varepsilon := 1/2$, $N := 1$, and $n := 2$.
b) For $n := 1$ the antecedent $n > N$ is false for all $N \in \mathbb{N}$, and in particular for, say, $N := 1$. Consequently, the statement as a whole is true.
c) The statement is true because for $\varepsilon := 2$ the consequent $1/n < \varepsilon$ is true for all $n \in \mathbb{N}$.

1.6.14 a) The antecedent $\forall_{n\in\mathbb{N}}\, n > N$ is false for all $N \in \mathbb{N}$, and therefore, it is false in particular for, say, $N := 1$. Thus the statement as a whole is true.
b) The consequent $\exists_{n\in\mathbb{N}}\, 1/n < \varepsilon$ is true for all $\varepsilon > 0$, and therefore, the statement as a whole is true as well.
c) For $N := 2$ the antecedent $\forall_{n\in\mathbb{N}}\, n \geq N$ is false, and therefore, the statement as a whole is true.

1.7.11 Using b) and d) in conjunction with the fact that $(P \Rightarrow Q) \Leftrightarrow (\bar{P} \vee Q)$ is a tautology, it follows that

$$\exists_{x\in\mathcal{U}}(R \Rightarrow P(x)) \Leftrightarrow \exists_{x\in\mathcal{U}}(\bar{R} \vee P(x))$$
$$\Leftrightarrow (\bar{R} \vee \exists_{x\in\mathcal{U}}\, P(x))$$
$$\Leftrightarrow (R \Rightarrow \exists_{x\in\mathcal{U}}\, P(x))$$

and

$$\exists_{x\in\mathcal{U}}(P(x) \Rightarrow R) \Leftrightarrow \exists_{x\in\mathcal{U}}(\bar{P}(x) \vee R)$$
$$\Leftrightarrow ((\exists_{x\in\mathcal{U}}\bar{P}(x)) \vee R)$$
$$\Leftrightarrow ((\neg\forall_{x\in\mathcal{U}}\, P(x)) \vee R)$$
$$\Leftrightarrow ((\forall_{x\in\mathcal{U}}\, P(x)) \Rightarrow R),$$

as desired.

1.7.13 If R is true, then the two statements are equivalent, because they both are equivalent to $\forall_{x\in\mathcal{U}} P(x)$. By contrast, if R is false, then the statement $\forall_{x\in\mathcal{U}}(R \Leftrightarrow P(x))$ is equivalent to $\forall_{x\in\mathcal{U}} \bar{P}(x)$, and the statement $R \Leftrightarrow \forall_{x\in\mathcal{U}} P(x)$ is equivalent to $\exists_{x\in\mathcal{U}} \bar{P}(x)$. So in this case, the former statement implies the latter as long as $\mathcal{U}$ is nonempty. Consequently, we may conclude that in a nonempty universe the statement $\forall_{x\in\mathcal{U}}(R \Leftrightarrow P(x))$ implies the statement $R \Leftrightarrow \forall_{x\in\mathcal{U}} P(x)$. Moreover, the reverse implication is not true because the existence of an x for which $\bar{P}(x)$ is true does not in general imply that $\bar{P}(x)$ is true for all x (counterexamples are easy to find).

1.7.15 If R is true then both statements are true, and if R is false, then the statement $(\exists_{x\in\mathcal{U}} P(x)) \Rightarrow R$ is equivalent to $\forall_{x\in\mathcal{U}} \bar{P}(x)$, and the statement $\exists_{x\in\mathcal{U}}(P(x) \Rightarrow R)$ is equivalent to $\exists_{x\in\mathcal{U}} \bar{P}(x)$. This shows that $(\exists_{x\in\mathcal{U}} P(x)) \Rightarrow R$ implies $\exists_{x\in\mathcal{U}}(P(x) \Rightarrow R)$ as long as $\mathcal{U}$ is nonempty.

1.7.17 d) If there exists an x so that both $P(x)$ and $Q(x)$ are true, then, in particular, there exists an x so that $P(x)$ is true, and there exists an x so that $Q(x)$ is true. Thus $\exists_x(P(x) \wedge Q(x))$ does indeed imply $\exists_x P(x) \wedge \exists_x Q(x)$.
e) Using the fact that $(P \Rightarrow Q) \Leftrightarrow (\bar{P} \vee Q)$ is a tautology, it follows that

$$(\forall_x P(x) \Rightarrow \exists_x Q(x)) \Leftrightarrow (\exists_x \bar{P}(x) \vee \exists_x Q(x))$$
$$\Leftrightarrow \exists_x(\bar{P}(x) \vee Q(x)) \quad \text{(by Proposition 1.7.7c)}$$
$$\Leftrightarrow \exists_x(P(x) \Rightarrow Q(x)).$$

1.7.18 d) If $\mathcal{U} := \mathbb{R}$, $P(x) :\Leftrightarrow x < 0$, and $Q(x) :\Leftrightarrow x > 0$, then $\exists_x P(x) \wedge \exists_x Q(x)$ is true, but $\exists_x(P(x) \wedge Q(x))$ is false.
f) If $\mathcal{U} := \mathbb{R}$ and $P(x) :\Leftrightarrow Q(x) :\Leftrightarrow x < 0$, then $\forall_x(P(x) \Rightarrow Q(x))$ is true, but $(\exists_x P(x)) \Rightarrow \forall_x Q(x)$ is false.

1.7.20 If we set $P(x) :\Leftrightarrow x^2 < 0$ and $Q(x) :\Leftrightarrow x < 0$ for all $x \in \mathbb{R}$, then $\exists_x(P(x) \Leftrightarrow Q(x))$ is true (because $P(0) \Leftrightarrow Q(0)$ is true), but $\exists_x P(x) \Leftrightarrow \exists_x Q(x)$ is false. Conversely, if $P(x) :\Leftrightarrow x > 0$ and $Q(x) :\Leftrightarrow x \leq 0$ for all $x \in \mathbb{R}$, then $\exists_x P(x) \Leftrightarrow \exists_x Q(x)$ is true (because both existence statements in this biconditional are true) but $\exists_x(P(x) \Leftrightarrow Q(x))$ is false.

Chapter 2

2.1.5 If $x + y$ were rational, then $x = (x + y) - y$ would be rational as well, in contradiction to the assumption that x is irrational. Thus $x + y$ must be irrational.

2.1.6 If $\sqrt{2} + \sqrt{3}$ were rational, then so would be $(\sqrt{2} + \sqrt{3})^2 = 3 + 2\sqrt{6}$. By implication, $\sqrt{6}$ would be rational as well and thus there would exist integers $m.n \in \mathbb{N}$ such that $\sqrt{6} = m/n$ and $\gcd(m, n)) = 1$. Consequently, $6n^2$ would be equal to m^2 and m^2 would be even. Since this would be possible only if m itself were even, we would find that $m = 2k$ for some $k \in \mathbb{N}$ and that $3n^2 = 2k^2$. Hence n would be even as well and $\gcd(m, n)$ would be greater than one. Having thus arrived at the false conclusion that $1 = \gcd(m, n) > 1$, we may infer that $\sqrt{2} + \sqrt{3}$ is indeed irrational.

2.1.8 Assume that $2^{1/n}$ is rational for some integer $n > 1$. Then $2^{1/n} = p/q$ for some $p, q \in \mathbb{N}$ with $\gcd(p, q) = 1$. Hence $2q^n = p^n$, and therefore, p must be even. Thus $p = 2k$ for some $k \in \mathbb{N}$, and, by implication, $q^n = 2^{n-1}k^n$. Since $n > 1$, it follows that q must be even as well, and therefore, $\gcd(p, q) > 1$. Having thus arrived at a contradiction, the proof is complete.

2.2.5 If both m and n are odd, then there are integers $i, j \in \mathbb{N}$ such that $m = 2i - 1$ and $n = 2j - 1$. Hence $mn = 4ij - 2i - 2j + 1 = 2(ij - i - j) + 1$ is odd, as desired. In order to show that the assumption that mn is odd implies that both m and n are odd, we will show, by way of contraposition, that the assumption that either m or n is even implies that mn is even. Assuming w.l.o.g. that m is even, it follows that there is a $k \in \mathbb{N}$ such that $n = 2k$. Hence $mn = 2mk$ is even as well.

2.3.10 a) The statement is true because the sequence $(1/n)_{n \in \mathbb{N}}$ is convergent (its limit is zero) and because every convergent sequence is also a Cauchy sequence (by Theorem 2.3.7).

b) The statement is false because its negation

$$\forall_{N \in \mathbb{N}} \exists_{\varepsilon > 0} \exists_{m \in \mathbb{N}} \exists_{n \in \mathbb{N}} ((m > N \wedge n > N) \wedge |a_n - a_m| \geq \varepsilon)$$

is true. For let $N \in \mathbb{N}$ be given and let $\varepsilon := 1/(N+1) - 1/(N+2)$, $n := N+1$, and $m := N+2$, then $n > N \wedge m > N$ and $|a_n - a_m| = 1/(N+1) - 1/(N+2) = \varepsilon$.

c) The statement is true because for given integers $n, m \in \mathbb{N}$ the antecedent $n > N \wedge m > N$ is false for $N := n$.

2.3.11 For all $n \in \mathbb{N}$ it is the case that $\sqrt{2}/n \in \mathbb{R} \setminus \mathbb{Q}$ because if $\sqrt{2}/n$ were rational for some $n \in \mathbb{N}$ then so would be $\sqrt{2}$. Since $\lim_{n \to \infty} \sqrt{2}/n = 0 \in \mathbb{Q}$, it follows that $(\sqrt{2}/n)_{n \in \mathbb{N}}$ is a sequence of irrational numbers that converges to a rational limit.

2.3.12 Since $(a_n)_{n \in \mathbb{N}}$ is convergent, there exists an $L \in \mathbb{R}$ and an $M \in \mathbb{N}$ such that $|a_n - L| < 1/2$ for all $n > M$. Hence

$$|a_n - a_m| \leq |a_n - L| + |a_m - L| < \frac{1}{2} + \frac{1}{2} = 1$$

for all $n, m > M$. Furthermore, the assumption $\forall_{n \in \mathbb{N}} a_n \in \mathbb{N}$ implies that $|a_n - a_m| \geq 1$ whenever $a_n \neq a_m$. Consequently, for all $n, m > M$ it is the case that $a_n = a_m$, and therefore, $a_n = a_N$ for all $n \geq N := M + 1$.

2.4.10 If $T > 0$ is the well defined period of f and if $n \in \mathbb{N}$, then

$$f(t + nT) = f(t + (n - 1)T) = \cdots = f(t + T) = f(t),$$

and therefore $S = nT > 0$ satisfies the periodicity condition. Conversely, if $S > 0$ satisfies the periodicity condition, then $S \geq T$ and $n := \min\{k \in \mathbb{Z} \mid$

$kT \geq S$} is a well defined positive integer for which it is the case that $nT \geq S$ and $(n-1)T < S$. Hence $0 \leq R := nT - S < nT - (n-1)T = T$ and

$$f(t+R) = f(t+nT-S) = f(t-S) = f(t-S+S) = f(t)$$

for all $t \in \mathbb{R}$. Consequently, R is a non-negative number that satisfies the periodicity condition and is less than T. Since T is the smallest strictly positive number that satisfies the periodicity, it must be the case that $0 = R$ and therefore, $S = nT$, as desired.

2.4.11 The function

$$f(x) := \begin{cases} 0 & \text{if } x < 0, \\ 1 & \text{if } x \geq 0 \end{cases}$$

is evidently non-periodic, and since $f(t+T) = f(t) = 1$ for all $T > t \geq 0$ and $f(t+T) = f(t) = 0$ for all $t < 0$ and $0 < T < |t|$, it is easy to see that f satisfies the given property.

2.4.12 One possible function that possesses the given properties is

$$f(x) := \begin{cases} 1 & \text{if } x \in \mathbb{N}, \\ 0 & \text{otherwise.} \end{cases}$$

We leave it to the reader to verify this claim.

Chapter 3

3.2.4 a) $\{(x,y) \mid x,y \in \mathbb{R} \wedge x^2 + y^2 = 1\}$, **b)** $\{x \in \mathbb{R} \mid 2 < x \leq 5\}$,
d) $\{n \in \mathbb{N} \mid \exists_{k \in \mathbb{N}} \, n = 2k - 1\}$.

3.3.13 $\mathcal{P}(S) = \{\emptyset, \{0\}, \{1\}, \{2\}, \{1,2\}, \{1,3\}, \{2,3\}, \{1,2,3\}\}$

3.3.14 a) This statement is true, because the empty set is a subset of every set.
b) This statement is false because $\mathcal{P}(\emptyset) = \{\emptyset\} \neq \emptyset$.

3.3.15 a) This statement is true, because the empty set is a subset of every set.
b) This statement is true because $\emptyset \subset S$, and therefore, $\emptyset \in \mathcal{P}(S)$.
c) This statement is also true because $\{\emptyset\} \subset \mathcal{P}(S)$ (by b)).

3.4.5 e)

$$\begin{aligned}
x \in A \cap (B \cup C) &\Leftrightarrow (x \in A \wedge x \in B \cup C) \\
&\Leftrightarrow (x \in A \wedge (x \in B \vee x \in C)) \\
&\Leftrightarrow ((x \in A \wedge x \in B) \vee (x \in A \wedge x \in C)) \\
&\quad \text{(by Proposition 1.4.5i)} \\
&\Leftrightarrow (x \in A \cap B \vee x \in A \cap C) \\
&\Leftrightarrow x \in (A \cap B) \cup (A \cap C).
\end{aligned}$$

m)

$$x \in (A \cap B)^c \Leftrightarrow \neg(x \in A \cap B)$$
$$\Leftrightarrow \neg(x \in A \wedge x \in B)$$
$$\Leftrightarrow x \notin A \vee x \notin B \quad \text{(by Proposition 1.4.5a)}$$
$$\Leftrightarrow x \in A^c \cup B^c.$$

r) According to p), we need to show that $A \cup B = (A \cap B) \cup ((A \cup B) \smallsetminus (A \cap B))$, and since $A \cap B \subset A \cup B$ (by h) and g)), it suffices to prove that $M = N \cup (M \smallsetminus N)$ whenever $N \subset M$. So let $N \subset M$ and let $P(x) :\Leftrightarrow x \in M$ and $Q(x) :\Leftrightarrow x \in N$. Given these definitions, it remains for us to show that

$$(\forall_x Q(x) \Rightarrow P(x)) \Rightarrow \forall_x(P(x) \Leftrightarrow (Q(x) \vee (P(x) \wedge \bar{Q}(x))))$$

or, equivalently, that

$$(\forall_x Q(x) \Rightarrow P(x)) \Rightarrow \forall_x(P(x) \Leftrightarrow (Q(x) \vee P(x))).$$

Moreover, since $P \Rightarrow (Q \vee P)$ is easily seen to be a tautology, we only need to prove that

$$(\forall_x Q(x) \Rightarrow P(x)) \Rightarrow \forall_x(P(x) \Leftarrow (Q(x) \vee P(x))),$$

and the validity of this latter statement immediately follows from the fact that $(Q \Rightarrow P) \Leftrightarrow (P \Leftarrow (Q \vee P))$ is a tautology.

3.4.6 a) Setting $P(x) :\Leftrightarrow x \in A$, $Q(x) :\Leftrightarrow x \in B$, and $R(x) :\Leftrightarrow x \in C$, it follows that

$$x \in (A \triangle B) \triangle C \Leftrightarrow (x \in (A \triangle B) \smallsetminus C \vee x \in C \smallsetminus (A \triangle B))$$
$$\Leftrightarrow (((P(x) \wedge \bar{Q}(x)) \vee (Q(x) \wedge \bar{P}(x))) \wedge \bar{R}(x))$$
$$\vee (R(x) \wedge \neg((P(x) \wedge \bar{Q}(x)) \vee (Q(x) \wedge \bar{P}(x))))$$
$$\Leftrightarrow ((P(x) \vee Q(x)) \wedge (\bar{P}(x) \vee \bar{Q}(x)) \wedge \bar{R}(x))$$
$$\vee (R(x) \wedge (\bar{P}(x) \vee Q(x)) \wedge (P(x) \vee \bar{Q}(x)))$$
$$\Leftrightarrow ((P(x) \vee Q(x) \vee R(x)) \wedge (\bar{P}(x) \vee \bar{Q}(x) \vee R(x))$$
$$\wedge (\bar{P}(x) \vee Q(x) \vee \bar{R}(x)) \wedge (P(x) \vee \bar{Q}(x) \vee \bar{R}(x)),$$

and similarly,

$$x \in A \triangle (B \triangle C) \Leftrightarrow ((R(x) \vee P(x) \vee Q(x)) \wedge (\bar{R}(x) \vee \bar{P}(x) \vee Q(x))$$
$$\wedge (\bar{R}(x) \vee P(x) \vee \bar{Q}(x)) \wedge (R(x) \vee \bar{P}(x) \vee \bar{Q}(x)).$$

Hence $\forall_x x \in (A \triangle B) \triangle C \Leftrightarrow x \in A \triangle (B \triangle C)$, as desired.

3.4.7 a) This statement is in general not true, because if $A := \{0\}$ and $B := \{1\}$, then $\{0,1\} \in \mathcal{P}(A \cup B)$ but $\{0,1\} \notin \mathcal{P}(A) \cup \mathcal{P}(B)$.

b) This statement is true because if $M \in \mathcal{P}(A) \cup \mathcal{P}(B)$, then $M \subset A \vee M \subset B$, and therefore, $M \subset A \cup B$.

3.5.20 If $n|m$ and $m|n$, then there are integers $i, j \in \mathbb{N}$ such that $m = jn$ and $n = im$. Hence $n = ijn$, and therefore, $ij = 1$. Since this is possible only if $i = 1$ and $j = 1$ (because $i, j \geq 1$), it follows that $m = n$, and this shows that r is antisymmetric. To prove that r is transitive, we assume that $n|m$ and $m|k$. Then there are integers $i, j \in \mathbb{N}$ such that $m = jn$ and $k = im$, and therefore, $k = ijn$. Thus $n|k$, as desired. Finally, r is not linear, because 2 does not divide 3, 3 does not divide 2, and $2 \neq 3$.

3.5.21 The maximal number of elements is one. If the number of elements is zero, then $S = \emptyset$, $\mathcal{P}(S) = \{\emptyset\}$, and $r = \{(\emptyset, \emptyset)\}$ is linear. If the number of elements is one, then $S = \{x\}$, $\mathcal{P}(S) = \{\emptyset, \{x\}\}$, and

$$r = \{(\emptyset, \emptyset), (\emptyset, \{1\}), (\{1\}, \{1\})\}$$

is again linear. However, if S contains two elements x and y such that $x \neq y$, then $\{x\}, \{y\} \in \mathcal{P}(S)$, but $\{x\} \not\subset \{y\}$, $\{y\} \not\subset \{x\}$, and $\{x\} \neq \{y\}$. Thus r is not linear if the number of elements in S is greater than one.

3.5.23 The number of pairs $(i, j) \in S \times S$ for which $i < j$ is $\sum_{k=1}^{n-1} k = n(n-1)/2$. Thus the number of subsets of S that consist of pairs of this form is $2^{n(n-1)/2}$. Since r is linear and antisymmetric, it follows that $(i, j) \in r \Leftrightarrow (j, i) \notin r$ whenever $i < j$. In other words, if $r \cap \{(i, j) \in S \times S \mid i < j\}$ is known, then so is $r \cap \{(i, j) \in S \times S \mid i > j\}$. Consequently, since the set $\{(i, j) \in S \times S \mid i = j\}$ contains n elements that can be chosen to be in r or not, it follows that the total number of antisymmetric linear relations on S is $2^{n(n-1)/2+n} = 2^{n(n+1)/2}$.

3.5.25 Since r is reflexive, it follows that $S = \{x \in S \mid (x, x) \in r\} \subset \{x \in S \mid \exists_{y \in S}(x, y) \in r\} = D(r) \subset S$, and therefore, $D(r) = S$.

3.5.26 For all $A \in \mathcal{P}(S)$ it is the case that $A \triangle A = \emptyset \subset T$ (by Exercise 3.4.6). Thus r is reflexive. Furthermore, according to Exercise 3.4.6, we have $A \triangle B = B \triangle A$ for all $A, B \in \mathcal{P}(S)$, and therefore, it readily follows that r is symmetric. Finally, if $A \triangle B \subset T$ and $B \triangle C \subset T$, then $(A \triangle B) \cup (B \triangle C) \subset T$, and therefore, $A \triangle C \subset T$ because, in general, $A \triangle C \subset (A \triangle B) \cup (B \triangle C)$. Consequently, r is transitive as well. Note: the proof of the fact that $A \triangle C \subset (A \triangle B) \cup (B \triangle C)$ is an easy exercise in formal logic.

3.5.27 Since r is not reflexive, there is an $x \in S$ such that $(x, x) \notin r$. Furthermore, since $D(r) = S$, we can find a $y \in S$ such that $(x, y) \in r$. If it were the case that $(y, x) \in r$, then the conjunction $(x, y) \in r \wedge (y, x) \in r$ would be true, and (x, x) would be in r by transitivity. However, since (x, x) is not in r, the assumption $(y, x) \in r$ must have been false, and therefore, r is not symmetric (because $(x, y) \in r$).

3.5.29 Assume that $(x, y), (u, v) \in r$. By argumentative symmetry, we only need to show that $(x, u) \in r \Rightarrow (y, v) \in r$. So let $(x, u) \in r$. Then $(y, x) \in r$

(by symmetry because $(x, y) \in r$), and therefore, the conjunction $(y, x) \in r \wedge (x, u) \in r$ is true. Thus $(y, u) \in r$ (by transitivity), and since (u, v) is also in r, the transitivity of r further implies that $(y, v) \in r$, as desired.

3.5.31 We may assume w.l.o.g. that $\forall_{z \in M} P(x, z) \wedge \forall_{z \in M} P(y, z)$. Since r is linear, it follows that $(x, y) \in r \vee (y, x) \in r \vee x = y$. If $(x, y) \in r$, then $x = y$ because $P(x, y)$ is true, and if $(y, x) \in r$, then we also find that $x = y$ because $P(y, x)$ is true.

3.6.8

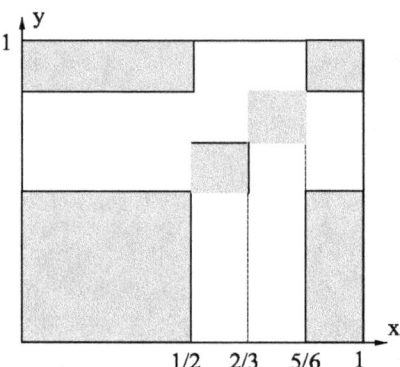

$$\text{1/2} \quad \text{2/3} \quad \text{5/6} \quad \text{1}$$

3.6.10 For all $A \subset S$ it is the case that $[A] = \{B \subset S \mid A \triangle B \subset T\}$, but in order to better understand this equation, it is helpful to observe that

$$(A \triangle B \subset T) \Leftrightarrow (A \setminus T = B \setminus T) \quad \text{(see below)}.$$

For in the light of this equivalence, we readily find that

$$[A] = \{B \subset S \mid \exists_{N \subset T} B = N \cup (A \setminus T)\}.$$

In order to prove the equivalence above, we notice that

$$A \triangle B \subset T \Leftrightarrow \forall_x (((x \in A \wedge x \notin B) \vee (x \in B \wedge x \notin A)) \Rightarrow x \in T)$$

and that

$$A \setminus T = B \setminus T \Leftrightarrow \forall_x ((x \in A \wedge x \notin T) \Leftrightarrow (x \in B \wedge x \notin T)).$$

Consequently, the equivalence above is valid because

$$(((P \wedge \bar{Q}) \vee (Q \wedge \bar{P})) \Rightarrow R) \Leftrightarrow ((P \wedge \bar{R}) \Leftrightarrow (Q \wedge \bar{R}))$$

is easily seen to be a tautology.

3.6.12 Since $x/x = 1$ for all $x \in \mathbb{R} \setminus \{0\}$, it follows that r is reflexive. Furthermore, r is symmetric because $x/y = y/x$ whenever $x/y \in \{-1, 1\}$, and r is transitive because if $x/y, y/z \in \{-1, 1\}$, then $x/z = (x/y)(y/z) \in \{-1, 1\}$.

Finally, since $x/y = \pm 1 \Leftrightarrow x = \pm y$ for all $x, y \in \mathbb{R} \setminus \{0\}$, it follows that for all $x \in \mathbb{R} \setminus \{0\}$ it is the case that $[x] = \{x, -x\}$. Hence $\pi = \{\{x, -x\} \mid x > 0\}$.

3.7.11 Since B is nonempty, we can choose y_1 and y_2 such that $y_1 = y_2$. In other words, we can choose y_1 and y_2 so that the consequent is true, and therefore, the statement is true.

3.7.12 The statement is true for b) and c). In b) we can choose $x = \pm 1$, and in c) we can choose all $x \in \mathbb{R}$.

3.8.27 If $x \in A$, then $(x, f(x)) \in f$, and therefore, $(f(x), x) \in f^{-1}$. Thus $f^{-1}(f(x)) = x$, as desired.

3.8.30 If f is not surjective, then there exists a $y \in B$ such that $f^{-1}(\{y\}) = \emptyset = f^{-1}(\emptyset)$. Thus $f^{-1}(\{y\}) \subset f^{-1}(\emptyset)$, but $\{y\} \not\subset \emptyset$.

3.8.31 a) $\emptyset$, **b)** $[0, 1]$.

3.8.32

$$x \in \bigcap_{i \in I} R_i \cup \bigcap_{i \in I} S_i \Leftrightarrow ((\forall_{i \in I} \, x \in R_i) \vee (\forall_{i \in I} \, x \in S_i))$$

$$\Rightarrow \forall_{i \in I} (x \in R_i \vee x \in S_i) \quad \text{(by Proposition 1.7.7b)}$$
$$\Leftrightarrow \forall_{i \in I} (x \in R_i \cup S_i)$$
$$\Leftrightarrow x \in \bigcap_{i \in I} (R_i \cup S_i),$$

as desired. Moreover, the reverse inclusion is not valid as the following example illustrates: if $I := \{1, 2\}$, $R_1 := \{1\}$, $R_2 := \{2\}$, $S_1 := \{2\}$, and $S_2 := \{1\}$, then $(R_1 \cup S_1) \cap (R_2 \cup S_2) = \{1, 2\} \not\subset \emptyset = (R_1 \cap R_2) \cup (S_1 \cap S_2)$.

3.8.35 Since every rational number is a quotient of the form m/n with $m \in \mathbb{Z}$ and $n \in \mathbb{N}$, it follows that the rational numbers can be listed in a two-dimensional array with m as the x-coordinate and n as the y-coordinate. Consequently, the rational numbers can be enumerated in the sense that there exists a sequence $(r_k)_{k \in \mathbb{N}}$ such that $\mathbb{Q} = \{r_k \mid k \in \mathbb{N}\}$ ($r_1 = 0/1$, $r_2 = 1/1$, $r_3 = 1/2$, $r_4 = 0/2$, $r_5 = -1/2$, $r_6 = -1/1$, $r_7 = -2/1$, $r_8 = -2/2$, $r_9 = -2/3$, $r_{10} = -1/3$, $r_{11} = 0/3$, $r_{12} = 1/3$, $r_{13} = 2/3$, $r_{14} = 2/2$, $r_{15} = 2/1$, etc.). Setting $I_k := (r_k - 1/2^k, r_k + 1/2^k)$, it follows that $r_k \in I_k$ for all $k \in \mathbb{N}$ and that the length of each I_k is equal to $1/2^{k-1}$. Since the sum $\sum_{k=1}^{n} 1/2^{k-1} = 1 + 1/2 + 1/4 + \cdots + 1/2^{n-1}$ is easily seen to be equal to $2 - 1/2^{n-1}$ for all $n \in \mathbb{N}$, it follows that the total length of the union of the intervals I_k is less than or equal to 2, and therefore, $\bigcup_{k=1}^{\infty} I_k \neq \mathbb{R}$.

3.8.37 If $f(x) := 1$ for $x \in \{1, 2\}$, $M_1 := \{1\}$, and $M_2 := \{2\}$, then $f(M_1) \cap f(M_2) = \{1\} \not\subset \emptyset = f(\emptyset) = f(M_1 \cap M_2)$.

3.8.39 a)

$$f(M_1) \triangle f(M_2) = (f(M_1) \smallsetminus f(M_2)) \cup (f(M_2) \smallsetminus f(M_1))$$
$$\subset f(M_1 \smallsetminus M_2) \cup f(M_2 \smallsetminus M_1) \quad \text{(by Proposition 3.8.22)}$$
$$= f((M_1 \smallsetminus M_2) \cup (M_2 \smallsetminus M_1)) \quad \text{(by Theorem 3.8.19a)}$$
$$= f(M_1 \triangle M_2).$$

b) If $f(x) := 1$ for all $x \in \{1, 2, 3\}$, $M_1 := \{1.2\}$, $M_2 := \{1.3\}$, and $M_3 := \{2, 3\}$, then

$$f(M_1) \triangle f(M_2) \triangle f(M_3) = \{1\} \not\subset \emptyset = f(\emptyset) = f(M_1 \triangle M_2 \triangle M_3).$$

c) Assume that $f(M_1 \triangle M_2) = f(M_1) \triangle f(M_2)$ for all $M_1, M_2 \subset A$ and that $f(x_1) = f(x_2)$ for some $x_1, x_2 \in A$. Setting $M_1 := \{x_1\}$ and $M_2 := \{x_2\}$, it follows that $\emptyset = f(M_1) \triangle f(M_2) = f(M_1 \triangle M_2)$, and therefore, $\emptyset = M_1 \triangle M_2 = \{x_1\} \triangle \{x_2\}$. But since the latter equation can be true only if $x_1 = x_2$, we may infer that f is injective, as desired.

Chapter 4

4.2.12 In order to show that s is antisymmetric, it is sufficient to prove that $(x, y) \in s \wedge (y, x) \in s$ is false for all $x, y \in \mathbb{S}$. Since

$$((x, y) \in s \wedge (y, x) \in s) \Leftrightarrow ((x, y) \in r \wedge (y, x) \in r \wedge x \neq y$$
$$\Rightarrow (x = y \wedge x \neq y) \quad \text{(because } r \text{ is antisymmetric)},$$

it follows that $(x, y) \in s \wedge (y, x) \in s$ is indeed false because $x = y \wedge x \neq y$ is obviously false as well. To prove that s is transitive, we assume that $(x, y) \in s \wedge (y, z) \in s$. Then it is the case in particular that $(x, y) \in r \wedge (y, z) \in r$, and therefore, $(x, z) \in r$ because r is transitive. Thus in order to show that $(x, z) \in s$, we only need to demonstrate that $x \neq z$. If x were equal to z, then $(x, y) \in r \wedge (y, x) = (y, z) \in r$, and therefore, $x = y$ (because r is antisymmetric) in contradiction to the assumption that $(x, y) \in s$. Thus $x \neq z$, as desired. Finally, since the linearity condition $(x, y) \in s \vee (y, x) \in s \vee x = y$ is equivalent to

$$((x, y) \in r \wedge x \neq y) \vee ((y, x) \in r \wedge y \neq x) \vee x = y$$

and since this latter condition is easily seen to be equivalent to

$$(x, y) \in r \vee (y, x) \in r \vee x = y,$$

it follows that s is linear because r is linear.

4.3.16 Since, trivially, $f(x) \in f(C)$ for all $x \in C$, it is the case that $C \subset f^{-1}(f(C))$. To prove the reverse inclusion, we assume that $x \in f^{-1}(f(C))$. Then $f(x) \in f(C)$, and therefore, there exists a $y \in C$ such that $f(y) = f(x)$. Since f is injective, we may infer that $x = y \in C$, as desired.

4.4.4 Setting $A := \{n \in \mathbb{N} \mid \forall_{m \in \mathbb{N}} \, n \leq nm\}$, it follows that $1 \in A$ because $1 \leq m = \mu(1, m) = 1 \cdot m$ for all $m \in \mathbb{N}$ (by Proposition 4.2.10a and Theorem 4.3.15). If $n \in A$, then $n \leq nm$, and therefore, $\varphi(n) = n + 1 \leq nm + 1$ (by Proposition 4.2.10c), and since $1 \leq m$, Proposition 4.2.10c also implies that $nm + 1 \leq nm + m = \mu(n, m) + m = \mu(\varphi(n), m) = (n + 1)m$. Thus $n + 1 \leq (n + 1)m$, and therefore, $n + 1 \in A$, as desired.

4.5.12 Using the results of Exercise 4.5.11 and Example 4.5.1, we find that

$$(n + 1)^4 - 1 = \sum_{k=1}^{n} ((k + 1)^4 - k^4) = 4 \sum_{k=1}^{n} k^3 + 6 \sum_{k=1}^{n} k^2 + 4 \sum_{k=1}^{n} k + \sum_{k=1}^{n} 1$$

$$= 4 \sum_{k=1}^{n} k^3 + \frac{6n(n + 1)(2n + 1)}{6} + \frac{4n(n + 1)}{2} + n,$$

and therefore,

$$\sum_{k=1}^{n} k^3 = \frac{1}{4} \left((n + 1)^4 - n(n + 1)(2n + 1) - 2n(n + 1) - (n + 1) \right)$$

$$= \frac{n + 1}{4} \left((n + 1)^3 - n(2n + 1) - (2n + 1) \right)$$

$$= \frac{(n + 1)^2}{4} \left((n + 1)^2 - (2n + 1) \right) = \frac{n^2(n + 1)^2}{4}.$$

In order to prove this formula by means of induction, we notice that $\sum_{k=1}^{1} k^3 = 1 = 1^2(1+1)^2/4$ and assume that the formula is satisfied for some $n \in \mathbb{N}$. Then

$$\sum_{k=1}^{n+1} k^3 = (n + 1)^3 + \sum_{k=1}^{n} k^3 = (n + 1)^3 + \frac{n^2(n + 1)^2}{4}$$

$$= (n + 1)^2 \left(n + 1 + \frac{n^2}{4} \right) = \frac{(n + 1)^2(n + 2)^2}{4},$$

as desired.

4.5.15 Since the given inequality is evidently satisfied for $n = 1$, we proceed to assume that it is satisfied for some $n \in \mathbb{N}$. Then

$$\sum_{k=1}^{n+1} \frac{1}{k^2} = \frac{1}{(n + 1)^2} + \sum_{k=1}^{n} \frac{1}{k^2} \leq \frac{1}{(n + 1)^2} + 2 - \frac{1}{n} < \frac{1}{n(n + 1)} + 2 - \frac{1}{n} = 2 - \frac{1}{n + 1},$$

as desired.

4.5.16 For $n = 1$, the given inequality reduces to the true assertion that $1 \geq 1$.

Assuming that the inequality is true for some $n \in \mathbb{N}$, it follows that

$$\frac{x_1}{x_{n+1}} + \sum_{k=1}^{n} \frac{x_{k+1}}{x_k} = \frac{x_1}{x_{n+1}} - \frac{x_1}{x_n} + \frac{x_1}{x_n} + \frac{x_{n+1}}{x_n} + \sum_{k=1}^{n-1} \frac{x_{k+1}}{x_k}$$

$$\geq \frac{x_1}{x_{n+1}} - \frac{x_1}{x_n} + \frac{x_{n+1}}{x_n} + n$$

$$= \frac{x_1 x_n - x_1 x_{n+1} + x_{n+1}^2 - x_n x_{n+1}}{x_n x_{n+1}} + n + 1$$

$$= \frac{(x_1 - x_{n+1})(x_n - x_{n+1})}{x_n x_{n+1}} + n + 1$$

$$\geq n + 1$$

because $x_1 \geq \cdots \geq x_n \geq x_{n+1}$.

4.5.17 For $n = 1$ and $k = 0$, the given equation reduces to the equation $1 = 1$ and is therefore true. Assuming that the equation is true for a given $n \in \mathbb{N}$ and all $k \in \{0, \ldots, n-1\}$, we need to show that it is true as well for $n + 1$ and all $k \in \{0, \ldots, n\}$. Considering first the case $k = n$, we may use the binomial formula to infer that

$$\sum_{i=0}^{n} \binom{n+1}{i}(-1)^i = -\binom{n+1}{n+1}(-1)^{n+1} + \sum_{i=0}^{n+1} \binom{n+1}{i}(-1)^i$$

$$= (-1)^n \binom{n}{n} + (1-1)^{n+1} = (-1)^n \binom{n}{n},$$

as desired. Furthermore, using the inductive assumption, we find that for any $k \in \{1, \ldots, n-1\}$ it is the case that

$$\sum_{i=0}^{k} \binom{n+1}{i}(-1)^i = 1 + \sum_{i=1}^{k} \left(\binom{n}{i} + \binom{n}{i-1} \right)(-1)^i$$

$$= \sum_{i=0}^{k} \binom{n}{i}(-1)^i + \sum_{i=0}^{k-1} \binom{n}{i}(-1)^{i+1}$$

$$= (-1)^k \binom{n-1}{k} - (-1)^{k-1} \binom{n-1}{k-1} = (-1)^k \binom{n}{k},$$

and finally, to complete the proof, we observe that for $k = 0$ we have

$$\sum_{i=0}^{k} \binom{n+1}{i}(-1)^i = 1 = (-1)^k \binom{n}{k}.$$

4.5.20 Setting $b := f(0)$ and $m := f(1) - f(0)$, we wish to show that $f(x) = mx + b$ for all $x \in \mathbb{R}$. If $x \in [0, 1]$, then we set $\lambda := x$ and conclude that

$$f(x) = f(x \cdot 1 + (1-x) \cdot 0) = x f(1) + (1-x) f(0) = mx + b,$$

as claimed. If $x > 1$, then we define $\lambda := 1/x$ and find that

$$f(1) = f\left(\frac{1}{x} \cdot x + \left(1 - \frac{1}{x}\right) \cdot 0\right) = \frac{f(x)}{x} + \left(1 - \frac{1}{x}\right)f(0).$$

Hence $f(x) = mx + b$, again as claimed. Finally, if $x < 0$, then we set $\lambda := 1/2$. This yields

$$f(0) = f\left(\frac{x}{2} - \frac{x}{2}\right) = \frac{f(x)}{2} + \frac{f(-x)}{2},$$

and therefore, $f(x) = 2f(0) - f(-x) = 2b - (m(-x) + b) = mx + b$.

4.5.21 Assuming the existence of the set B, as described in the hint, there exists for any $x \in \mathbb{R} \setminus \{0\}$ a unique value $n(x) \in \mathbb{N}$, unique pairwise distinct values $b_1(x), \ldots, b_{n(x)}(x) \in B$, and unique coefficients $r_1(x), \ldots, r_{n(x)}(x) \in \mathbb{Q} \setminus \{0\}$ such that $x = \sum_{k=1}^{n(x)} r_k(x)b_k(x)$. Consequently, for any function $g : B \to \mathbb{R}$, the function

$$f(x) := \begin{cases} \sum_{k=1}^{n(x)} r_k(x)g(b_k(x)) & \text{if } x \neq 0, \\ 0 & \text{if } x = 0 \end{cases}$$

is well defined (because the values $n(x)$, $b_k(x)$, and $r_k(x)$ are unique). Given this definition and given two values $x \in \mathbb{R} \setminus \{0\}$ and $r \in \mathbb{Q} \setminus \{0\}$, it follows that

$$f(rx) = f\left(r\sum_{k=1}^{n(x)} r_k(x)b_k(x)\right) = \sum_{k=1}^{n(x)} r \cdot r_k(x)g(b_k(x)) = rf(x)$$

(again by the stated uniqueness property). (Note: the equation $f(rx) = rf(x)$ is trivially valid as well when either x or r are zero, but this fact will not be needed for the present exercise.) Next we wish to show that $f(x + y) = f(x) + f(y)$ for all $x, y \in \mathbb{R}$. To do so, we observe to begin with that this equation is trivially satisfied if $x = 0$ or $y = 0$. Furthermore, if $x + y = 0$, then $f(x) + f(y) = f(x) + f((-1)x) = f(x) + (-1)f(x) = 0 = f(0) = f(x + y)$. Consequently, we may assume that x, y, and $x + y$ are all different from zero. Given this assumption, it follows that

$$\sum_{k=1}^{n(x)} r_k(x)b_k(x) + \sum_{k=1}^{n(y)} r_k(y)b_k(y) = x + y = \sum_{k=1}^{n(x+y)} r_k(x+y)b_k(x+y),$$

and since the representation of $x + y$ is unique, we may infer that

$$f(x) + f(y) = \sum_{k=1}^{n(x)} r_k(x)g(b_k(x)) + \sum_{k=1}^{n(y)} r_k(y)g(b_k(y))$$

$$= \sum_{k=1}^{n(x+y)} r_k(x+y)g(b_k(x+y)) = f(x+y),$$

as desired. (Note: it is possible that the set

$$\{b_1(x), \ldots, b_{n(x)}(x)\} \cup \{b_1(y), \ldots, b_{n(y)}(y)\}$$

contains values that the set $\{b_1(x + y), \ldots, b_{n(x+y)}(x + y)\}$ does not contain because there can be indices i and j such that $b_i(x) = b_j(y)$ and $r_i(x) = -r_j(y)$, but in that case the corresponding values $r_i(x)g(b_i(x))$ and $r_j(y)g(b_j(y))$ cancel each other out, and the validity of the equation $f(x) + f(y) = f(x + y)$ is not affected.) Consequently, it is the case for all $x, y \in \mathbb{R}$ that $f((x + y)/2) = f(x + y)/2 = (f(x) + f(y))/2$. However, if we pick a value $b_0 \in B$ and set

$$g(b) := \begin{cases} 1 & \text{if } b = b_0, \\ 0 & \text{otherwise}, \end{cases}$$

then the corresponding function f is not affine because $f(b) = 0$ for all $b \in B \setminus \{b_0\}$ and $f(b_0) = 1$ (and because an affine function that is equal to zero at more than one point is evidently equal to zero at all points).

4.6.10 a) $f(2) = \sqrt{2 + \sqrt{2}}$, $f(3) = \sqrt{2 + \sqrt{2 + \sqrt{2}}}$, and

$$f(4) = \sqrt{2 + \sqrt{2 + \sqrt{2 + \sqrt{2}}}}.$$

b) Since the definition of $f(n)$ clearly implies that $f(n) > 0$ for all $n \in \mathbb{N}$, it follows that

$$\begin{aligned}
f(n + 1) > f(n) &\Leftrightarrow f(n + 1)^2 > f(n)^2 \\
&\Leftrightarrow 2 + f(n) > f(n)^2 \\
&\Leftrightarrow f(n)^2 - f(n) - 2 < 0 \\
&\Leftrightarrow (f(n) - 2)(f(n) + 1) < 0 \\
&\Leftrightarrow -1 < f(n) < 2
\end{aligned}$$

for all $n \in \mathbb{N}$. Consequently, the inequality $f(n + 1) > f(n)$ is valid for all $n \in \mathbb{N}$ because it was shown in Eample 4.6.4 that $f(n) < 2$ for all $n \in \mathbb{N}$.

c) Again using the fact that $0 < f(n) < 2$ for all $n \in \mathbb{N}$, it follows that

$$\frac{2 - f(n)}{2} = \frac{4 - (2 + f(n))}{2} = \frac{4 - f(n + 1)^2}{2} = \frac{(2 + f(n + 1))(2 - f(n + 1))}{2}$$
$$> 2 - f(n + 1).$$

d) To begin with, we will use c) to show by induction that $2 - f(n) < 1/2^{n-1}$ for all $n \in \mathbb{N}$. For $n = 1$ we find that $2 - f(n) = 2 - \sqrt{2} < 1 = 1/2^{n-1}$, and assuming that $2 - f(n) < 1/2^{n-1}$ for some $n \in \mathbb{N}$, we may apply c) to infer that $2 - f(n+1) < (2 - f(n))/2 < 1/2^n$, as desired. Consequently, if $\varepsilon > 0$ and $N > 1/\varepsilon$, then

$$|f(n) - 2| = 2 - f(n) < \frac{1}{2^{n-1}} \leq \frac{1}{n} < \frac{1}{N} < \varepsilon$$

for all $n > N$, and therefore, $\lim_{n \to \infty} f(n) = 2$.

4.6.12 $f(n) = \prod_{k=1}^{n-1}(1 + 1/k^2)$ for all $n > 1$, and therefore, the tangent line estimate $\ln(1+x) \leq x$ implies that

$$f(n) = e^{\sum_{k=1}^{n-1} \ln(1+1/k^2)} \leq e^{\sum_{k=1}^{n-1} 1/k^2} \leq e^{1 + \int_1^{n-1} 1/x^2\, dx} = e^{2 - 1/(n-1)} < e^2.$$

4.6.14 a) The existence of the sequence $(a_n)_{n \in \mathbb{N}}$ follows from Theorem 4.6.2 by setting $g(n,x) := x/2 + 1$ for all $n \in \mathbb{N}$ and all $x \in \mathbb{R}$.
b) $a_1 = 1 = (2^1 - 1)/2^{1-1}$, and if $a_n = (2^n - 1)/2^{n-1}$ for some $n \in \mathbb{N}$, then

$$a_{n+1} = \frac{a_n}{2} + 1 = \frac{2^n - 1}{2^n} + 1 = \frac{2^{n+1} - 1}{2^n},$$

as desired.
c) Using b), we find that

$$|a_n - 2| = 2 - \frac{2^n - 1}{2^{n-1}} = \frac{1}{2^{n-1}},$$

and therefore it follows, as in Exercise 4.6.10d, that $\lim_{n \to \infty} a_n = 2$.

4.6.16 According to the uniqueness assertion of Theorem 4.6.7, we only need to show that the sequence $(F_n)_{n \in \mathbb{N}}$ that is defined by the equation

$$F_n := \frac{1}{\sqrt{5}}\left(\left(\frac{1+\sqrt{5}}{2}\right)^n - \left(\frac{1-\sqrt{5}}{2}\right)^n\right)$$

for all $n \in \mathbb{N}$ satisfies the defining equations of the Fibonacci sequence: $F_1 = F_2 = 1$, and $F_{n+2} = F_{n+1} + F_n$ for all $n \in \mathbb{N}$. By inspection, it is easy to see that this is indeed the case.

Chapter 5

5.1.11 The number of bijections from S to S is simply the number of permutations of the elements of S and is therefore equal to $|S|!$.

5.1.12 By Theorem 5.1.9, the answer to this exercise is the same as to Exercise 5.1.11.

5.1.13 There are n possible values for $f(1)$. Having chosen $f(1)$ and taking into account the requirement that f be injective, there are $n-1$ possible values that we can choose for $f(2)$, and so forth. Thus the number of injections is $n \cdot (n-1)\ldots(n-m+1) = n!/(n-m)!$.

5.1.14 It is sufficient to determine for a given $n \in \mathbb{N}$ the number P_n of bijective maps f from M_n into itself that satisfy the condition $f(x) \neq x$ for all $x \in M_n$, where $P_0 := 1$ for convenience. Given an integer $n \geq 2$, we notice that there are $\binom{n}{k}$ subsets of M_n that contain exactly k elements. Consequently, for any $k \in \{0, \ldots, n-2\}$ the number of bijective maps that leave exactly k elements in M_n invariant (i.e., $f(x) = x$) is $\binom{n}{k} P_{n-k}$. Observing further that the total number of bijective maps is $n!$, that there is exactly one map that leaves all elements in S invariant (note: a map that leaves all but one element invariant necessarily leaves all elements invariant), and that $P_1 = 0$, it follows that

$$n! = 1 + \sum_{k=0}^{n-2} \binom{n}{k} P_{n-k} = 1 + \sum_{k=0}^{n-2} \binom{n}{n-k} P_{n-k} = \sum_{k=0}^{n} \binom{n}{k} P_k$$

for all $n \geq 2$ (but it is easy to see that the formula also holds for $n \in \{0, 1\}$). To proceed, we will use this observation to prove, by way of induction, that P_n satisfies the equation

$$P_n = n P_{n-1} + (-1)^n$$

for all $n \in \mathbb{N}$. Since $P_0 = 1$ and $P_1 = 0$, the equation above is true for $n = 1$. Assuming that it is true for all $k \in \{1, \ldots, n\}$, we need to show that it is true as well for $n+1$. Since $\sum_{k=0}^{n+1} \binom{n+1}{k} (-1)^k = (1-1)^{n+1} = 0$, we find that

$$(n+1)! = \sum_{k=0}^{n+1} \binom{n+1}{k} P_k = P_{n+1} + \sum_{k=0}^{n} \binom{n+1}{k} P_k$$

$$= 1 + P_{n+1} + \sum_{k=1}^{n} \binom{n+1}{k} (k P_{k-1} + (-1)^k)$$

$$= P_{n+1} + (n+1) \sum_{k=1}^{n} \binom{n}{k-1} P_{k-1} + \sum_{k=0}^{n} \binom{n+1}{k} (-1)^k$$

$$= P_{n+1} + (n+1) \sum_{k=0}^{n-1} \binom{n}{k} P_k - (-1)^{n+1}$$

$$= P_{n+1} + (n+1)(n! - P_n) - (-1)^{n+1},$$

and therefore, $P_{n+1} = (n+1)P_n + (-1)^{n+1}$, as desired. To proceed, we will show, again by induction, that

$$P_n = \sum_{k=0}^{n} \frac{n!}{k!} (-1)^k$$

for all $n \geq 0$. For $n = 0$ we find that $P_0 = 1 = (-1)^0 0! / 0!$, and assuming that the equation is valid for some $n \in \mathbb{N} \cup \{0\}$, it follows that

$$P_{n+1} = (n+1)P_n + (-1)^{n+1} = (-1)^{n+1} + (n+1)\sum_{k=0}^{n} \frac{n!}{k!}(-1)^k$$

$$= (-1)^{n+1} + \sum_{k=0}^{n} \frac{(n+1)!}{k!}(-1)^k = \sum_{k=0}^{n+1} \frac{(n+1)!}{k!}(-1)^k.$$

Having thus established the above summation formula for P_n, we further notice that

$$\lim_{n \to \infty} \frac{P_n}{n!} = \sum_{k=0}^{\infty} \frac{(-1)^k}{k!} = \frac{1}{e}.$$

5.2.9 If $|A| \leq |M_1| = |\{1\}|$, then there is an injective map f from A to $\{1\}$, and since $|\{1\}| = 1$, this implies that f is actually bijective. Thus A is finite and $|A| = 1$. Having thus established the inductive base, we proceed to assume that we are given an integer $n \in \mathbb{N}$ such that a nonempty set A is finite whenever $|A| \leq |M_n|$. To complete the proof, we need to show that a nonempty set A is finite whenever $|A| \leq |M_{n+1}|$. To do so, we assume that $f : A \to M_{n+1}$ is an injection. If f happens to be a bijection, then A is finite because $|A| = |M_{n+1}| = n+1$. If f is not a bijection, then f is not a surjection (because f is injective), and therefore, there exists a $k_0 \in M_{n+1}$ such that $f(x) \neq k_0$ for all $x \in A$. Consequently, in setting

$$g(x) := \begin{cases} f(x) & \text{if } f(x) < k_0, \\ f(x) - 1 & \text{if } f(x) > k_0 \end{cases}$$

for all $x \in A$, we find that $g : A \to M_n$ is well defined and injective. Hence $|A| \leq |M_n|$, and therefore, A is finite (by the inductive assumption).

5.2.11 If $B \subset A$, then trivially $|A \cup B| = |A| = |\mathbb{N}|$. So we may assume that $B \setminus A \neq \emptyset$. In that case it follows that $B \setminus A$ is finite (because B is finite), and therefore, there is an $n \in \mathbb{N}$ for which there exists a bijection $g : M_n \to B \setminus A$. Furthermore, since $|A| = |\mathbb{N}|$, there exists a bijection $f : \mathbb{N} \to A$. Consequently, in setting

$$h(k) := \begin{cases} g(k) & \text{if } k \in M_n, \\ f(k - n) & \text{if } k > n, \end{cases}$$

we find that $h : \mathbb{N} \to A \cup B$ is a bijection as well (note: h is surjective because f and g are surjective, and h is injective because f and g are injective and $A \cap (B \setminus A) = \emptyset$). Thus $|A \cup B| = |\mathbb{N}|$, as desired.

5.3.20 We will prove only that $\bigcup_{i \in I} C_i \leq A$ and leave the proof of the inequality $\bigcap_{i \in I} C_i \leq A$ to the reader. Since $C_i \leq A$ for all $i \in I$, it follows in particular that $C_i \subset A$ for all $i \in I$, and therefore, trivially, $\bigcup_{i \in I} C_i \subset A$. To complete the proof, we assume that $y \leq x$ for some $x \in \bigcup_{i \in I} C_i$ and $y \in A$. Then there exists an $i_0 \in I$ such that $x \in C_{i_0}$, and since $C_{i_0} \leq A$, it follows that $y \in C_{i_0} \subset \bigcup_{i \in I} C_i$, as desired.

5.3.22 Let S be an infinite set. We wish to show that S is a disjoint union of subsets of S that are each cardinally equivalent to $\mathbb{N}$. To do so, we pick a non-empty subset M of S and agree to call a function $f : M \to \mathcal{P}(S)$ an M-function, if

$$\forall_{x \in M}(x \in f(x) \wedge |f(x)| = |\mathbb{N}| \wedge \forall_{y \in M}(y \neq x \Rightarrow f(y) \cap f(x) = \emptyset)).$$

In order to show that the set F of all M-functions is nonempty, we apply Theorem 5.2.6 to infer that there exists an injection $g : N \to S$ (because S is infinite). Setting $M := \{g(1)\}$ and $f(g(1)) := R(g)$, it follows that f is an M-function, and therefore, $f \in F$, as desired. In order to define a reflexive partial order on F (see Example 3.5.12), we assume that we are given M-functions $f_1 : M_1 \to \mathcal{P}(S)$ and $f_2 : M_2 \to \mathcal{P}(S)$ and define

$$f_1 \leq f_2 :\Leftrightarrow f_1 \subset f_2$$
$$\Leftrightarrow f_1 = f_2 \restriction_{M_1}$$
$$\Leftrightarrow (M_1 \subset M_2 \wedge \forall_{x \in M_1} f_1(x) = f_2(x)).$$

To proceed, we assume that G is a linearly subset of F and claim that $f := \bigcup_{g \in G} g$ is an M-function, where $M := \bigcup_{g \in G} D(g)$. In order to show that f is a function, we assume that $(x, y_1), (x, y_2) \in f$. Then there are functions $g_1, g_2 \in G$ such that $(x, y_1) \in g_1$ and $(x, y_2) \in g_2$. Since G is linearly ordered, we may assume w.l.o.g. that $g_1 \subset g_2$. Given this assumption, it follows that $(x, y_1) \in g_2 \wedge (x, y_2) \in g_2$, and therefore, $y_1 = y_2$ because g_2 is a function. If $x \in M = \bigcup_{g \in G} D(g)$, then $x \in D(g)$ for some $g \in G$. Hence $x \in g(x) = f(x)$ and $|f(x)| = |g(x)| = |\mathbb{N}|$ because g is a $D(g)$-function and because the definition of f trivially implies that $g \subset f$. Finally, if $y \in M$ such that $y \neq x$, then there exists an $h \in G$ such that $y \in D(h)$. Assuming w.l.o.g. that $h \subset g$, we may infer that $x, y \in D(g)$, and therefore, $f(y) \cap f(x) = g(y) \cap g(x) = \emptyset$ (again because g is a $D(g)$-function). Consequently, f is an M-function, and since $g \subset f$ for all $g \in G$, it follows that f is an upper bound of G. Using Zorn's lemma, we may thus conclude that F has a maximal element $f_0 : M_0 \to \mathcal{P}(S)$. Given this maximal element, we set

$$T := \bigcup_{x \in M_0} f_0(x).$$

If $T = S$, then S is a disjoint union of the sets $f_0(x)$ (which are each cardinally equivalent to $\mathbb{N}$), and we are done. So we only need to consider the case where $S \smallsetminus T \neq \emptyset$. In that case there are two possibilities: $S \smallsetminus T$ is finite or $S \smallsetminus T$ is infinite. In the former case, we pick an element $x_0 \in M_0$ and observe that

$\{f_0(x_0) \cup S \smallsetminus T\} \cup \{f_0(x) \mid x \in M_0 \wedge x \neq x_0\}$ is a collection of disjoint sunsets of S that are each cardinally equivalent to $\mathbb{N}$ (by Exercise 5.2.11) and whose union is all of S. Consequently, all we have left to do is to consider that case where $S \smallsetminus T$ is infinite. In that case we may apply Theorem 5.2.6 to infer that there exists an injection $p : \mathbb{N} \rightarrow S \smallsetminus T$. Given this injection, we define $N := M_0 \cup \{p(1)\}$ and set

$$q(x) := \begin{cases} f_0(x) & \text{if } x \in M_0, \\ R(p) & \text{if } x = p(1) \end{cases}$$

for all $x \in N$. Then q is an N-function (this is easy to see) that is strictly larger than f_0 (because $f_0 \subset q$ and $f_0 \neq q$). Since this contradicts the fact that f_0 is maximal, it follows that the assumption $|S \smallsetminus T| = \infty$ was false, and therefore, the proof of the assertion that S is a disjoint union of subsets of S that are each cardinally equivalent to $\mathbb{N}$ is complete.

In order to complete the proof of the present exercise, we assume that A and B are cardinally equivalent and that A is infinite. Then there is a collection of pairwise disjoint subsets $\{A_i\}_{i \in I}$ of A such that $\bigcup_{i \in I} A_i = A$ and $|A_i| = |\mathbb{N}|$ for all $i \in I$. So for every $i \in I$ there exists a bijection $f_i : \mathbb{N} \rightarrow A_i$, and therefore, the function $f : I \times \mathbb{N} \rightarrow A$, $(i, n) \mapsto f_i(n)$ is a bijection (note: f is injective because each f_i is injective and the sets A_i are pairwise disjoint, and f is surjective because each f_i is surjective and $\bigcup_{i \in I} A_i = A$). Furthermore, since $|A| = |B|$, there exists a bijection $g : A \rightarrow B$, and therefore, the function $h : I \times \mathbb{N} \rightarrow A \cup B$ that is defined by the equation

$$h(i, n) := \begin{cases} f(i, n/2) & \text{if } n = 2k \text{ for some } k \in \mathbb{N}, \\ g(f(i, (n+1)/2)) & \text{if } n = 2k - 1 \text{ for some } k \in \mathbb{N} \end{cases}$$

for all $(i, n) \in I \times \mathbb{N}$ is surjective because

$$\mathbb{N} = \{n/2 \mid \exists_{k \in \mathbb{N}} \ n = 2k\} = \{(n+1)/2 \mid \exists_{k \in \mathbb{N}} \ n = 2k - 1\}$$

and because f and g are surjective as well. Hence the function $h \circ f^{-1} : A \rightarrow A \cup B$ is also a surjection, and since $i : A \rightarrow A \cup B$, $x \mapsto x$ is evidently an injection, Theorem 5.3.2 implies that $|A| = |A \cup B|$, as desired.

Chapter 6

6.1.7 a) If $x = [(m, n)]_\mathbb{Z}$, $y = [(k, l)]_\mathbb{Z}$, and $z = [(i, j)]_\mathbb{Z}$, then, according to Theorem 4.2.2a, it is the case that

$$\begin{aligned} x + (y + z) &= [(m, n)]_\mathbb{Z} + ([(k, l)]_\mathbb{Z} + [(i, j)]_\mathbb{Z}) = [(m, n)]_\mathbb{Z} + [(k + i, l + j)]_\mathbb{Z} \\ &= [(m + (k + i), n + (l + j)]_\mathbb{Z} = [(m + k) + i, (n + l) + j)]_\mathbb{Z} \\ &= [(m + k, n + l)]_\mathbb{Z} + [(i, j)]_\mathbb{Z} = (x + y) + z, \end{aligned}$$

as desired.

b) $x + 0 = [(m, n)]_\mathbb{Z} + [(1, 1)]_\mathbb{Z} = [(m + 1, n + 1)]_\mathbb{Z} = [(m, n)]_\mathbb{Z} = x$ (by Lemma 6.1.1).

6.1.9 a) Using Lemma 6.1.1, it follows that

$$i(m + n) = [(m + n + 1, 1)]_\mathbb{Z} = [(m + 1, n + 1, 1 + 1)]_\mathbb{Z}$$
$$= [(m + 1, 1)]_\mathbb{Z} + [(n + 1, 1)]_\mathbb{N} = i(m) + i(n).$$

b) Again using Lemma 6.1.1, we find that

$$i(m)i(n) = [(m + 1, 1)]_\mathbb{Z}[(n + 1, 1)]_\mathbb{Z}$$
$$= [((m + 1)(n + 1) + 1, m + 1 + n + 1)]_\mathbb{Z}$$
$$= [(mn + m + n + 2, m + n + 2)]_\mathbb{Z}$$
$$= [(mn + 1, 1)]_\mathbb{Z} = i(mn).$$

6.1.11 f) It is sufficient to show that $y < x < y + 1$ is false for all integers $x = [(m, n)]_\mathbb{Z}$ and $y = [(k, l)]_\mathbb{Z}$. Assuming that $y < x < y + 1$ is true, it follows that there are integers $[i, j]_\mathbb{N}$ and $[(p, q)]_\mathbb{Z}$ such that $i > j$, $p > q$, $y + [i, j]_\mathbb{N} = x$, and $x + [p, q]_\mathbb{N} = y + 1 = [(k + 2, l + 1)]_\mathbb{Z}$. Hence $[(k + i + p, l + j + q)]_\mathbb{Z} = [(k + 2, l + 1)]_\mathbb{Z}$, and therefore, Lemma 6.1.1 implies that

$$(k - l) + (i - j) + (p - q) = (k - l) = 1,$$

or equivalently,

$$(i - j) + (p - q) = 1.$$

But this is impossible, because $i - j, p - q \in \mathbb{N}$ implies that $(i - j) + (p - q) > 1$ (note: no element in $\mathbb{N}$ can be greater than one and equal to one simultaneously because $n + 1 = \varphi(n) \neq 1$ for all $n \in \mathbb{N}$).

6.2.9

$$i(-m) = \frac{-m}{1} = -\frac{m}{1} = -i(m),$$

$$i(m + n) = \frac{m + n}{1} = \frac{m}{1} + \frac{n}{1} = i(m) + i(n)$$

and

$$i(mn) = \frac{mn}{1} = \frac{m}{1} \cdot \frac{n}{1} = i(m)i(n).$$

Furthermore, i is injective, because if $m/1 = i(m) = i(n) = n/1$, then $m = m \cdot 1 = 1 \cdot n = n$. (We leave it to the reader to prove the remaining order-related statement.)

6.2.11 Using a) in conjunction with Proposition 6.2.7h, it readily follows that

$$xy < 0 \Leftrightarrow 0 < -xy \Leftrightarrow 0 < (-x)y \Leftrightarrow ((0 < -x \wedge 0 < y) \vee (-x < 0 \wedge y < 0))$$
$$\Leftrightarrow ((x < 0 \wedge 0 < y) \vee (0 < x \wedge y < 0)),$$

as claimed.

6.2.13 We will prove only that $|xy| = |x||y|$ and leave it to the reader to use this equation to establish (6.13). If $x = 0$ or $y = 0$, then $|xy| = 0 = |x||y|$ (by Theorem 6.2.5i and by the definition of the absolute value). So we may assume that both x and y are different from zero. Then there are the following four cases: $x > 0 \wedge y > 0$, $x < 0 \wedge y < 0$, $x > 0 \wedge y < 0$, and $x < 0 \wedge y > 0$. Using Proposition 6.2.6 in conjunction with Proposition 6.2.7h, we may handle these cases as follows: if $x > 0 \wedge y > 0$, then $|xy| = xy = |x||y|$; if $x < 0 \wedge y < 0$, then $|xy| = xy = (-x)(-y) = |x||y|$; if $x > 0 \wedge y < 0$, then $|xy| = -xy = x(-y) = |x||y|$, and the fourth case is analogous to the third.

6.3.11 If $x = [(a_n)_{n \in \mathbb{N}}]_\mathbb{R} = [(a'_n)_{n \mathbb{N}}]_\mathbb{R}$, then $0 = \lim_{n \to \infty}(a_n - a'_n) = \lim_{n \to \infty}(a'_n - a_n)$, and therefore $[(-a_n)_{n \in \mathbb{N}}]_\mathbb{R} = [(-a'_n)_{n \in \mathbb{N}}]_\mathbb{R}$. Furthermore, since the operations of addition and sign inversion are well defined, it follows that subtraction is well defined as well because $x - y = x + (-y)$.

6.3.13 d) $x + y = [(a_n)_{n \in \mathbb{N}}]_\mathbb{R} + [(b_n)_{n \in \mathbb{N}}]_\mathbb{R} = [(a_n + b_n)_{n \in \mathbb{N}}]_\mathbb{R} = [(b_n + a_n)_{n \in \mathbb{N}}]_\mathbb{R} = [(b_n)_{n \in \mathbb{N}}]_\mathbb{R} + [(a_n)_{n \in \mathbb{N}}]_\mathbb{R} = y + x$ (by Theorem 6.2.5d).
j) $x(y + z) = [(a_n(b_n + c_n))_{n \in \mathbb{N}}]_\mathbb{R} = [(a_n b_n + a_n c_n)_{n \in \mathbb{N}}]_\mathbb{R} = xy + xz$ (by Theorem 6.2.5j)

6.3.15 c) If $x + z < y + z$ for some $z \in \mathbb{R}$, then, by definition, $0 < (y + z) - (x + z) = y + z - x - z = y - x$, and therefore, $x < y$.
s) If $1 < x = [(a_n)_{n \in \mathbb{N}}]_\mathbb{R}$, then there exists an $N \in \mathbb{N}$ such that $1 < a_n$ for all $n > N$, and since $(a_n)_{n \in \mathbb{N}}$ is bounded (by Proposition 6.3.4), there exists an $M \in \mathbb{N}$ such that $a_n < M$ for all $n \in \mathbb{N}$. Thus $1/a_n > 1/M > 0$ for all $n > N$, and therefore, $0 < 1/x$. Consequently, we may apply i) to infer that $1/x = (1/x) \cdot 1 < (1/x)x = 1$. Conversely, if $0 < 1/x < 1$, then $x = 1/(1/x) > 0$ (by the argument just given), and therefore, $1 = (1/x)x < 1x = x$.

6.3.17 We will show that

$$\lim_{n \to \infty} \frac{n+1}{n} = \lim_{n \to \infty} \frac{n^2 + 3n + 1}{n^2 + 6} = 1$$

and leave the proof of the remaining limit equation as an exercise to the reader. Using Theorem 6.3.1b in conjunction with the obvious fact that $\lim_{n \to \infty} 1/n = 0$, we find that

$$\lim_{n \to \infty} \frac{n+1}{n} = \lim_{n \to \infty}\left(1 + \frac{1}{n}\right) = 1 + \lim_{n \to \infty}\frac{1}{n} = 1,$$

and

$$\lim_{n \to \infty} \frac{n^2 + 3n + 1}{n^2 + 6} = \lim_{n \to \infty}\left(1 + \frac{3n - 5}{n^2 + 6}\right) = 1 + \lim_{n \to \infty}\frac{3n - 5}{n^2 + 6}.$$

So we only need to show that $\lim_{n \to \infty}(3n - 5)/(n^2 + 6) = 0$. To do so, we pick an $\varepsilon > 0$ and an $N \in \mathbb{N}$ such that $N > 3/\varepsilon$. Then for any integer $n > N \geq 1$

it is the case that $3n - 5 > 0$, and therefore,

$$\left|\frac{3n-5}{n^2+6}\right| = \frac{3n-5}{n^2+6} < \frac{3n}{n^2} = \frac{3}{n} < \frac{3}{N} < \varepsilon,$$

as desired.

6.4.13 Since the map $i : [0,1] \to \mathbb{R}$, $x \mapsto x$ is evidently injective, it is sufficient to show (by Theorem 5.3.2) that the function $g : (0,1) \to \mathbb{R}$ that is defined by the equation

$$g(x) := \frac{2x-1}{1-(2x-1)^2}$$

for all $x \in (0,1)$ is surjective (note: the fact that g is not defined at 0 and 1 is irrelevant, because we can simply assign to 0 and 1 some arbitrary output value such as 0, that is, we can set $g(0) := g(1) := 0$). In order to prove that g is surjective, we pick a value $y \in \mathbb{R}$ and solve for x the equation

$$y = g(x) = \frac{2x-1}{1-(2x-1)^2}.$$

Using the quadratic formula, we easily find that

$$x = \frac{1}{2}\left(1 - \frac{1}{2y} \pm \sqrt{1 + \frac{1}{4y^2}}\right).$$

(Note: since $g(1/2) = 0 \,(= g(0) = g(1))$, we may here assume that $y \neq 0$.) Furthermore, since the inequality

$$0 < \frac{1}{2}\left(1 - \frac{1}{2y} \pm \sqrt{1 + \frac{1}{4y^2}}\right) < 1$$

is readily seen to be equivalent to the inequality

$$\frac{1}{2y^2} < \pm\frac{1}{y}\sqrt{1 + \frac{1}{4y^2}},$$

it is sufficient to show that

$$\frac{1}{2y^2} < \frac{1}{y}\sqrt{1 + \frac{1}{4y^2}}$$

whenever $y > 0$, and

$$\frac{1}{2y^2} < -\frac{1}{y}\sqrt{1 + \frac{1}{4y^2}}$$

whenever $y < 0$: if $y > 0$, then

$$\frac{1}{2y^2} = \frac{1}{y}\sqrt{\frac{1}{4y^2}} < \frac{1}{y}\sqrt{1 + \frac{1}{4y^2}},$$

and if $y < 0$, then

$$\frac{1}{2y^2} = -\frac{1}{y}\sqrt{\frac{1}{4y^2}} < \frac{1}{y}\sqrt{1 + \frac{1}{4y^2}},$$

as desired. Thus g is surjective and $|\mathbb{R}| = |[0,1]|$.

6.4.16 Using the sequence $(r_n)_{n \in \mathbb{N}}$ in Exercise 3.8.35, we find that the function $g : \mathbb{N} \to \mathbb{Q}$, $n \mapsto r_n$ is surjective. Since the function $i : \mathbb{N} \to \mathbb{Q}$, $x \mapsto x$ is evidently injective, Theorem 5.3.2 implies that $|\mathbb{N}| = |\mathbb{Q}|$, as claimed.

6.4.17 Since the map $i : \mathbb{Z} \to \mathbb{Q}$, $x \mapsto x$ is injective, it follows that $|\mathbb{Z}| \le |\mathbb{Q}|$, and since $|\mathbb{Q}| = |\mathbb{N}|$ (by Exercise 6.4.16), we may apply Theorem 5.2.4d to infer that $|\mathbb{Z}| \le |\mathbb{N}|$. Furthermore, since the map $j : \mathbb{N} \to \mathbb{Z}$, $x \mapsto x$ is injective, we find that $|\mathbb{N}| \le |\mathbb{Z}|$, and therefore, Theorem 5.2.3 in conjunction with Theorem 5.3.2 implies that $|\mathbb{Z}| = |\mathbb{N}|$.

6.5.4 According to Lemma 6.1.1, the equation

$$\left(\left[(([((n^2, 1)]_{\mathbb{Z}}, [(n^2+2, 1)]_{\mathbb{Z}})]_{\mathbb{Q}})_{n \in \mathbb{N}} \right]_{\mathbb{R}}, \left[(([((n, 1)]_{\mathbb{Z}}, [(n^2+2.1)]_{\mathbb{Z}})]_{\mathbb{Q}})_{n \in \mathbb{N}} \right]_{\mathbb{R}} \right)$$
$$+ \left(\left[(([((n+4, 1)]_{\mathbb{Z}}, [(n, 1)]_{\mathbb{Z}})]_{\mathbb{Q}})_{n \in \mathbb{N}} \right]_{\mathbb{R}}, \left[(([((n, 1)]_{\mathbb{Z}}, [(n^2, 2)]_{\mathbb{Z}})]_{\mathbb{Q}})_{n \in \mathbb{N}} \right]_{\mathbb{R}} \right)$$
$$= \left(\left[(([((2n^2, 1)]_{\mathbb{Z}}, [(n^2, 8)]_{\mathbb{Z}})]_{\mathbb{Q}})_{n \in \mathbb{N}} \right]_{\mathbb{R}}, \left[(([((n^2, 1)]_{\mathbb{Z}}, [(n^3, 1)]_{\mathbb{Z}})]_{\mathbb{Q}})_{n \in \mathbb{N}} \right]_{\mathbb{R}} \right).$$

is equivalent to the equation

$$\left(\left[([(n^2-1, n^2+1)]_{\mathbb{Q}})_{n \in \mathbb{N}} \right]_{\mathbb{R}}, \left[([(n-1, n^2+1)]_{\mathbb{Q}})_{n \in \mathbb{N}} \right]_{\mathbb{R}} \right)$$
$$+ \left(\left[([(n+3, n-1)]_{\mathbb{Q}})_{n \in \mathbb{N}} \right]_{\mathbb{R}}, \left[([(n-1, n^2-2)]_{\mathbb{Q}})_{n \in \mathbb{N}} \right]_{\mathbb{R}} \right)$$
$$= \left(\left[([(2n^2-1, n^2-8)]_{\mathbb{Q}})_{n \in \mathbb{N}} \right]_{\mathbb{R}}, \left[([(n^2-1, n^3-1)]_{\mathbb{Q}})_{n \in \mathbb{N}} \right]_{\mathbb{R}} \right),$$

and using standard quotient notation, we find that

$$\left(\left[\left(\frac{n^2-1}{n^2+1} \right)_{n \in \mathbb{N}} \right]_{\mathbb{R}}, \left[\left(\frac{n-1}{n^2+1} \right)_{n \in \mathbb{N}} \right]_{\mathbb{R}} \right)$$
$$+ \left(\left[\left(\frac{n+3}{n-1} \right)_{n \in \mathbb{N}} \right]_{\mathbb{R}}, \left[\left(\frac{n-1}{n^2-2} \right)_{n \in \mathbb{N}} \right]_{\mathbb{R}} \right)$$

$$= \left(\left[\left(\frac{2n^2 - 1}{n^2 - 8} \right) \right]_{n \in \mathbb{N}}\Big|_\mathbb{R}, \left[\left(\frac{n^2 - 1}{n^3 - 1} \right) \right]_{n \in \mathbb{N}}\Big|_\mathbb{R} \right).$$

Since the limits of the sequences that appear in his equation are 1, 0, 1, 0, 2, and 0, respectively (and in this order), it follows that the latter equation above is equivalent to

$$1 + 0i + 1 + 0i = (1,0) + (1,0) = (2,0) = 2 + 0i,$$

which in turn is equivalent to $1 + 1 = 2$.

6.5.5 b)

$$\overline{uv} = \overline{(a + ib)(c + id)} = \overline{ac - bd + i(ad + bc)} = ac - bd - i(ad + bc)$$
$$= (a - ib)(c - id) = \overline{u} \cdot \overline{v}.$$

6.5.6 a) We need to show that $|u+v|^2 \le |u|^2 + 2|u||v| + |v|^2$. Setting $u := a + ib$ and $v := c + id$, it follows that

$$|u + v|^2 = (a + c)^2 + (b + d)^2 = a^2 + c^2 + 2ac + 2bd + b^2 + d^2$$

and

$$|u|^2 + 2|u||v| + |v|^2 = a^2 + b^2 + 2\sqrt{a^2 + b^2}\sqrt{c^2 + d^2} + c^2 + d^2.$$

Consequently, we only need to show that

$$ac + bd \le \sqrt{a^2 + b^2}\sqrt{c^2 + d^2} = \sqrt{a^2c^2 + a^2d^2 + b^2c^2 + b^2d^2}$$

or that

$$a^2c^2 + 2acbd + b^2d^2 \le a^2c^2 + a^2d^2 + b^2c^2 + b^2d^2.$$

But this latter inequality is equivalent to the inequality $0 \le (ad - bc)^2$ which evidently is true.

b)

$$|uv|^2 = |ac - bd + i(ad + bc)|^2 = (ac - bd)^2 + (ad + bc)^2$$
$$= a^2c^2 - 2acbd + b^2d^2 + a^2d^2 + 2adbc + b^2c^2$$
$$= (a^2 + b^2)(c^2 + d^2) = |u|^2|v|^2.$$

Bibliography

[B]   Blume, Frank, *Applied Calculus for Scientists and Engineers*, Jones and Bartlett, Sudbury, 2005.

[DG] Daepp, U., Gorkin, P., *Reading, Writing, and Proving*, Springer, New York, 2003.

[JW] Just, Winfried, and Weese, Martin, *Discovering Modern Set Theory. I*, American Mathematical Society, 1996.

[Q]   Quine, W.V., *Methods of Logic*, Harvard University Press, Cambridge, MA, 1982.

Index

A

absolute value
    on $\mathbb{C}$ 199
    on $\mathbb{Q}$ 179
    on $\mathbb{R}$ 185
addition map on $\mathbb{N}$ 113
addition
    associative law 100
    commutative law 100
    in $\mathbb{C}$ 196
    in $\mathbb{N}$ 106, 97
    in $\mathbb{Q}$ 171
    in $\mathbb{R}$ 180
    in $\mathbb{Z}$ 158
alternation 5
antecedent 7
antisymmetry 71, 144
associativity 15, 100
Axiom of Choice 86
Axioms of Peano 18, 95

B

biconditional 7
bijection 82

C

Cantor, Georg 57
cardinal equiv. of $\mathbb{R}$ and $\mathcal{P}(N)$ 190
cardinality 133
    and linearity 154
    of a power set 143

of a set 135
Cartesian product 70
    of equivalence relations 74
Cauchy sequence 49, 176
    and boundedness 178
    products of 178
    sums of 178
closed set 150
complement of a set 64
completeness of $\mathbb{R}$ 50, 186
complex conjugate 197
complex numbers 195
    absolute value of 199
    addition of 196
    definition of 196
    division of 198
    multiplication of 196
    properties of 198
composition of functions 84
conditional 6
conjunction 4
consequent 7
consistency 12
Continuum Hypothesis 194
convergence
    of a sequence 31
    of a series 189

D

DeMorgan's Laws 15
density of $\mathbb{Q}$ in $\mathbb{R}$ 186
difference of sets 64
disjunction 5